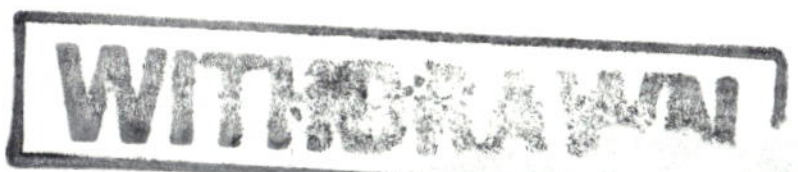

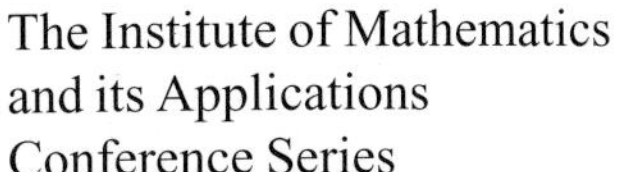

The Institute of Mathematics
and its Applications
Conference Series

Volumes in the previous series were published by Academic Press to whom all enquiries should be addressed. The following and all forthcoming titles are published by Oxford University Press throughout the world.

NEW SERIES

1. *Supercomputers and parallel computation* Edited by D. J. Paddon
2. *The mathematical basis of finite element methods* Edited by David F. Griffiths
3. *Multigrid methods for integral and differential equations* Edited by D. J. Paddon and H. Holstein
4. *Turbulence and diffusion in stable environments* Edited by J. C. R. Hunt
5. *Wave propagation and scattering* Edited by B. J. Uscinski
6. *The mathematics of surfaces* Edited by J. A. Gregory
7. *Numerical methods for fluid dynamics II* Edited by K. W. Morton and M. J. Baines
8. *Analysing conflict and its resolution* Edited by P. G. Bennett
9. *The state of the art in numerical analysis* Edited by A. Iserles and M. J. D. Powell
10. *Algorithms for approximation* Edited by J. C. Mason and M. G. Cox
11. *The mathematics of surfaces II* Edited by R. R. Martin
12. *Mathematics in signal processing* Edited by T. S. Durrani, J. B. Abbiss, J. E. Hudson, R. N. Madan, J. G. McWhirter, and T. A. Moore
13. *Simulation and optimization of large systems* Edited by Andrzej J. Osiadacz
14. *Computers in mathematical research* Edited by N. M. Stephens and M. P. Thorne
15. *Stably stratified flow and dense gas dispersion* Edited by J. S. Puttock
16. *Mathematical modelling in non-destructive testing* Edited by Michael Blakemore and George A. Georgiou
17. *Numerical methods for fluid dynamics III* Edited by K. W. Morton and M. J. Baines
18. *Mathematics in oil production* Edited by Sir Sam Edwards and P. R. King
19. *Mathematics in major accident risk assessment* Edited by R. A. Cox
20. *Cryptography and coding* Edited by Henry J. Beker and F. C. Piper
21. *Mathematics in remote sensing* Edited by S. R. Brooks
22. *Applications of matrix theory* Edited by M. J. C. Gover and S. Barnett
23. *The mathematics of surfaces III* Edited by D. C. Handscomb
24. *The interface of mathematics and particle physics* Edited by D. G. Quillen, G. B. Segal, and Tsou S. T.
25. *Computational methods in aeronautical fluid dynamics* Edited by P. Stow
26. *Mathematics in signal processing II* Edited by J. G. McWhirter
27. *Mathematical structures for software engineering* Edited by Bernard de Neumann, Dan Simpson, and Gil Slater
28. *Computer modelling in the environmental sciences* Edited by D. G. Farmer and M. J. Rycroft
29. *Statistics in medicine* Edited by F. Dunstan and J. Pickles
30. *The mathematical revolution inspired by computing* Edited by J. H. Johnson and M. J. Loomes
31. *The mathematics of oil recovery* Edited by P. R. King
32. *Credit scoring and credit control* Edited by L. C. Thomas, J. N. Crook, and D. B. Edelman
33. *Cryptography and coding* Edited by Chris Mitchell
34. *The dynamics of numerics and numerics of dynamics* Edited by D. S. Broomhead and A. Iserles
35. *The unified computation laboratory: modelling, specifications, and tools* Edited by Charles Rattray and Robert G. Clark
36. *Mathematics in the automotive industry* Edited by James R. Smith
37. *The mathematics of control theory* Edited by N. K. Nichols and D. H. Owens
38. *Mathematics in transport planning and control* Edited by J. D. Griffiths
39. *Computational ordinary differential equations* Edited by J. R. Cash and I. Gladwell

Continued overleaf

40. *Waves and turbulence in stably stratified flows* Edited by S. D. Mobbs and J. C. King
41. *Robotics: applied mathematics and computational aspects* Edited by Kevin Warwick
42. *Mathematical modelling for materials processing* Edited by Mark Cross, J. F. T. Pittman, and Richard D. Wood
43. *Wavelets, fractals, and Fourier transforms* Edited by M. Farge, J. C. R. Hunt, and J. C. Vassilicos
44. *Computational aeronautical fluid dynamics* Edited by J. Fezoui, J. Hunt, and J. Periaux
45. *Cryptography and coding III* Edited by M. J. Ganley
46. *Parallel computation* Edited by A. E. Fincham and B. Ford
47. *Aerospace vehicle dynamics* Edited by M. V. Cook and M. J. Rycroft
48. *Computer-aided surface geometry and design: mathematics of surfaces IV* Edited by A. Bowyer
49. *Mathematics in signal processing III* Edited by J. G. McWhirter
50. *Design and application of curves and surfaces: mathematics of surfaces V* Edited by R. B. Fisher
51. *Artificial intelligence in mathematics* Edited by J. H. Johnson, S. McKee, and A. Vella
52. *Stably stratified flows: flow and dispersion over topography* Edited by I. P. Castro and N. Rockliffe
53. *Computational learning theory: EuroCOLT 93* Edited by J. S. Shawe-Taylor and M. Antony
54. *Complex stochastic systems and engineering* Edited by D. M. Titterington
55. *Mathematics of dependable systems* Edited by C. J. Mitchell and V. Stavridou
56. *The mathematics of deforming surfaces* Edited by D. G. Dritschel and R. J. Perkins
57. *Mathematical education of engineers* Edited by L. R. Mustoe and S. Hibberd
58. *The mathematics of surfaces VI* Edited by G. Mullineux
59. *Applications of finite fields* Edited by Dieter Gollmann
60. *Applications of combinatorial mathematics* Edited by C. J. Mitchell
61. *Image processing: mathematical methods and applications* Edited by J. M. Blackledge
62. *Flow and dispersion through groups of obstacles* Edited by R. J. Perkins and S. E. Belcher
63. *The state of the art in numerical analysis* Edited by I. S. Duff and G. A. Watson
64. *Mathematics of dependable systems II* Edited by V. Stavridou
65. *Approximations and numerical methods for the solution of Maxwell's equations* Edited by F. El Dabaghi, K. Morgan, A. K. Parrott and J. Periaux
66. *Mathematics of heat transfer* Edited by G. E. Tupholme and A. S. Wood
67. *Mathematics in signal processing IV* Edited by J. G. McWhirter and I. K. Proudler
68. *Mixing and dispersion in stably stratified flows* Edited by Peter A. Davies
69. *Wind-over-wave couplings: perspectives and prospects* Edited by S. G. Sajjadi, N. H. Thomas, and J. C. R. Hunt
70. *Cardiovascular flow modelling and measurement with application to clinical medicine* Edited by S. J. Sajjadi, G. B. Nash, and M. W. Rampling
71. *Mathematics in signal processing V* Edited by J. G. McWhirter and I. K. Proudler

Mathematics in Signal Processing V

Based on the proceedings of a conference on Mathematics in Signal Processing. Organized by the Institute of Mathematics and its Applications and held at the University of Warwick in December 2000

Edited by

J. G. McWHIRTER

Senior Fellow, Defence Evaluation and Research Agency, Malvern

and

I. K. PROUDLER

Fellow, Defence Evaluation and Research Agency, Malvern

CLARENDON PRESS • OXFORD • 2002

OXFORD
UNIVERSITY PRESS

Great Clarendon Street, Oxford OX2 6DP

Oxford University Press is a department of the University of Oxford. It furthers the University's objective of excellence in research, scholarship, and education by publishing worldwide in

Oxford New York

Auckland Bangkok Buenos Aires Cape Town Chennai Dar es Salaam Delhi Hong Kong Istanbul Karachi Kolkata Kuala Lumpur Madrid Melbourne Mexico City Mumbai Nairobi São Paulo Shanghai Taipei Tokyo Toronto

with an associated company in Berlin

Oxford is a registered trade mark of Oxford University Press in the UK and in certain other countries

Published in the United States by Oxford University Press Inc., New York

First published 2002

A catalogue record for this book is available from the British Library

Library of Congress Cataloging in Publication Data

(Data available)

ISBN 0 19 850734 8 (Hbk)

10 9 8 7 6 5 4 3 2 1

Typeset by HK Typesetting Ltd, Fleetwood
Printed in Great Britain
on acid-free paper by
Biddles Ltd., Guildford and King's Lynn

PREFACE

The fifth IMA conference on "Mathematics in Signal Processing" was held at the University of Warwick from 18th to 20th December 2000. The aim of the conference was to bring together mathematicians and engineers with a view to exploring recent developments and identifying fruitful avenues for further research.

This book comprises a selection of the papers presented at that meeting and provides a representative cross-section of the many topics discussed. Unfortunately, it was not possible to include all papers presented at the conference or submitted for this book. Many good papers, worthy of publication had to be omitted. This selection was made by the editors giving priority to papers with strong mathematical content, relevance to signal processing and tutorial value. We would like to offer our thanks and sincere apologies to the authors of papers which were not included in this book.

A strong theme of the conference was nonlinear, non-Gaussian signal processing with particular emphasis on blind signal separation and the use of higher order statistics (HOS). This has been a particularly significant area for research in signal processing over the last few years and is now at a very exciting stage of development. It is likely to be of great importance for many years to come.

At the conference, this theme was introduced with great insight and enthusuiasm by Pierre Comon in his keynote talk. It is highly appropriate that the content of that talk constitutes the opening paper for our book. It deals with the topic of tensor decomposition which arises quite naturally in the context of HOS as applied to blind signal separation. This is a very interesting and challenging mathematical problem which lies at the heart of extending numerical linear algebra to more than two dimensions. It is the epitomy of a topic requiring inter-disciplinary research between mathematicians and engineers and must be ripe for collaborative funding. The application to blind signal separation and blind deconvolution is discussed in the next two papers by Regalia and by De Lathauwer *et al.* These help to set the scene for a number of the papers which follow.

The book covers a wide range of other interesting topics too numerous to list here. They range from the statistics of impulse noise and chaotic dynamical systems to biomedical signal processing, importance sampling and evidential reasoning. This rich blend of theoretical ideas and novel application was very much the hallmark of our conference. We hope that the reader will enjoy the selection as much as we did.

Finally, we would like to thank all authors for their contribution to this book and for helping to make the conference such a success.

Editors:
J.G. McWhirter and I.K. Proudler
Defence Evaluation and Research Agency, Malvern

ACKNOWLEDGEMENTS

The Institute of Mathematics and its Applications would like to thank the following who contributed to the organisation of the conference and the production of this book.

Professor John McWhirter FRS FREng, Chairman of the Conference Organising Committee and editor of this book; Dr Ian Proudler, co-editor of this book; members of the conference organising committee; Dr Pierre Comon, keynote speaker at the conference; the authors of the papers; Mrs Pamela Bye (IMA Conference Officer) and Mr Terry Edwards (IMA Services Officer) for help with all aspects of the conference; Ms Kim Roberts (HK Typesetting) for assistance in producing the final copy of this book.

Members of the Conference Organising Committee:
Professor J.G. McWhirter FRS FREng (Chairman) (DERA, Malvern); Professor O.R. Hinton (University of Newcastle); Dr M. MacLeod (University of Cambridge); Professor M. Sandler (King's College, London); Dr S. McLaughlin (University of Edinburgh); Dr I.K. Proudler (DERA, Malvern).

CONTENTS

CONTRIBUTORS

M. ANDERLE; Department of Mathematics, Colorado State University, Fort Collins, CO 80523, USA.

P. ASHWIN; School of Mathematical Sciences, Laver Building, University of Exeter, Exeter EX4 4QE.

N. AYDIN; Department of Electronics and Electrical Engineering, The University of Edinburgh, Edinburgh.

C.R. BAKER; Department of Statistics, University of North Carolina, Chapel Hill, NC 27599-3260, USA.

B. BARKAT; Nanyang Technological University, School of Electrical and Electronic Engineering, Block S2, Nanyang Avenue, Singapore 639798, Singapore.

E. BIGLIERI; Dipartimento di Elettronica, Politecnico di Torino, Turin 10129, Italy.

D.S. BROOMHEAD; Department of Mathematics, University of Manchester Institute of Science and Technology, PO Box 88, Manchester M60 1QD.

A.G. BROWN; Defence Evaluation Research Agency, St. Andrews Road, Great Malvern, WR14 3PS.

P. COMON; Lab. I3S, CNRS, BP121, F-06903 Sophia-Antipolis cedex, France.

J. CHAMBERS; Signal and Image Processing Group, Department of Electronic and Electrical Engineering, University of Bath, Bath, BA2 7AY.

C.-C. CHEN; Electrical Engineering Department, UCLA, Los Angeles, CA 90095-1594, USA.

S. CHEN; Department of Electronics and Computer Science, University of Southampton, Southampton SO17 1BJ.

A. CLIMESCU-HAULICA; Communications Research Center Canada, 3701 Carling Ave.,Ottawa, Ontario K2H 8S2, Canada.

M. DAVIES; Signal Processing Laboratory, King's College London, The Strand, London WC2R 2LS.

J.-P. DELMAS; Institut National des Télécommunications, 9 rue Charles Fourier, 91011 Evry Cedex, France.

L. DE LATHAUWER; Equipe Traitement des Images et du Signal, Ecole Nationale Supérieure de l'Electronique et de ses Applications and Université de Cergy-Pontoise, France.

B. DE MOOR; Department of Electrical Engineering, Katholieke Universiteit Leuven, Belgium.

T.R. FIELD; DERA Malvern, St. Andrews Road, Malvern WR14 3PS.

X.-C. FU; School of Mathematical Sciences, Laver Building, University of Exeter, Exeter EX4 4QE.

P.M. GRANT; Department of Electronics and Electrical Engineering, University of Edinburgh, The King's Buildings, Edinburgh EH9 3JL.

A.F. GUALTIEROTTI; IDHEAP, 21 Route de la Maldière, CH-1022 Chavannes-près-Lausanne, Switzerland.

L. HANZO; Department of Electronics and Computer Science, University of Southampton, Southampton SO17 1BJ.

S.D. HAYWARD; Defence Evaluation and Research Agency, St. Andrews Road, Malvern, WR14 3PS.

E. HEYVAERT; Department of Electrical Engineering, Katholieke Universiteit Leuven Kasteelpark Arenberg 10, 3001 Leuven-Heverlee, Belgium.

J.P. HUKE; Department of Mathematics, University of Manchester Institute of Science and Technology, PO Box 88, Manchester M60 1QD.

M. KIRBY; Department of Mathematics, Colorado State University, Fort Collins, CO 80523, USA.

M. KLAJMAN; Communications and Signal Processing Research Group, Imperial College of Science, Technology and Medicine, London SW7 2BT.

H. KOEPPL; Infineon Technologies Austria, Design Centre Villach, Austria.

F.T. LUK; Department of Computer Science and Engineering, Chinese University of Hong Kong, Shatin, NT, Hong Kong.

Y. LUO; Signal and Image Processing Group, Department of Electronic and Electrical Engineering, University of Bath, Bath, BA2 7AY.

S.P. LUTTRELL; Signal Processing and Imagery Department, Defence Evaluation and Research Agency, St. Andrew's Rd, Malvern, WR14 3PS.

M.D. MACLEOD; Department of Engineering, University of Cambridge, Trumpington Street, Cambridge, CB2 1PZ.

I. MANN; Department of Electronics and Electrical Engineering, University of Edinburgh, Edinburgh, EH9 3JL.

S. McLAUGHLIN; Department of Electronics and Electrical Engineering, University of Edinburgh, Edinburgh, EH9 3JL.

Y. MEURISSE; Institut National des Télécommunications, 9 rue Charles Fourier, 91011 Evry Cedex, France.

M.R. MULDOON; Department of Mathematics, University of Manchester Institute of Science and Technology, PO Box 88, Manchester M60 1QD.

B. MULGREW; Department of Electronics and Electrical Engineering, University of Edinburgh, The King's Buildings, Edinburgh EH9 3JL.

G. PAOLI; Infineon Technologies Austria, Design Center Villach, Austria.

S. QIAO; Department of Computing and Software, McMaster University, Hamilton, Ontario L8S 4L7, Canada.

J.F. RALPH; Department of Electrical Engineering and Electronics, The University of Liverpool, Brownlow Hill, Liverpool, L69 3GJ.

P.A. REGALIA; Institut National des Télécommunications, 9, rue Charles Fourier, F-91011 Evry cedex, France.

C.Z.W. HASSELL SWEATMAN; Department of Electronics and Electrical Engineering, University of Edinburgh, The King's Buildings, Edinburgh EH9 3JL.

J.S. THOMPSON; Department of Electronics and Electrical Engineering, University of Edinburgh, The King's Buildings, Edinburgh EH9 3JL.

K. UMENO; CRL, Ministry of Post and Telecommunications, Koganei, Tokyo 184-9795, Japan.

D. VANDEVOORDE; Edison Design Group, Belmont, CA 94002-3112 USA.

J. VANDEWALLE; Department of Electrical Engineering, Katholieke Universiteit Leuven, Belgium.

L. VANHAMME; Department of Electrical Engineering, Katholieke Universiteit Leuven Kasteelpark Arenberg 10, 3001 Leuven-Heverlee, Belgium.

P. VAN HECKE; Biomedical NMR Unit, Faculty of Medicine, Katholieke Universiteit Leuven Kasteelpark Arenberg 10, 3001 Leuven-Heverlee, Belgium.

S. VAN HUFFEL; Department of Electrical Engineering, Katholieke Universiteit Leuven Kasteelpark Arenberg 10, 3001 Leuven-Heverlee, Belgium.

Y. WANG; Department of Electrical Engineering, Katholieke Universiteit Leuven Kasteelpark Arenberg 10, 3001 Leuven-Heverlee, Belgium.

S. WEISS; Department of Electronics and Computer Science, University of Southampton SO17 1BJ.

R. WILSON; University of Warwick, Coventry CV4 7AL.

K. YAO; Electrical Engineering Department, UCLA, Los Angeles CA 98095-1594, USA.

P. YUVAPOOSITANON; Department of Electronic Engineering, Mahanakorn University of Technology, Bangkok, Thailand 10530.

Tensor Decompositions
State of the Art and Applications

Pierre Comon
Lab. I3S, CNRS, BP121, F-06903 Sophia-Antipolis cedex, France

Abstract

In this paper, we present a partial survey of the tools borrowed from tensor algebra, which have been utilized recently in Statistics and Signal Processing. It is shown why the decompositions well known in linear algebra can hardly be extended to tensors. The concept of rank is itself difficult to define, and its calculation raises difficulties. Numerical algorithms have nevertheless been developed, and some are reported here, but their limitations are emphasized. These reports hopefully open research perspectives for enterprising readers.

1 Introduction

Applications. The decomposition of arrays of order higher than 2 has proven to be useful in a number of applications. The most striking case is perhaps *Factor Analysis*, where statisticians early identified difficult problems, tackling the limits of linear algebra. The difficulty lies in the fact that such arrays may have more factors than their dimensions. Next, data are often arranged in many-way arrays, and the reduction to 2-way arrays sometimes results in a loss of information. Lastly, the solution of some problems, including the so-called *Blind Source Separation* (BSS) generally requires the use of High-Order Statistics (HOS), which are intrinsically tensor objects [59]. When second-order statistics suffice to establish identifiability, the corresponding algorithms are quite sensitive to model uncertainties [57], so that the complementary use of HOS statistics is often still recommended.

BSS finds applications in Sonar, Radar [13], Electrocardiography [32], Speech [61, 52, 26], and Telecommunications [35, 36, 70, 12, 39], among others. In particular, the surveillance of radio-communications in the civil context, or interception and classification in military applications, resort to BSS. Moreover, in *Mobile Communications*, the mitigation of interfering users and the compensation for channel fading effects are now devised with the help of BSS; this is closely related to the general problem of *Blind Deconvolution.*

High-Order Factor Analysis is applied in many areas including Economy, Psychology [10, 11], Chemometrics [37, 4], and Sensor Array Processing [18, 70, 65]. Other fields where array decompositions can turn out to be useful include Exploratory Analysis [45], Complexity Analysis [50, 43], and Sparse Coding [44].

Bibliographical survey. Bergman [1] (1969) and Harshman [41] were the first to notice that the concept of *rank* was difficult to extend from matrices to higher-order arrays. Caroll and Chang [10] provided the first *canonical decomposition* algorithm of a three-way array, later referred to as *Candecomp* model. Several years later, Kruskal [50] conducted a detailed analysis of uniqueness, and related several definitions of rank. The algorithm *Candelinc* was devised by Caroll and others in the 1980s [11]; it allowed computation of a canonical decomposition subject to *a priori* linear constraints.

Leurgans *et al.* [54] derived sufficient identifiability conditions for the 3-way array decomposition; as opposed to Kruskal, their proof was constructive and yielded a numerical algorithm running several matrix SVDs.

Instead of finding an exact decomposition of a d-way array, which requires more than d terms (as we shall subsequently see), Comon proposed [15] to approximately decompose it into d terms. The problem was then reduced to finding an invertible linear transform (change of coordinates): see [16, 9] and references therein. This decomposition is now referred to as *Independent Component Analysis* (ICA), whereas the exact *Canonical Decomposition* is sometimes referred to as *underdetermined* or *over-complete* ICA [52, 18].

The terminology of ICA is meaningful in the context of Signal Processing and BSS [46, 16, 8]. Constructive algorithms for ICA either proceed by sweeping the pairs of indices [14, 16, 9], or are of iterative nature, like power methods [28, 49], gradient descents [58], or Robbins-Monro algorithms [46, 61]. Some less efficient early methods were based on contracted versions of the array [6] or on noiseless observations [14].

A solid account on decompositions of 3-way arrays can also be found in DeLathauwer's PhD thesis [26]; an interesting tool defined therein is the *HOSVD* [30], generalizing the concept of SVD to arrays of order 3, in a different manner compared to Caroll, but quite similar to the *Tuckals* decomposition [55, 68, 37]. A good survey of rank issues can also be found in [49]. An account on identifiability issues can be found in [5].

2 Tensors

2.1 Terminology

The *order* of an array refers to the number of its *ways*; the entries of an array of order d are accessed via d indices, say $i_1 \dots i_d$, with every index i_a ranging from 1 to n_a. The integer n_a is one of the d *dimensions* of the array. For instance, a matrix is a 2-way array (order 2), and thus has two dimensions. A vector is an array of order 1, and a scalar is of order 0.

Throughout this paper, and unless otherwise specified, variables take their values in the real field, although all the statements hold true in the complex field with more complicated notations; boldface lowercase letters, like $\boldsymbol{u}$, will denote single-way arrays, that is, vectors, whereas boldface uppercase letters, like $\boldsymbol{G}$, will denote arrays with more than one way, that is, matrices or many-way arrays.

The entries of arrays are scalar quantities and are denoted with plain letters, such as u_i or $G_{ijk\ell}$.

A tensor of order d is a d-way array that enjoys the *multilinearity property* after a change of coordinate system. In order to fix the ideas, consider a third order tensor $\boldsymbol{T}$ with entries T_{ijk}, and a change of coordinates defined by three square invertible matrices, $\boldsymbol{A}$, $\boldsymbol{B}$, and $\boldsymbol{C}$. Then, in the new coordinate system, the tensor $\boldsymbol{T}'$ can be written as a function of tensor $\boldsymbol{T}$ as

$$T'_{ijk} = \sum_{abc} A_{ia} B_{jb} C_{kc} T_{abc}. \tag{2.1}$$

In particular, moments and cumulants of random variables may be treated as tensors [59]. This product is sometimes referred to as the Tucker product [49] between matrices $\boldsymbol{A}$, $\boldsymbol{B}$, and $\boldsymbol{C}$, weighted by $\boldsymbol{T}$. Note that tensors enjoy property (2.1) even if the above matrices are not invertible; only linearity is required.

Tensor algebra is a well identified framework; in particular, two kinds of indices are distinguished, covariant or contravariant, depending on the role they play in the application under consideration: an array can indeed be seen as an *operator* from one space to another. For the sake of simplicity, we *shall not* pay too much attention to this distinction, although it turns out to be important in contexts other than the present one.

2.2 Notation

Given two arrays of order m and n, one defines their *outer product* $\boldsymbol{C} = \boldsymbol{A} \circ \boldsymbol{B}$ as the array of order $m + n$:

$$C_{ij\ldots\ell\, ab\ldots d} = A_{ij\ldots\ell}\, B_{ab\ldots d}. \tag{2.2}$$

For instance, the outer product of two vectors, $\boldsymbol{u} \circ \boldsymbol{v}$, is a matrix.

Given two arrays, $\boldsymbol{A} = \{A_{ij\ldots\ell}\}$ and $\boldsymbol{B} = \{B_{i'j'\ldots\ell'}\}$ of orders d_A and d_B respectively, having the same first dimension, one can define the *mode*-1 contraction product

$$(A \bullet B)_{j\ldots\ell j'\ldots\ell'} = \sum_{i=1}^{n_1} A_{ij\ldots\ell} B_{ij'\ldots\ell'}.$$

For instance, the standard matrix-vector product is $\boldsymbol{A}\boldsymbol{u} = \boldsymbol{A}^{\mathrm{T}} \bullet \boldsymbol{u}$. Similarly, one defines the *mode-p* inner product when arrays $\boldsymbol{A}$ and $\boldsymbol{B}$ have the same pth dimension, by summing over the pth index; the product is denoted as

$$\boldsymbol{A} \underset{p}{\bullet} \boldsymbol{B}.$$

If unspecified, the contraction applies by default to the first index. Some authors denote this product as $\boldsymbol{A} \times_p \boldsymbol{B}$, but we find it less readable.

We define the Kronecker product $\boldsymbol{u} \otimes \boldsymbol{v}$ between two vectors $\boldsymbol{u}$ and $\boldsymbol{v}$ as the vector containing all the possible cross-products [3]. If $\boldsymbol{u}$ and $\boldsymbol{v}$ are of dimension J and K, then $\boldsymbol{u} \otimes \boldsymbol{v}$ is of dimension JK.

Lastly, we define the symmetric Kronecker product of a K-dimensional vector $\boldsymbol{w}$ by itself, denoted $\boldsymbol{w} \oslash \boldsymbol{w} = \boldsymbol{w}^{\oslash 2}$, as the $K((K+1)/2$-dimensional vector containing all the distinct products, with an appropriate weighting of the cross-terms so that $||\boldsymbol{w} \oslash \boldsymbol{w}|| = ||\boldsymbol{w} \otimes \boldsymbol{w}||$. The product $\boldsymbol{w}^{\oslash d}$ is defined in a similar manner for $d > 2$. For instance, $\boldsymbol{w}^{\oslash 3}$ is of size $K(K+1)(K+2)/6$.

The $\mathbf{vecs}\{\cdot\}$ operator puts a $K \times K$ symmetric matrix in the form of a $K(K+1)/2$-dimensional vector; conversely, $\mathbf{Unvecs}\{\cdot\}$ puts it back in matrix form. For instance, $\mathbf{Unvecs}\{\boldsymbol{f}^{\oslash 2}\} = \boldsymbol{f}\,\boldsymbol{f}^{\mathrm{T}}$: see [39, 70] for a more detailed description.

2.3 Homogeneous polynomials

d-way arrays can be written in two different manners, as pointed out by several authors [59], related to each other by a bijective mapping, $\boldsymbol{f}$. Assume the notations $\boldsymbol{x}^{\boldsymbol{j}} \stackrel{\text{def}}{=} \prod_{k=1}^{K} x_k^{j_k}$ and $|\boldsymbol{j}| \stackrel{\text{def}}{=} \sum_k j_k$. Then for homogeneous monomials of degree d, $\boldsymbol{x}^{\boldsymbol{j}}$, we have $|\boldsymbol{j}| = d$.

To start with, take the example of $(K, d) = (4, 3)$: one can associate every entry T_{ijk} to a monomial $T_{ijk}\; x_i x_j x_k$. For instance, T_{114} is associated with $T_{114}\, x_1^2 x_4$, and thus to $T_{114}\, \boldsymbol{x}^{[2,0,0,1]}$; this means that $f([1,1,4]) = [2,0,0,1]$.

More generally, the d-dimensional vector index $\boldsymbol{i} = [i, j, k]$ can be associated with a K-dimensional vector index $\boldsymbol{f}(\boldsymbol{i})$ containing the number of times each variable x_k appears in the associated monomial. Whereas the d entries of $\boldsymbol{i}$ take their values in $\{1, \ldots, K\}$, the K entries of $\boldsymbol{f}(\boldsymbol{i})$ take their values in $\{1, \ldots, d\}$ with the constraint that $\sum_k f_k(\boldsymbol{i}) = d, \forall \boldsymbol{i}$.

As a consequence, the linear space of symmetric tensors can be bijectively associated with the linear space of homogeneous polynomials. To see this, it suffices to associate every polynomial $p(\boldsymbol{x})$ with the symmetric tensor $\boldsymbol{G}$ as

$$p(\boldsymbol{x}) = \sum_{|\boldsymbol{f}(\boldsymbol{i})|=d} G_{\boldsymbol{i}}\, \boldsymbol{x}^{\boldsymbol{f}(\boldsymbol{i})} \tag{2.3}$$

where $G_{\boldsymbol{i}}$ are the entries of $\boldsymbol{G}$. The dimension of these spaces is $S = \binom{K+d-1}{d}$, and one can choose as a basis the set of monomials: $\mathcal{B}(K; d) = \{\boldsymbol{x}^{\boldsymbol{j}},\, |\boldsymbol{j}| = d\}$.

> *Example.* Let p and q be two homogeneous polynomials in K variables, associated with tensors $\boldsymbol{P}$ and $\boldsymbol{Q}$, possibly of different orders. Then, polynomial pq is associated with $\boldsymbol{P} \circ \boldsymbol{Q}$:
>
> $$p(\boldsymbol{x})q(\boldsymbol{x}) = \sum_{\boldsymbol{i}} \sum_{\boldsymbol{j}} P_{\boldsymbol{i}} Q_{\boldsymbol{j}} \boldsymbol{x}^{\boldsymbol{f}(\boldsymbol{i})+\boldsymbol{f}(\boldsymbol{j})} = \sum_{[\boldsymbol{i}\,\boldsymbol{j}]} [\boldsymbol{P} \circ \boldsymbol{Q}]_{[\boldsymbol{i}\,\boldsymbol{j}]}\, \boldsymbol{x}^{\boldsymbol{f}([\boldsymbol{i}\,\boldsymbol{j}])}.$$

In practice, it is convenient to take into account the symmetry of the tensor, by defining $c(\boldsymbol{j})$ as the number of times the entry $T_{\boldsymbol{f}^{-1}(\boldsymbol{j})}$ appears in the array $\boldsymbol{T}$: $c(\boldsymbol{j}) = |\boldsymbol{j}|!/(\boldsymbol{j})!$, where $(\boldsymbol{j})! \stackrel{\text{def}}{=} \prod_k j_k!$. For binary quantics, $K = 2$ and $c(\boldsymbol{j}) = \binom{d}{j_1} = \binom{d}{j_2}$; for instance, for $d = 4$, $c([3,1]) = 4$, and $c([2,2]) = 6$.

Coefficients $\gamma(\boldsymbol{j},p)$ of a polynomial p in basis $\mathcal{B}$ are chosen so as satisfy the relation

$$p(\boldsymbol{x}) = \sum_{|\boldsymbol{j}|=d} \gamma(\boldsymbol{j},p)\, c(\boldsymbol{j})\, \boldsymbol{x}^{\boldsymbol{j}}.$$

Now, both spaces can be provided with a scalar product. For d-way arrays of dimension K, define the Froebenius scalar product

$$\langle \boldsymbol{G}, \boldsymbol{H} \rangle = \sum_{\boldsymbol{i}} \boldsymbol{G}_{\boldsymbol{i}}\, \boldsymbol{H}_{\boldsymbol{i}}$$

and the induced Euclidian norm. For homogeneous polynomials of degree d in K variables, define the scalar product as

$$\langle p, q \rangle = \sum_{|\boldsymbol{j}|=d} c(\boldsymbol{j})\, \gamma(\boldsymbol{j},p)\, \gamma(\boldsymbol{j},q).$$

In particular, monomials in $\mathcal{B}$ satisfy $||\boldsymbol{x}^{\boldsymbol{j}}||^2 = (\boldsymbol{j})!/d! = 1/c(\boldsymbol{j})$. In the case of binary quantics ($K = 2$), this was called the *apolar* scalar product [51]. The latter definition has several advantages [24]. For instance, if $a(\boldsymbol{x})$ is a homogeneous linear form with coefficient vector $\boldsymbol{a}$, then the scalar product $\langle a^d, q \rangle$ with any homogeneous polynomial q of degree d turns out to be the value of q at $\boldsymbol{a}$:

$$\langle a(\boldsymbol{x})^d, q(\boldsymbol{x}) \rangle = \sum_{\boldsymbol{i}} c(\boldsymbol{i})\, \gamma(\boldsymbol{i},q)\, \boldsymbol{a}^{\boldsymbol{i}} = q(\boldsymbol{a}).$$

The interest in establishing a link between tensors and polynomials lies in the fact that polynomials have been studied rather deeply during the last hundred and more years [63]. Some of the results obtained will be useful in this chapter, and in particular the classification of cubics.

2.4 Genericity

A property will be referred to as *generic* if it is true on a dense algebraic subset. The topology used is the standard one for homogeneous polynomials, namely that of Zariski [64]. Recall that in this topology, the closed subsets are defined by algebraic equations of the form $p(\boldsymbol{x}) = 0$, where p is a polynomial. Its particularity is that two open non-empty subsets always intersect; in other words, the topology is not separated.

For instance, a symmetric matrix is generically of full rank. In fact, the set of singular matrices is defined by the polynomial relation $\det(\boldsymbol{A}) = 0$, which is associated with a closed subset, whose complementary is dense. We shall subsequently see that this does not hold true any more for tensors of order higher than 2. For instance, the $2 \times 2 \times 2$ tensor $\boldsymbol{T}$ such that $T_{122} = T_{212} = T_{221} = 1$, and zero elsewhere (see Fig. 1) is known to be of rank 3. However, the generic rank is 2 in that case, as will be discussed in Section 4.2.

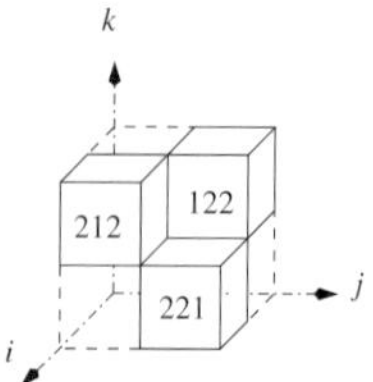

FIG. 1. Non-generic example of a binary cubic of maximal rank: position of non-zero entries in the $2 \times 2 \times 2$ associated symmetric tensor.

2.5 Array ranks

Let $\boldsymbol{T}$ be a tensor, not necessarily symmetric, of dimensions $n_1 \times \cdots \times n_d$. One defines the *tensor rank* ω of $\boldsymbol{T}$ as the minimal number of rank 1 tensors whose linear combination yields $\boldsymbol{T}$. The properties of tensor rank will be extensively discussed in Section 4.2. For completeness, let us also mention the definition of *mode-n ranks.*

The mode-n vectors of $\boldsymbol{T}$ are obtained by varying index i_n and keeping the others fixed; there are thus as many mode$-n$ vectors as possibilities of fixing indices i_k, $k \neq n$. The mode$-n$ rank, R_n, is defined as the dimension of the linear space spanned by all mode$-n$ vectors of $\boldsymbol{T}$.

Bounds. Howell [43] showed that the tensor rank can be bounded as $\omega \leq \max_{i \neq j}\{n_i n_j\}$. On the other hand, mode$-n$ ranks and tensor rank are related by the inequality $R_n \leq \omega$, $\forall k$.

In the symmetric case, Reznick showed that the tensor rank can be bounded as a function of the dimension K and the order, d:

$$\omega \leq \begin{pmatrix} K+d-2 \\ d-1 \end{pmatrix} \tag{2.4}$$

but this bound is rather loose, except in some very particular cases, as will be commented in Section 4.2.

3 Cumulants

3.1 Definitions

Let $\boldsymbol{z}$ be a random variable of dimension K, with components z_i. Then one defines its moment and cumulant tensors of order d as

$$\begin{aligned} \mathcal{M}^{\boldsymbol{z}}_{i_1 i_2 \dots i_d} &= \mathrm{E}\{z_{i_1} z_{i_2} \dots z_{i_d}\} \\ \mathcal{C}^{\boldsymbol{z}}_{i_1 i_2 \dots i_d} &= \mathrm{Cum}\{z_{i_1}, z_{i_2}, \dots, z_{i_d}\} \end{aligned}$$

When the moment tensors of order less than or equal to d exist and are finite, the cumulant tensor of order d exists and is finite. Whereas moments are the coefficients of the expansion of the first characteristic function $\Phi^{\boldsymbol{z}}(\boldsymbol{u}) = \mathrm{E}\{exp(\jmath \boldsymbol{u}^{\mathrm{T}} \boldsymbol{z})\}$ about the origin, where the dotless $\jmath$ denotes $\sqrt{-1}$, cumulants are those of the second characteristic function, $\Psi^{\boldsymbol{z}}(\boldsymbol{u}) = \log(\Phi^{\boldsymbol{z}}(\boldsymbol{u}))$; for complex random variables, it suffices to consider the joint distribution of their real and imaginary parts. Moments and cumulants enjoy the multilinearity property (2.1) and may be considered as tensors [59].

One important property of cumulant tensors is the following: if at least two variables, or groups of variables, among $\{z_1, \dots z_K\}$ are statistically independent, then all cumulants involving these variables are null. For instance, if all the z_i are mutually independent, then $\mathcal{C}^{\boldsymbol{z}}_{ij\dots\ell} = \delta(i, j, \dots \ell)\, \mathcal{C}^{\boldsymbol{z}}_{ii\dots i}$ [48], where the Kronecker δ is null unless *all* its arguments are equal. This property is not enjoyed by moments, hence the interest in cumulants.

The reverse is not true. In fact, unless the random variable $\boldsymbol{z}$ is Gaussian, an infinite number of cumulants must vanish in order to ensure their strict sense independence. Therefore, when a cumulant tensor $\mathcal{C}^{\boldsymbol{z}}$ of order d is diagonal, we shall say that the random variables z_i are *independent at order* d.

Gaussian variables play a particular role, since they are the only random variables that have a finite set of non-zero cumulants [47, 34]. The cumulants of order 1 and 2 are better known under the names of statistical *mean* and *covariance*. To illustrate this property, I looked for a long time for a simple non-Gaussian random variable having null cumulants of order 3 and 4. Take a random variable with values in the complex plane. If its distribution is 0 with probability one-half, and uniformly distributed on the unit circle with probability one-half, then its mean $\mathrm{E}\{z\}$ is zero, its variance $\mathrm{E}\{zz^*\} = 1$, its second-order non-circular cumulant $\mathrm{E}\{z^2\} = 0$, its two marginal cumulants of order 3, $\mathrm{E}\{z^3\}$ and $\mathrm{E}\{z^2z^*\}$, are zero, as well as its three marginal cumulants of order 4, $\mathrm{Cum}\{z, z, z, z\}$, $\mathrm{Cum}\{z, z, z, z^*\}$, $\mathrm{Cum}\{z, z, z^*, z^*\}$. Yet, this variable is obviously not Gaussian. This is a very striking example, that one encounters in secondary radar applications.

3.2 Blind source separation

Consider the linear statistical model

$$\boldsymbol{y} = \boldsymbol{A}\,\boldsymbol{x} + \boldsymbol{v} \tag{3.1}$$

where $\boldsymbol{y}$ is an observed random variable of dimension K, $\boldsymbol{x}$ is a random vector of dimension P referred to as the *source vector*, $\boldsymbol{A}$ is a mixing matrix, and $\boldsymbol{v}$ stands for background noise, possible interferers, and measurement errors, independent of $\boldsymbol{x}$.

The BSS problem consists of estimating the mixing matrix, $\boldsymbol{A}$, and possibly the corresponding estimates of $\boldsymbol{x}$, solely from measurements of $\boldsymbol{y}$. In the classical BSS framework, the components x_i of $\boldsymbol{x}$ are assumed to be statistically independent (generally not in the strict sense because a weaker independence is sufficient) [16]. In some cases, however, sources x_i may be correlated [70, 39, 21], as we shall subsequently see, and this is not necessarily an obstacle to their separation.

To fix the ideas and simplify the notation, assume sources x_i are independent at order 4. Then, from the properties of cumulants we just described, we have

$$\mathcal{C}^{\boldsymbol{y}}_{ijk\ell} = \sum_{p=1}^{P} A_{ip} A_{jp} A_{kp} A_{\ell p} \mathcal{C}^{\boldsymbol{x}}_{pppp} \tag{3.2}$$

up to an additive noise term, $\mathcal{C}^{\boldsymbol{v}}$. From measurements of $\boldsymbol{y}$, it is possible to estimate the cumulant tensor $\mathcal{C}^{\boldsymbol{y}}$. Estimating $\boldsymbol{A}$ then amounts to finding the decomposition (3.2). In practice, because of the noise $\boldsymbol{v}$, this decomposition is not exact. In addition, since the only property utilized is the source independence at a given order, matrix $\boldsymbol{A}$ can only be identified up to a multiplicative factor $\boldsymbol{P\Lambda}$, where $\boldsymbol{P}$ is a permutation and $\boldsymbol{\Lambda}$ is diagonal invertible; see identifiability issues in [16, 5].

Blind Deconvolution is related to the above BSS modeling in two respects. First, a convolution with a finite impulse response can always be written as the product with a Töplitz matrix, which means that the modelling (3.1) still holds valid, provided matrix $\boldsymbol{A}$ is subject to the Töplitz structure [16, 40]. Second, if the source process is linear, then extracting the sources is equivalent to computing the linear prediction residue [17]. Then, the problem reduces to an unstructured static separation as in (3.1).

4 Array decompositions

4.1 Diagonalization by change of coordinates

Preprocessing for square mixtures. If the mixing matrix $\boldsymbol{A}$ is square and invertible, which means that the number of sources, P, is equal to the observation dimension, K, then the BSS problem may be seen as a bijective congruent transformation (ICA).

Denote $\boldsymbol{R}_y$ the covariance matrix of the observation. The goal is to find an estimate $\boldsymbol{z}$ of $\boldsymbol{x}$ such that its components z_i are statistically independent. The first idea is thus to build a vector $\tilde{\boldsymbol{y}} = \boldsymbol{T}\,\boldsymbol{y}$ that has a diagonal covariance, yielding decorrelated components. This can be easily done by searching for a (non-unique) square root factor of $\boldsymbol{R}_y$; it can be obtained by a Cholesky factorization or by an eigen-value decomposition (EVD) of $\boldsymbol{R}_y$. We then define $\boldsymbol{T}$ as the inverse of this factor, so that $\boldsymbol{T}\boldsymbol{R}_y\boldsymbol{T}^{\mathrm{T}} = \boldsymbol{I}$.

With this preprocessing $\boldsymbol{T}$, the obtained random variable $\tilde{\boldsymbol{y}}$ has a covariance equal to identity. We say that this variable is *standardized.*

Now, it may be more appropriate, when the noise covariance $\boldsymbol{R}_v$ (or conversely the signal covariance $\boldsymbol{R}_s = \boldsymbol{R}_y - \boldsymbol{R}_v$) is known, to build $\boldsymbol{T}$ as the inverse of a square root of the signal covariance: $\boldsymbol{T}\boldsymbol{R}_s\boldsymbol{T}^{\mathrm{T}} = \boldsymbol{I}$. In fact, this yields an unbiased solution in the presence of noise. Unfortunately, neither $\boldsymbol{R}_v$ nor $\boldsymbol{R}_s$ are known in general, hence the former procedure based on $\boldsymbol{R}_y$.

Preprocessing for rectangular mixtures. In practice, one can always reduce the problem to the latter when the number of sources, P, is smaller than the observation dimension, K, in the absence of noise, or when the noise covariance is known. This is now explained below.

If noise is present, denote $\boldsymbol{T}_v$ the inverse of a square root of $\boldsymbol{R}_v$, such that we have $\boldsymbol{T}_v\boldsymbol{R}_v\boldsymbol{T}_v{}^{\mathrm{T}} = \sigma_v\boldsymbol{I}$; if noise is absent, set $\boldsymbol{T}_v = \boldsymbol{I}$ and $\sigma_v = 0$. Now consider the matrix $\boldsymbol{T}_v\boldsymbol{R}_y\boldsymbol{T}_v = \boldsymbol{T}_v\boldsymbol{R}_s\boldsymbol{T}_v + \sigma\boldsymbol{I}$. Its EVD allows one to detect the number of non-zero eigenvalues in $\boldsymbol{R}_s$ [2], equal to P by definition, as well as to estimate the source space spanned by the associated eigenvectors: $\boldsymbol{T}_v\boldsymbol{R}_y\boldsymbol{T}_v = \boldsymbol{U}\boldsymbol{\Sigma}\boldsymbol{U}^{\mathrm{T}} + \sigma\boldsymbol{I}$. The matrix $\boldsymbol{U}$ is here of dimension $K \times P$ and of full rank. The preprocessing defined as $\boldsymbol{T} = \boldsymbol{U}^{\mathrm{T}}\boldsymbol{T}_v$ eventually yields a P-dimensional standardized vector $\tilde{\boldsymbol{y}} = \boldsymbol{T}\boldsymbol{y}$ whose noiseless part has a unit covariance, as in the previous paragraph.

Lastly, if the mixture is rectangular, but with more sources than sensors, that is, $P > K$, the mixture cannot be linearly inverted. Such mixtures are referred to as *underdetermined* or *over-complete*, as already pointed out in the bibliographical survey, and their identification will be addressed separately in Section 4.2. In such a case, the preprocessing is unuseful, and not recommended.

Orthogonal change of coordinates. In the preprocessing, we have done only part of the job. In fact, we have constructed a matrix $\boldsymbol{T}$ such that ideally $\boldsymbol{T}\boldsymbol{A}\boldsymbol{A}^{\mathrm{T}}\boldsymbol{T}^{\mathrm{T}} = \boldsymbol{I}$, but this only implies that $\boldsymbol{T}\boldsymbol{A} = \boldsymbol{Q}^{\mathrm{T}}$, for some $P \times P$ orthogonal matrix $\boldsymbol{Q}$. This $\boldsymbol{Q}$ factor still remains undetermined. It is thus necessary to resort statistics of order higher than 2, namely 3 or 4, unless other hypotheses can be assumed. The choice between these two possibilities depends on the conditioning of the problem, directly linked to the value of the diagonal tensor $\mathcal{C}^{\boldsymbol{x}}$. At order 3, this tensor vanishes for all symmetrically distributed sources, which strongly limits its use. At order 4, this tensor is generally non-zero, except in some exceptional pathological cases, as that mentioned in Section 3.1.

In order to find $\boldsymbol{Q}$, one can attempt to diagonalize (approximately) the cumulant tensor of $\boldsymbol{z} = \boldsymbol{Q}\,\tilde{\boldsymbol{y}}$, $\mathcal{C}^{\boldsymbol{z}}_{ijk\ell} = \sum_{pqrs} Q_{ip}Q_{jq}Q_{kr}Q_{\ell s}\mathcal{C}^{\tilde{\boldsymbol{y}}}_{pqrs}$. The random variable $\boldsymbol{z}$ is eventually an estimate of the source vector $\boldsymbol{x}$; in the absence of noise, we have $\boldsymbol{z} = \boldsymbol{P}\boldsymbol{\Lambda}\boldsymbol{x}$. Because $\boldsymbol{Q}$ is orthogonal, minimizing the non-diagonal entries is equivalent to maximizing the diagonal ones [15], so that $\boldsymbol{Q}$ can be determined by

$$\boldsymbol{Q} = \operatorname{Arg}\operatorname*{Max}_{\boldsymbol{Q}} \Upsilon_{\alpha,4}\,;\; \Upsilon_{\alpha,4} = \sum_i |\mathcal{C}^{\boldsymbol{z}}_{iiii}|^{\alpha} \tag{4.1}$$

where $\alpha \geq 1$. Several optimization criteria of this type, called *contrasts*, have been proposed [16, 60, 26, 8, 20] and are justified by Information Theory arguments. Contrary to the matrix case [38], it is generally impossible to exactly null the non-diagonal entries of a symmetric tensor of order higher than 2, by just rotating the coordinate axes. In other words, the class of decompositions presented in this section lead to rank-K *approximations* of K-dimensional symmetric tensors. More will be said in the next section. Numerical ICA algorithms are surveyed in Section 5.

Table 1 Generic rank ω of symmetric tensors as a function of the dimension K and the order d

ω	K	**2**	**3**	**4**	**5**	**6**	**7**	**8**
d	**3**	2	4	5	8	10	12	15
	4	3	6	10	15	22	30	42

Table 2 Generic dimension D of the manifold of solutions

D	K	**2**	**3**	**4**	**5**	**6**	**7**	**8**
d	**3**	0	2	0	5	4	0	0
	4	1	3	5	5	6	0	6

4.2 Decomposition into a sum of rank-1 arrays

When the number of sources, P, is strictly larger than the observation dimension K, the previous approach does not apply. In fact, the matrix $\boldsymbol{A}$ now has fewer rows than columns, and the noiseless relation $\boldsymbol{y} = \boldsymbol{A}\,\boldsymbol{x}$ cannot be linearly inverted. In other words, $\boldsymbol{A}$ must be identified without attempting to extract the sources x_p. A symmetric tensor of order d can be expressed via a Canonical Decomposition (CAND) of the form

$$\mathcal{C}^{\boldsymbol{y}} = \sum_{p=1}^{\omega} \gamma(p)\, \boldsymbol{a}(p) \circ \boldsymbol{a}(p) \circ \boldsymbol{a}(p) \circ \boldsymbol{a}(p). \tag{4.2}$$

The number of terms, ω, reaches a minimum when it equals the *tensor rank*. This CAND decomposition allows the identification of matrix $\boldsymbol{A}$ if (i) it is unique up to $\boldsymbol{\Lambda P}$-indeterminations, and (ii) the tensor rank ω is larger than or equal to the number of sources, P.

Generic rank. We report in the tables below the generic value of the *tensor rank* as a function of the dimension K and the order d [24]. We also report the dimension D of the manifold of solutions; when it is zero, it means that there are a finite number of CAND (at most d^S), and there is a chance of identifying matrix $\boldsymbol{A}$ this way.

> *Example.* For matrices ($d = 2$), it is known that a quadratic form cannot be uniquely decomposed into a sum of squares. The manifold of solutions is of dimension $D = K(K-1)/2$).

The first striking fact that appears in Table 1 is that *the rank can exceed the dimension*, which is not true for matrices. For instance, it can be seen that $P = 5$ sources can be identified in dimension $K = 4$ with a third order cumulant tensor, whereas this number increases to $P = 10$ with a 4th order tensor.

One can also deduce from Table 2 that third order tensors have a finite number of CAND for even dimensions. For fourth order tensors, this is satisfied for dimension seven, but not for lower ones. This is unfortunate, for fourth order cumulants are very often better conditioned than third order ones. Furthermore,

Table 3 Equivalence classes of binary cubics: orbits under the action of $\mathcal{GI}$, the group of invertible two-dimensional changes of coordinates

$\mathcal{GI}$–orbit	$\omega(p)$
x^3	1
$x^2y + xy^2$	2 (**generic**)
x^2y	3

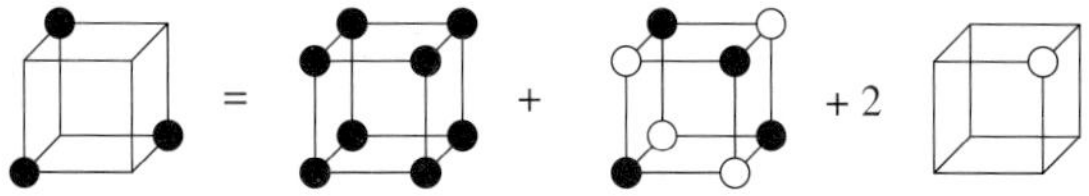

FIG. 2. Explicit decomposition of the non-generic example of binary cubic of maximal rank. Black bullets represent $+1$ and white bullets -1

most of the proofs leading to these tables are not constructive. The only known constructive result is given by the Sylvester theorem (Section 5.3).

Non-generic rank. In addition, these results are only valid in generic cases. And it turns out that, contrary to matrices (that is, second-order tensors), *the generic rank is not always maximal.* In other words, the rank can exceed its generic value. Unfortunately, the maximal achievable rank is not known for all pairs (P, K). But we can still illustrate this odd fact with particular values.

Example. For instance, for $K = 2$ and $d = 3$, the maximal rank is 3. The symmetric tensors having rank 3 are associated with polynomials in the orbit of x^2y. The tensor associated with the latter homogeneous polynomial is represented in Fig. 1, where only three entries are equal to 1, the others being null. As reported in Table 3 there is a single class associated with every value of the tensor rank. Now to make it more explicit, the polynomial x^2y can be written as

$$6\,x^2y = (x+y)^3 + (-x+y)^3 - 2y^3.$$

This relation can be rewritten in tensor form as

$$\boldsymbol{T} = \begin{pmatrix} 1 \\ 1 \end{pmatrix}^{\circ 3} + \begin{pmatrix} -1 \\ 1 \end{pmatrix}^{\circ 3} - 2\begin{pmatrix} 0 \\ 1 \end{pmatrix}^{\circ 3}$$

which is an explicit irreducible CAND. This decomposition is depicted in Fig. 2. Also note that, in this case, the Reznick bound (2.4) is reached: $\omega = \binom{3}{2} = 3$.

Example. Now take $K = 3$ and $d = 3$. We are thus handling $3 \times 3 \times 3$ symmetric tensors, or equivalently, ternary cubics. The generic rank is 4, but the maximal rank is 5, according to Table 4. The class of maximal rank is unique, and a representative is depicted in Fig. 3; the

Table 4 Equivalence classes for ternary cubics: orbits under the action of $\mathcal{GI}$, the group of invertible three-dimensional changes of coordinates

$\mathcal{GI}$-orbit	$\omega(p)$
x^3	1
$x^2y + xy^2$	2
x^2y	3
$x^3 + 3\,y^2z$	4
$x^3 + y^3 + 6\,xyz$	4
$x^3 + 6\,xyz$	4
$a\,(x^3 + y^3 + z^3) + 6b\,xyz$	4 (generic)
$x^2y + xz^2$	5

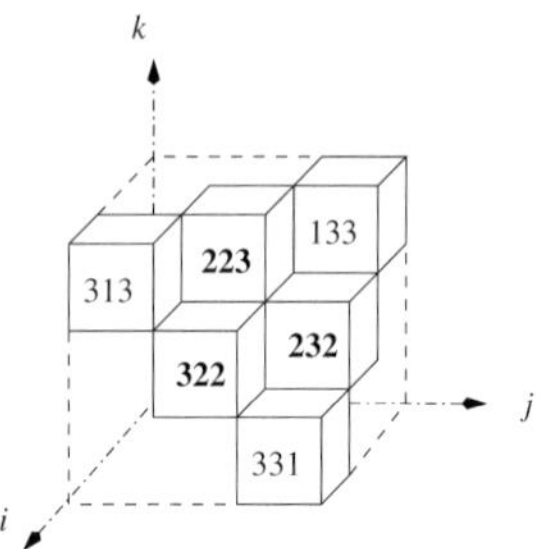

FIG. 3. Non-generic example of ternary cubic, proved to be of maximal rank: position of non-zero entries.

6 non-zero entries are all equal. Note that other non-generic classes occur with also a rank of 4, as pointed out in Table 4.

Example. Finally, consider ternary quartics, that is, $(K, d) = (3, 4)$. In this case, the number of free parameters in the tensor is $S = \binom{6}{4} = 15$. The number of free parameters in CAND exceeds 15 as soon as $\omega \geq 5$. So we could hope that we are lucky, because the number of free parameters is the same on both sides of CAND. Unfortunately, this is not the case, and Clebsch showed that the maximal achievable rank was 6 [33]. Of course, this maximal rank is not generic, as reported in Table 1.

4.3 Rank-1 approximation

Approximating a tensor by another of rank 1 has at least two applications in the present context. The first one is encountered when $P \leq K$ and when the source extraction is performed one source at a time in model (3.1), contrary to Section 4.1; this is referred to as a *deflation* procedure.

The maximization of the contrast (4.1) then reduces to that of a single output standardized cumulant (here the *kurtosis*), because a single unit-norm vector is sought, instead of a whole orthogonal matrix:

$$\boldsymbol{w} = \mathrm{Arg} \max_{||\boldsymbol{w}||=1} \sum_{ijk\ell} w_i w_j w_k w_\ell \, \mathcal{C}^{\boldsymbol{y}}_{ijk\ell}. \tag{4.3}$$

Yet, it has been shown [28, 18, 49] that this maximization problem is equivalent to minimizing $||\mathcal{C}^{\boldsymbol{y}} - \sigma\, \boldsymbol{w} \circ \boldsymbol{w} \circ \boldsymbol{w} \circ \boldsymbol{w}||$, which is simply finding the best rank-1 approximate of tensor $\mathcal{C}^{\boldsymbol{y}}$.

The second application is found in analytical BSS when sources are discrete [39] or of constant modulus [70]. In this problem, we have to solve a system of N equations of the form $(\boldsymbol{f}^{\mathsf{T}} \boldsymbol{y}_n)^d = 1$, $1 \le n \le N$. This is equivalent to solving a larger linear system $\boldsymbol{Y}\, \boldsymbol{f}^{\oslash d} = \mathbf{1}$, under the constraint of $\mathbf{Unvecs}\{\boldsymbol{f}^{\oslash d}\}$ being a rank-1 tensor. Denote $\{\boldsymbol{u}_q^{\oslash d}, 1 \le q \le \bar{P}\}$ a basis of $\mathrm{Ker}\{\boldsymbol{Y}\}$. The solution to this system takes the form

$$\boldsymbol{f}^{\oslash d} = \boldsymbol{f}^{\oslash d}_{\min} + \sum_{p=1}^{\bar{P}} \lambda_p \, \boldsymbol{u}_p^{\oslash d}$$

where $\boldsymbol{f}^{\oslash d}_{\min}$ is the minimum norm solution. Unfolding these vectors in tensor form leads to the relation

$$\boldsymbol{F} = \boldsymbol{F}_{\min} + \sum_{p=1}^{\bar{P}} \lambda_p \, \boldsymbol{U}_p. \tag{4.4}$$

This problem can be shown to be related to the rank-1 combination problem that we describe below.

4.4 Rank-1 combination

The rank-1 combination problem consists of finding the numbers λ_p so that, given matrices $\boldsymbol{U}_p$, matrix $\sum_p \lambda_p \boldsymbol{U}_p$ has a rank of 1. Up to now, this problem has spawned solutions that are not entirely satisfactory. As a consequence, so are the solutions to (4.4).

Incidentally, we can restate the Joint Approximate Diagonalization (JAD) problem addressed in [9] for the BSS into rank-1 combinations.

The Joint Approximate Diagonalization of $\bar{P}$ matrices $\boldsymbol{N}_p$ consists of finding a square matrix $\boldsymbol{T}$ such that $\boldsymbol{N}_p \approx \boldsymbol{T} \boldsymbol{\Lambda}_p \boldsymbol{T}^{\mathsf{T}}$, for all p, where $\boldsymbol{\Lambda}_p$ are diagonal matrices. From a property recalled in Section 2.2, this relation can be rewritten in vector form as $\mathbf{vecs}\{\boldsymbol{N}_p\} \stackrel{\mathrm{def}}{=} \boldsymbol{n}_p \approx \sum_i \lambda_{pi} \boldsymbol{t}_i^{\oslash 2}$, $\boldsymbol{t}_i$ denoting the ith column of $\boldsymbol{T}$. If the matrix $[\lambda_{pi}]$ is full rank and has more columns than rows, then there exists a matrix $\boldsymbol{B}$ such that $\boldsymbol{t}_j^{\oslash 2} \approx \sum_p B_{pj} \boldsymbol{n}_p$. Thus, given matrices $\boldsymbol{N}_p$, the problem is to find for every j, scalar coefficients β_p such that $\sum_p \beta_p \boldsymbol{N}_p$ is a rank-1 matrix, and hence the link with the rank-1 combination problem.

However, the two problems are not equivalent, for matrix $\boldsymbol{B}$ is not necessarily square.

5 Numerical algorithms

5.1 Contrast maximization

The ICA diagonalization of Section 4.1 (as well as the JAD briefly mentioned in Section 4.4) can be solved entirely analytically in dimension $K = 2$, in a number of instances. In order to exploit this property, Comon [15] proposed a sweeping of the pairs of indices, in a similar manner as in the Jacobi diagonalization algorithm for Hermitian matrices [38]. This idea has been later applied to JAD by Cardoso [9]. To see this in more detail, consider the Givens rotation

$$\boldsymbol{Q} = \begin{pmatrix} \cos\phi & \sin\phi\, \exp(\jmath\theta) \\ -\sin\phi\, \exp(-\jmath\theta) & \cos\phi \end{pmatrix}$$

where the angle ϕ is imposed to lie in the interval $(-\pi/2, \pi/2]$, because of inherent $\boldsymbol{\Lambda P}$-indeterminacies. Thus this matrix is entirely defined by the vector $\boldsymbol{u} = [\cos 2\phi,\ \sin 2\phi\, \cos\theta,\ \sin 2\phi\, \sin\theta]$. Now, as in (4.1), define the contrast $\Upsilon_{\alpha,d}$ as the sum of the d-th order tensor diagonal entries raised to the power α. Then it can be shown that $\Upsilon_{1,3}$ and $\Upsilon_{1,4}$ are real quadratic forms in $\boldsymbol{u}$, and can thus be easily maximized with respect to $\boldsymbol{u}$, and hence to (θ, ϕ) (by convention, if $\alpha = 1$, the absolute value is dropped in (4.1)). On the other hand, this holds true for $\Upsilon_{2,3}$ but not for $\Upsilon_{2,4}$, which can be shown to be a quartic [16, 20]. Nevertheless, polynomials of degree 4 can still be rooted analytically.

The procedure originally proposed by Comon [14] consisting of sweeping all the pairs, like in some numerical algorithms dedicated to matrices, has never been proved to always lead to one of the $\boldsymbol{\Lambda P}$-equivalent absolute maxima, even if this is always observed in practice. Counter-examples have never been found either. So we consider this convergence issue as an open problem, belonging to the general class of optimization problems over multiplicative groups. However, some elements of convergence are now reported below.

Convergence. For compactness, denote $\boldsymbol{G}$ the cumulant tensor of the standardized observation, $\bar{\boldsymbol{y}}$, which has been denoted $\mathcal{C}^{\bar{\boldsymbol{y}}}$ up to now. Also denote $\boldsymbol{Z} = \mathcal{C}^{\boldsymbol{z}}$ the cumulant tensor obtained after an orthogonal transformation $\boldsymbol{Q}$. According to the multi-linearity property, we have that

$$Z_{pq\ldots r} = \sum_{ij\ldots\ell} Q_{pi} Q_{qj} \ldots Q_{r\ell}\, G_{ij\ldots\ell}. \tag{5.1}$$

Consider first the matrix case (order 2) in order to fix the ideas. The contrast (4.1) can then be written as

$$\Upsilon_{2,2} = \sum_{p} |Z_{pp}|^2. \tag{5.2}$$

Because $\boldsymbol{Q}$ is orthogonal, its differential can be written as

$$d\boldsymbol{Q} = d\boldsymbol{S}\, \boldsymbol{Q} \tag{5.3}$$

where matrix $\boldsymbol{S}$ is skew-symmetric. This yields the relation characterizing stationary points, $\boldsymbol{Z}$: $\frac{1}{2}\,d\Upsilon_{2,2} = 2\sum_{p,t} Z_{pp}S_{pt}Z_{tp} = 0$. Yet, this is true for any skew-symmetric matrix $\boldsymbol{S}$, and hence for every skew-symmetric matrix having only two non-zero entries (one $+1$ and one -1); based on this argument, one concludes that

$$(Z_{qq} - Z_{rr})Z_{qr} = 0, \qquad \text{for } q \neq r. \tag{5.4}$$

Next, the local convexity can be examined with the help of the same tools, observing that

$$\frac{1}{4}d^2\Upsilon_{2,2} = 4Z_{qr}^2 - (Z_{qq} - Z_{rr})^2. \tag{5.5}$$

Thus, there are three kinds of stationary points: (i) those for which all diagonal entries are equal, which correspond to minima of $\Upsilon_{2,2}$, (ii) those for which all non-diagonal entries are null, which correspond to maxima, and (iii) saddle points, for which some diagonal entries are equal and some non-diagonal entries vanish. This result is well known, and proves that the only maxima are diagonal matrices, which can be deduced from each other by mere permutation within the diagonal.

Now let us develop the same calculations for tensors of order 3 and 4. Stationary values are given by the relations

$$\begin{aligned}
\frac{1}{2}\,d\Upsilon_{2,3} &= 3\sum_{p,t} Z_{ppp}\,dS_{pt}\,Z_{tpp} = 0,\\
\frac{1}{2}\,d\Upsilon_{2,4} &= 4\sum_{p,t} Z_{pppp}\,dS_{pt}\,Z_{tppp} = 0
\end{aligned}$$

or, on the basis of skew-symmetric matrices, for $q \neq r$:

$$Z_{qqq}Z_{qqr} - Z_{rrr}Z_{qrr} = 0, \tag{5.6}$$

$$Z_{qqqq}Z_{qqqr} - Z_{rrrr}Z_{qrrr} = 0, \tag{5.7}$$

whereas local convexity conditions are governed by [16]

$$\frac{1}{6}d^2\Upsilon_3 = 4Z_{qqr}^2 + 4Z_{qrr}^2 - (Z_{qqq} - Z_{qrr})^2 - (Z_{rrr} - Z_{qqr})^2 \tag{5.8}$$

$$\frac{1}{8}d^2\Upsilon_4 = \frac{9}{2}Z_{qqrr}^2 + 4Z_{qqqr}^2 + 4Z_{qrrr}^2 - (Z_{qqqq} - \frac{3}{2}Z_{qqrr})^2 - (Z_{rrrr} - \frac{3}{2}Z_{qqrr})^2. \tag{5.9}$$

The comparison of these results with (5.4) and (5.5) lead to two conclusions: (a) non-diagonal terms do not factorize any more in (5.6) and (5.7), so that stationary values are more difficult to characterize, and (b) diagonal tensors are still local maxima, but there are *a priori* others. This is another problem, linked to optimization in groups, that this author considers as open.

Sweeping strategies. We have presented several numerical algorithms aiming at separating $P = 2$ sources from $K = 2$ sensors in the presence of noise of unknown statistics. Inspired from the Jacobi cyclic-by rows sweeping strategy proposed for matrices, we can process all the $K(K-1)/2$ pairs one by one sequentially [14, 16]. However, as in the matrix case, the noise part (constituted by the actual background noise and all the other $K-2$ sources) changes at every step, so that a single sweeping is not sufficient. In practice, an order of $\sqrt{K}$ sweeps have been shown to be sufficient.

Other strategies have been analysed, and consist of processing first the pair of sensors that yields the maximal increase in the contrast criterion. This strategy has also been implemented successfully, but is not always numerically efficient.

When processing one pair (i, j), one can either recompute all the entries of the cumulant tensor that have been affected (that is, those whose indices contain i or j), or compute the rotated data instead. The two possibilities do not have the same numerical complexity, and the best choice depends on the number of sensors, K, and on the number of samples, N.

5.2 Parafac algorithm

In [54], SVD-based algorithms are proposed to compute CAND of third-order tensors in larger dimensions. However, these algorithms, called PARAFAC, need the number of sources, P, to be smaller than or equal to $\frac{3}{2}K-1$, in the symmetric case we are interested in. See also [50, 4] for more details. In view of Table 1 reported above, this value of P is strictly smaller than the generic rank, ω, except for $(d, K) = (3, 2)$ or $(d, K) = (3, 4)$. As a consequence, PARAFAC algorithms can only *approximate* d-way arrays, in general.

In the unsymmetric problem, the goal is to find three matrices, $\boldsymbol{A}$, $\boldsymbol{B}$, and $\boldsymbol{C}$, such that $G_{ijk} = \sum_p A_{ip}B_{jp}C_{kp}$. One possible numerical algorithm is based on alternating least squares, as explained below for 3-way arrays [10]:

- Start with $(\boldsymbol{A}(0), \boldsymbol{B}(0), \boldsymbol{C}(0))$
- Define matrices $\boldsymbol{G}^{(1)}$, $\boldsymbol{G}^{(2)}$, $\boldsymbol{G}^{(3)}$:
  $\boldsymbol{G}_{ijk} = \boldsymbol{G}^{(1)}_{ip} = \boldsymbol{G}^{(2)}_{jq} = \boldsymbol{G}^{(3)}_{kr}$; $p = (jk)$, $q = (ik)$, $r = (ij)$
- Estimate stage $t+1$ from stage t by pseudo-inversion:
  - ∗ Update mode 1:
    $\boldsymbol{A}(t+1) = \boldsymbol{G}^{(1)}\,[\boldsymbol{B}(t)^{\mathrm{T}}\,\boldsymbol{C}(t)^{\mathrm{T}}]^{-}$
  - ∗ Update mode 2:
    $\boldsymbol{B}(t+1) = \boldsymbol{G}^{(2)}\,[\boldsymbol{A}(t+1)^{\mathrm{T}}\,\boldsymbol{C}(t)^{\mathrm{T}}]^{-}$
  - ∗ Update mode 3:
    $\boldsymbol{C}(t+1) = \boldsymbol{G}^{(3)}\,[\boldsymbol{A}(t+1)^{\mathrm{T}}\,\boldsymbol{B}(t+1)^{\mathrm{T}}]^{-}$

where $\boldsymbol{M}^{-}$ denotes the Moore–Penrose pseudo-inverse of $\boldsymbol{M}$. See also [4, 26, 50] for more details on PARAFAC algorithms.

5.3 Sylvester theorem

As already pointed out earlier, a rank-1 tensor is associated with a linear form raised to the dth power. In terms of polynomials, the CAND decomposition can thus be rephrased: how can one decompose a quantic into a sum of dth powers of linear forms [24]? This is this topic that addresses this theorem, restricted to the binary case, however (that is, two variables).

Theorem 5.1 *A binary quantic* $p(x,y) = \sum_{i=0}^{d} \gamma_i\, c(i)\, x^i\, y^{d-i}$ *can be written as a sum of dth powers of* ω *distinct linear forms:*

$$p(x,y) = \sum_{j=1}^{\omega} \lambda_j\, (\alpha_j\, x + \beta_j\, y)^d,$$

if and only if (i) there exists a vector $\boldsymbol{g}$ *of dimension* $\omega+1$, *with components* g_ℓ, *such that*

$$\begin{bmatrix} \gamma_0 & \gamma_1 & \cdots & \gamma_\omega \\ \vdots & & & \vdots \\ \gamma_{d-\omega} & \cdots & \gamma_{d-1} & \gamma_d \end{bmatrix} \boldsymbol{g} = \boldsymbol{0}. \tag{5.10}$$

and (ii) the polynomial $q(x,y) \stackrel{\text{def}}{=} \sum_{\ell=0}^{\omega} g_\ell\, x^\ell\, y^{\omega-\ell}$ *admits* ω *distinct roots.*

Sylvester's theorem not only proves the existence of the ω forms (second column in the tables), but also gives a means to compute them [18, 24]. For odd values of d, we have thus a generic rank of $\omega = \frac{d+1}{2}$, whereas for even values of d, $\omega = \frac{d}{2}+1$. So when d is odd, there is generically a unique vector $\boldsymbol{g}$ satisfying (5.10), but there are two of them when d is even. This theorem shows that in column $K=2$ of Table 2 we have $D=0$ when d is odd, and $D=1$ when d is even.

In [27], several extensions to this theorem are proposed in the complex case. The basic idea remains the same, but the result becomes more complicated.

The disappointing fact is that Sylvester's theorem cannot be extended to dimensions higher than two. In fact, a key step in the proof [24, 18] is that for any polynomial p of degree d, and any monomial m of degree $d-\omega$, there exists a polynomial q of degree ω such that qm is orthogonal to p. Equation (5.10) expresses that orthogonality in terms of polynomial coefficients. It is clear that this holds true only when $d \geq \omega$, which is unfortunately satisfied only in the binary case, according to Table 1. Possibilities of extension to more than two variables is discussed in [24].

Simultaneous CAND. Let us go back to Table 2. Among others, this table reports that there are infinitely many CAND for even orders, d. In order to fix this indeterminacy in the case $(d,K,P)=(4,2,3)$ (the manifold of solutions is of dimension one in that situation), it is proposed in [18] to simultaneously diagonalize a second cumulant tensor of order 4.

The help of virtual sources. In [18, 21] an algorithm dedicated to discrete sources is proposed, and performs both the identification of $\boldsymbol{A}$ and the extraction of sources x_i, in the case $(d, K, P) = (2, 2, 3)$.

In a few words, assume three sources x_i are mixed and received on two sensors, and assume these sources are all distributed in $\{-1, +1\}$ (they are called BPSK in digital communications). One can prove, if sources x_i are statistically independent, that the 'virtual' source $x_1 x_2 x_3$ is also BPSK-distributed, but obviously statistically dependent of the three former ones. However, one can still prove that all its fourth-order pairwise cross-cumulants vanish. Yet, only *pairwise* cumulants are utilized in the sweeping strategies maximizing contrasts such as $\Upsilon_{2,4}$ in (4.1). As a consequence, viewed by the algorithm, sources are independent; one can thus build from $\boldsymbol{y}^{\mathrm{T}} = [y_1, y_2]$ virtual measurements y_1^3, $y_1^2 y_2$, $y_1 y_2^2$, and y_2^3, that can be modelled as linear mixtures of 4th order pairwise independent unknown sources. This allows the separation of the four sources (three actual and one virtual) from six sensors (two actual and four virtual).

5.4 Rank-one approximation

The rank-1 approximation problem (Section 4.3) has been partly solved by algorithms inspired from the matrix *power method* and devised for arrays of higher orders [28, 26, 49].

Criteria. Given tensor $\mathcal{C}^{\boldsymbol{y}}$, the goal is to find a vector $\boldsymbol{w}$ minimizing

$$\Omega_o = ||\mathcal{C}^{\boldsymbol{y}} - \sigma\, \boldsymbol{w} \circ \boldsymbol{w} \circ \boldsymbol{w} \circ \boldsymbol{w}|| \tag{5.11}$$

for some scalar number σ. One can prove that minimizing (5.11) is equivalent to maximizing [18]:

$$\Omega_d = ||\mathcal{C}^{\boldsymbol{y}} \bullet \boldsymbol{w} \bullet \boldsymbol{w} \bullet \cdots \bullet \boldsymbol{w}|| \tag{5.12}$$

or to minimizing

$$\Omega_{d-1} = ||\mathcal{C}^{\boldsymbol{y}} \bullet \boldsymbol{w} \bullet \cdots \bullet \boldsymbol{w} - \lambda\, \boldsymbol{w}||. \tag{5.13}$$

However, the other criteria Ω_r, $0 < r < d-1$, are generally not equivalent.

Stationary uplets. $(\boldsymbol{v}, \lambda)$ of Ω_o, Ω_{d-1} or Ω_d are the same and satisfy

$$\mathcal{C}^{\boldsymbol{y}} \bullet \underbrace{\boldsymbol{v} \bullet \cdots \bullet \boldsymbol{v}}_{d-1 \text{ times}} = \lambda\, \boldsymbol{v}$$

this suggests a Rayleigh-like iteration, that we can call the *tensor Rayleigh symmetric iteration*:

$$\begin{aligned} \boldsymbol{w} &\leftarrow \mathcal{C}^{\boldsymbol{y}} \bullet \underbrace{\boldsymbol{w} \bullet \cdots \bullet \boldsymbol{w}}_{d-1 \text{ times}} \\ \boldsymbol{w} &\leftarrow \boldsymbol{w}/||\boldsymbol{w}||. \end{aligned}$$

In [28], it is suggested to run a non-symmetric iteration, and to initialize the algorithm with the HOSVD.

The rank-1 combination problem. (Section 4.4) has been solved in a suboptimal way up to now in [70, 39] by solving a large unconstrained linear system, and trying to restore the structure afterwards. The optimal one-stage solving still remains to be devised.

6 Concluding remarks

In this chapter, we have partly surveyed the tools dedicated to tensor decompositions, mainly through the problem of source separation. Thus, this presentation has been restrictive, but hopefully still informative.

Many other source separation algorithms do not resort to tensor tools, and have not been reported here. It is worth noting that some of them do not need the sources to be statistically independent, so that the output cumulant tensor is not even aimed at being diagonal. Instead, other properties of the sources can be exploited, such as their discrete character, or their constant modulus [70, 66, 39]. When more sources than sensors are present, general results state that it is sometimes possible to identify the mixture, but source extraction requires more knowledge about the sources (*e.g.*, their distribution). These issues have been tackled herein. Let us now turn to research perspectives.

In the area of source separation, current hot research topics include (i) blind identification of under-determined mixtures, (ii) blind equalization of convolutive mixtures, (iii) the theoretical proof of convergence of pair-sweeping algorithms, and, in the context of telecommunications, (iv) handling properly carrier residuals when present in the measurements. In all cases, analytical block-algorithms are suitable when computer power is available and when the stationarity duration is short.

As far as tensors are concerned, open research directions include (i) the determination of the maximal achievable rank for arbitrary order and dimensions, (ii) the actual calculation of general Canonical Decompositions for $K > 2$, and (iii) efficient numerical algorithms for computing an approximate of given rank.

Bibliography

[1] Bergman, G.M. Ranks of tensors and change of base field. *J. Algebra*, **11**, 613–621, 1969.

[2] Bienvenu, G. and Kopp, L. Optimality of high-resolution array processing using the eigensystem approach. *IEEE Trans. ASSP*, **31**, 1235–1248, 1983.

[3] Brewer, J.W. Kronecker products and matrix calculus in system theory. *IEEE Trans. on Circuits and Systems*, **25**, 114–122, 1978.

[4] Bro, R. Parafac, tutorial and applications. *Chemom. Intell. Lab. Syst.*, **38**, 149–171, 1997.

[5] Cao, X.R. and Liu, R.W. General approach to blind source separation. *IEEE Trans. Signal Process*, **44**, 562–570, 1996.

[6] Cardoso, J.F. Source separation using higher order moments. In *Proc. ICASSP* Glasgow, pp. 2109–2112, 1989.

[7] Cardoso, J.F. A tetradic decomposition of fourth order tensors: application to the source separation problem. In Elsevier, editor, *SVD and Signal Processing III*, pp. 375–382, Leuven, Belgium, Aug. 22–25 1995.

[8] Cardoso, J.F. High-order contrasts for independent component analysis. *Neural Comput.*, **11**, 157–192, 1999.

[9] Cardoso, J.F. and Souloumiac, A. Blind beamforming for non-Gaussian signals. *IEE Proceedings - Part F*, **140**, 362–370, 1993. Special issue on Applications of High-Order Statistics.

[10] Caroll, J.D. and Chang, J.J. Analysis of individual differences in multidimensional scaling via n-way generalization of Eckart–Young decomposition. *Psychometrika*, **35**, 283–319, 1970.

[11] Caroll, J.D., Pruzansky, S., and Kruskal, J.B. CANDELINC: A general approach to multidimensional analysis of many-way arrays with linear constraints on parameters. *Psychometrika*, **45**, 3–24, 1980.

[12] Castedo, L. and Macchi, O. Maximizing the information transfer for adaptive unsupervised source separation. In *Proc. SPAWC 97 Conf.*, (Paris, France, April 1997), pp. 65–68.

[13] Chaumette, E., Comon, P., and Muller, D. An ICA-based technique for radiating sources estimation; application to airport surveillance. *IEE Proceedings - Part F*, **140**, 395–401, 1993. Special issue on Applications of High-Order Statistics.

[14] Comon, P. Separation of stochastic processes. In *Proc. Workshop on Higher-Order Spectral Analysis*, pp. (Vail, CO, USA, June 28-30 1989.) 174–179, IEEE-ONR-NSF.

[15] Comon, P. Independent component analysis. In *Proc. Int. Signal Process Workshop on Higher-Order Statistics*, pp. 111–120, (Chamrousse, France, July 10-12 1991.) Republished in *Higher-Order Statistics*, (ed., J. L. Lacoume) Elsevier, 1992, pp 29–38.

[16] Comon, P. Independent Component Analysis, a new concept? *Signal Processing*, **36**, 287–314, 1994. Special issue on Higher-Order Statistics.

[17] Comon, P. Tensor diagonalization, a useful tool in signal processing. In (eds, M. Blanke and T. Soderstrom), *IFAC-SYSID, 10th IFAC Symposium on System Identification*, volume 1, pp. 77–82, (Copenhagen, Denmark, July 4–6 1994.) Invited session.

[18] Comon, P. Blind channel identification and extraction of more sources than sensors. In *SPIE Conference*, pp. 2–13, (San Diego, July 19-24 1998.) Keynote address.

[19] Comon, P. Block methods for channel identification and source separation. In *IEEE Symposium on Adaptive Systems for Signal Process Commun. Control*, pp. 87–92, (Lake Louise, Alberta, Canada, Oct. 1–4 2000.) Invited plenary.

[20] Comon, P. From source separation to blind equalization, contrast-based ap-

proaches. In *Int. Conf. on Image and Signal Processing (ICISP'01)*, (Agadir, Morocco, May 3–5, 2001.) Invited plenary.

[21] Comon, P. and Grellier, O. Non linear inversion of underdetermined mixtures. In *ICA99, IEEE Workshop on Indep. Comput. Anal. and Blind Source Separation*, pp. 461–465, (Aussois, France, Jan 11–15 1999.)

[22] Comon, P. and Grellier, O. Analytical blind identification of a SISO communication channel. In *IEEE Workshop on Statistical Signal and Array Processing*, pp. 206–210, (Pocono Manor, PA, USA, Aug. 14–16, 2000.)

[23] Comon, P. and Grellier, O., and Mourrain, B. Closed-form blind channel identification with MSK inputs. In *Asilomar Conference*, pp. 1569–1573, (Pacific Grove, CA, USA, November 1–4 1998.) Invited session.

[24] Comon, P. and Mourrain, B. Decomposition of quantics in sums of powers of linear forms. *Signal Processing*, **53**, 93–107, 1996. Special issue on High-Order Statistics.

[25] Coppi, R. and Bolasco, S. (eds). *Multi-Way Data Analysis*. Amsterdam, Elsevier, 1889.

[26] DeLathauwer, L. *Signal Processing based on Multilinear Algebra*. Doctorate, Katholieke Universiteit Leuven, September 1997.

[27] DeLathauwer, L., Comon, P., and DeMoor, B. ICA algorithms for 3 sources and 2 sensors. In *6th Signal Process. Workshop on Higher Order Statistics*, pp. 116–120, (Caesarea, Israel, June 14–16 1999.)

[28] DeLathauwer, L., Comon, P., *et al.* Higher-order power method, application in Independent Component Analysis. In *NOLTA Conference*, Vol. 1, pp. 91–96, (Las Vegas, 10–14 Dec 1995.)

[29] DeLathauwer, L. and DeMoor, B. From matrix to tensor: Multilinear algebra and signal processing. In *Mathematics in Signal Process., IMA Conf. Series*, (Warwick, Dec 17–19 1996.) Oxford, Oxford University Press.

[30] DeLathauwer, L., DeMoor, B., and Vandewalle, J. A singular value decomposition for higher-order tensors. In *2nd ATHOS Workshop*, (Sophia-Antipolis, France, Sept 20–21 1993.)

[31] DeLathauwer, L., DeMoor, B., and Vandewalle, J. Fetal electrocardiogram extraction by source subspace separation. In *IEEE-ATHOS Workshop on Higher-Order Statistics*, pp. 134–138, (Begur, Spain, 12–14 June 1995.)

[32] DeLathauwer, L., DeMoor, B., and Vandewalle, J. Fetal electrocardiogram extraction by blind source subspace separation. *IEEE Trans. Biomedical Engineering*, **47**, 567–572, 2000. Special topic section on Advances in Statistical Signal Processing for Biomedicine.

[33] Ehrenborg, R. and Rota, G.C. Apolarity and canonical forms for homogeneous polynomials. Eur. J. Combinatorics, 1993.

[34] Feller, W. *An Introduction to Probability Theory and its Applications*. New York, Wiley, 1968.

[35] Ferreol, A. and Chevalier, P. On the behavior of current second and higher order blind source separation methods for cyclostationary sources. *IEEE Trans. Signal Process.*, **48**, 1712–1725, 2000.

[36] Gassiat, E. and Gamboa, F. Source separation when the input sources are discrete or have constant modulus. *IEEE Trans. Signal Process.*, **45**, 3062–3072, 1997.

[37] Geladi, P. Analysis of multi-way data. *Chemom. Intell. Lab. Syst.*, **7**, 11–30, 1989.

[38] Golub, G.H. and Van Loan, C.F. *Matrix computations.* Balitmore, MD, Johns Hopkins University Press, 1989.

[39] Grellier, O. and Comon, P. Analytical blind discrete source separation. In *Eusipco*, (Tampere, Finland, 5–8 Sept. 2000.)

[40] Grigorascu, V.S. and Regalia, P.A. Tensor displacement structures and polyspectral matching. In T. Kailath and A. H. Sayed, (eds), *Fast Reliable Algorithms for Matrices with Structure*, chapter 9. SIAM Publ., Philadelphia, PA, 1998.

[41] Harshman, R.A. Determination and proof of minimum uniqueness conditions of PARAFAC1. *UCLA Working Papers in Phonetics*, **22**, 111–117, 1972.

[42] Haykin, S. *Unsupervised Adaptive Filtering*, Vol. 1. New York, Wiley, 2000. Series in Adaptive and Learning Systems for Communications, Signal Processing, and Control.

[43] Howell, T.D. Global properties of tensor rank. *Linear Algebra and Applications*, **22**, 9–23, 1978.

[44] Hyvärinen, A., Hoyer, P., and Oja, E. Denoising of non gaussian data by Independent Component Analysis and sparse coding. In *ICA99, IEEE Workshop on Ind. Comput. Anal. and Blind Source Separation*, pp. 485–489, (Aussois, France, Jan 11–15 1999.)

[45] Jones, M.C. and Sibson, R. What is projection pursuit. *J. Roy. Stat. Soc.*, **150**, 1–36, 1987.

[46] Jutten, C. and Hérault, J. Blind separation of sources, part I: An adaptive algorithm based on neuromimetic architecture. *Signal Processing*, **24**, 1–20, 1991.

[47] Kagan, A.M., Linnik, Y.V., and Rao, C.R. *Characterization Problems in Mathematical Statistics.* New York, Wiley, 1973.

[48] Kendall, M. and Stuart, A. *The Advanced Theory of Statistics, Distribution Theory*, Vol. 1. C. Griffin, 1977.

[49] Kofidis, E. and Regalia P.A. Tensor approximation and signal processing applications. In *AMS Conf. on Structured Matrices in Operator theory, Numerical Analysis, Control, Signal and Image Processing*, 2000.

[50] Kruskal, J.B. Three-way arrays: Rank and uniqueness of trilinear decom-

positions. *Lin. Alg. Appl.*, **18**, 95–138, 1977.

[51] Kung, J.P. and Rota, G.C. The invariant theory of binary forms. *Bull. Am. Math. Soc.*, **10**, 27–85, 1984.

[52] Lee, T.W., Lewicki, M.S., *et al.* Blind source separation of more sources than mixtures using overcomplete representations. *IEEE Signal Process. Lett.*, **6**, 87–90, 1999.

[53] Leshem, A. and van der Veen, A.J. Direction of arrival estimation for constant modulus signals. *IEEE Trans. Signal. Process.*, **47**, 3125–3129, 1999.

[54] Leurgans, S., Ross, R.T., and Abel, R.B. A decomposition for three-way arrays. *SIAM J. Matrix Anal. Appl.*, **14**, 1064–1083, 1993.

[55] Levin, J. Three-mode factor analysis. *Psych. Bull.*, **64**, 442–452, 1965.

[56] Lewicki, M. and Sejnowski, T.J. Learning non-linear overcomplete representations for efficient coding. In *Advances in Neural Information Processing Systems*, pp. 815–821, 1998.

[57] Liavas, A.P., Regalia, P.A., and Delmas, J.P. Robustness of least-squares and subspace methods with respect to effective channel undermodeling/overmodeling. *Trans. Signal Process.*, **47**, 1636–1645, 1999.

[58] Macchi, O. and Moreau, E. Self-adaptive source separation. part I: Convergence analysis of a direct linear network controlled by the Herault–Jutten algorithm. *IEEE Trans. Signal Process.*, **45**, 918–926, 1997.

[59] McCullagh, P. *Tensor Methods in Statistics.* Monographs on Statistics and Applied Probability. London, Chapman and Hall, 1987.

[60] Moreau, E. and Pesquet, J.C. Generalized contrasts for multichannel blind deconvolution of linear systems. *IEEE Signal Process. Lett.*, **4**, 182–183, 1997.

[61] Nguyen-Thi, H.L., Jutten, C., Kabre, H., and Caelen, J. Separation of sources: A method for speech enhancement. *Applied Signal. Process.*, **3**, 177–190, 1996.

[62] Reznick, B. Sums of even powers of real linear forms. *Memoirs of the AMS*, **96**, 1992.

[63] Salmon, G. *Lessons introductory to the Modern Higher Algebra.* Chesla., New York, 1885.

[64] Shafarevitch, I.R. *Basic Algebraic Geometry.* Berlin, Springer, 1977.

[65] Sidiropoulos, N.D. Bro, R., and Giannakis, G.B. Parallel factor analysis in sensor array processing. *IEEE Trans. Signal Procoess.*, **48**, 2377–2388, 2000.

[66] Talwar, S., Viberg, M., and Paulraj, A. Blind estimation of multiple co-channel digital signals arriving at an antenna array: Part I, algorithms. *IEEE Trans. Signal Process.*, pp. 1184–1197, 1996.

[67] Tong, L. Identification of multichannel MA parameters using higher-order

statistics. *Signal Processing*, **53**, 195–209, 1996. Special issue on High-Order Statistics.

[68] Tucker, L.R. Some mathematical notes for three-mode factor analysis. *Psychometrika*, **31**, 279–311, 1965.

[69] van der Veen, A.J. Analytical method for blind binary signal separation. *IEEE Trans. Signal Process.*, **45**, 1078–1082, 1997.

[70] van der Veen, A.J. and Paulraj, A. An analytical constant modulus algorithm. *IEEE Trans. Signal Process.*, **44**, 1136–1155, 1996.

Blind Deconvolution and Source Separation: Recent Results and Open Issues

Phillip A. Regalia
Institut National des Télécommunications, 9, rue Charles Fourier, F-91011 Evry cedex, France

Abstract

We provide a tutorial overview of recent results in signal restoration applied to convolutional mixtures, which may be specialized to blind deconvolution and blind signal separation. We review in particular an equivalence between the Godard and Shalvi–Weinstein criteria, characterizations of solutions when ideal restoration is unattainable, a stepsize bound for convergence of a gradient scheme, and relations to low-rank tensor approximation. Some open issues are highlighted along the way.

1 Introduction

The problems of blind deconvolution and blind signal separation have met with fruitful interplay in recent years. The two problems share a common setting: One is given a mixture of independent random variables or 'sources,' where the mixture may be temporal (as assumed in deconvolution) or spatial (as assumed in signal separation), and the goal is to restore one of the sources. A multi-source convolutional mixture setting combines elements of both problems, and can be specialized to either one depending on the spatial and temporal characteristics of the mixture. It is not surprising, therefore, to learn that algorithms originally devised in a deconvolution setting may be adapted for separation, and vice versa.

The intent of this paper is to provide a tutorial overview of recent results in blind signal restoration. We treat in particular an equivalence between two popularly used criteria, a characterization of stationary points and extrema when ideal solutions are unattainable, stepsize bounds which ensure convergence of a gradient search procedure, connections with low-rank tensor approximation, and generalized contrast functions used in independent component analysis. The results so reviewed all apply to the case where the equalizer length is insufficient to perfectly restore a source, as occurs in practice. In addition, the results concerning equivalence between criteria, tensor approximation, and generalized contrasts remain valid even with nonlinear channels and dependent and/or correlated sources. In the interest of clarity, we restrict our attention to real signals and channels.

2 Problem setting

We consider a multi-source communication channel, in which the received signal $\{\mathbf{u}_n\}$, comprising P components, is a convolutional mixture:

$$(P \updownarrow) \quad \mathbf{u}_n = \sum_{k=0}^{\infty} \mathbf{H}_k \, \mathbf{s}_{n-k} \, . \tag{2.1}$$

Here the vector $\mathbf{s}_n$ collects Q source signals, which are assumed i.i.d., mutually independent, and scaled to unit variance, while $\{\mathbf{H}_k\}$ is the impulse response of the Q-input/P-output channel. We suppose (realistically) that the number of sources Q exceeds the number of sensors (or oversampling factor) P, giving thus a rectangular channel which will not admit a left inverse.

A seeming generalization to the above model is to include channel noise $\mathbf{b}_n$, leading to the form

$$\mathbf{u}_n = \sum_{k=0}^{\infty} \mathbf{H}_k \, \mathbf{s}_{n-k} + \mathbf{b}_n. \tag{2.2}$$

If the noise $\mathbf{b}_n$ is Gaussian, however, then we may always develop an innovations model of the form

$$\mathbf{b}_n = \sum_{k=0}^{\infty} \mathbf{F}_k \alpha_{n-k} \, , \tag{2.3}$$

in which the components of α_n are white, Gaussian, mutually uncorrelated (and hence, i.i.d.), and scaled to unit variance, while $\{\mathbf{F}_k\}$ is the (matrix-valued) impulse response of a modelling filter for the Gaussian process $\mathbf{b}_n$. This, in effect, expresses the noise term as the output of a virtual channel with impulse response $\{\mathbf{F}_k\}$, driven by virtual sources contained in α_n. We may therefore combine (2.3) with (2.2) to write

$$\mathbf{u}_n = \sum_{k=0}^{\infty} \begin{bmatrix} \mathbf{H}_k & \mathbf{F}_k \end{bmatrix} \begin{bmatrix} \mathbf{s}_{n-k} \\ \alpha_{n-k} \end{bmatrix}.$$

This assumes the same form as our original model (2.1) modulo some basic notational adjustments. (This will generally lead to $Q > P$, that is, a channel with more inputs than outputs, because the vector α_n will, in general, contain P innovation sources). By assimilating background noise and interfering sources, we return to our original model (2.1).

The equalizer output y_n comes from a P-input/single-output FIR filter with ($1 \times P$ vector) coefficients $\{\mathbf{g}_k\}_{k=0}^{M}$:

$$y_n = \sum_{k=0}^{M} \mathbf{g}_k \, \mathbf{u}_{n-k} \, .$$

Convolving the equalizer inpulse response $\{\mathbf{g}_k\}$ with the channel impulse response $\{\mathbf{H}_k\}$ gives the combined response $\{\mathbf{c}_k\}$:

$$\underbrace{\begin{bmatrix} \mathbf{c}_0^t \\ \mathbf{c}_1^t \\ \vdots \\ \mathbf{c}_M^t \\ \mathbf{c}_{M+1}^t \\ \vdots \end{bmatrix}}_{\triangleq \mathbf{c}} = \underbrace{\begin{bmatrix} \mathbf{H}_0^t & \mathbf{0} & \cdots & \mathbf{0} \\ \mathbf{H}_1^t & \mathbf{H}_0^t & \ddots & \vdots \\ \vdots & \ddots & \ddots & \mathbf{0} \\ \mathbf{H}_M^t & \cdots & \mathbf{H}_1^t & \mathbf{H}_0^t \\ \mathbf{H}_{M+1}^t & \mathbf{H}_M^t & \cdots & \mathbf{H}_1^t \\ \vdots & \ddots & \ddots & \vdots \end{bmatrix}}_{\triangleq \mathcal{H}} \begin{bmatrix} \mathbf{g}_0^t \\ \mathbf{g}_1^t \\ \vdots \\ \mathbf{g}_M^t \end{bmatrix}. \tag{2.4}$$

This allows the equalizer output to be written in terms of the source sequences as the mixture

$$y_n = \sum_{k=0}^{\infty} \mathbf{c}_k \, \mathbf{s}_{n-k}.$$

No matter how the coefficients $\{\mathbf{g}_k\}_{k=0}^{M}$ are chosen in (2.4), the combined response vector $\mathbf{c}$ is restricted to the column space of $\mathcal{H}$; this subspace is called the set of *attainable combined responses* [7, 9]. We introduce the projection matrix onto the column space of $\mathcal{H}$, *viz.*

$$\mathcal{P} = \mathcal{H}(\mathcal{H}^t\mathcal{H})^{-1}\mathcal{H}^t,$$

and ask whether $\mathcal{P}$ coincides (or not) with the identity. If $\mathcal{P} = \mathbf{I}$ (called the *sufficient order case* in [9, 10]), then an arbitrary candidate combined response becomes attainable, including thus any ideal equalizer which would restore perfectly one of the sources. If, on the other hand, $\mathcal{P} \neq \mathbf{I}$ (called the *undermodeled case* in [9, 10], then only a subset of combined responses may be attained, often excluding any ideal equalizer. We note that candidate combined response vector $\mathbf{c}$ is attainable if and only if $\mathcal{P}\mathbf{c} = \mathbf{c}$.

3 Equivalence of criteria

Two commonly used criteria for blind equalization are the Godard (or constant modulus) and Shalvi–Weinstein cost functions, given respectively as

$$\begin{aligned} J_{\text{CMA}}(y_n) &= E[(1 - y_n^2)^2]; \\ J_{\text{SW}}(y_n) &= \frac{\text{cum}_4(y_n)}{[E(y_n^2)]^2} = \frac{E(y_n^4)}{[E(y_n^2)]^2} - 3; \end{aligned}$$

where $\text{cum}_4(y_n) = E(y_n^4) - 3[E(y_n^2)]^2$ is the fourth-order cumulant of the equalizer output. Observe that J_{SW} is radially invariant (that is, $J_{\text{SW}}(\beta y_n) = J_{\text{SW}}(y_n)$ for all $\beta \neq 0$), but that J_{CMA} is not.

As such, if we parametrize the equalizer in polar form as

$$\mathbf{g} = \rho\,\overline{\mathbf{g}}, \quad \text{with} \quad \begin{cases} \|\overline{\mathbf{g}}\|_2 = 1; \\ \rho > 0; \end{cases}$$

and let $\overline{y}_n$ denote the output of the unit-norm filter $\overline{\mathbf{g}}$, then J_{SW} is insensitive to the value of the radial factor ρ: $J_{\mathrm{SW}}(y_n) = J_{\mathrm{SW}}(\rho\overline{y}_n) = J_{\mathrm{SW}}(\overline{y}_n)$.

Now, optimizing the Godard criterion $J_{\mathrm{CMA}}(\rho\overline{y}_n)$ with respect to ρ leads to (see, e.g., [8])

$$\rho^2_{\mathrm{opt}} = \frac{E(\overline{y}_n^2)}{E(\overline{y}_n^4)},$$

which results in $E(y_n^2) = E(y_n^4)$. The cost function then simplifies to

$$\begin{aligned} \overline{J}_{\mathrm{CMA}}(\overline{y}_n) &\triangleq J_{\mathrm{CMA}}(\rho\overline{y}_n)\Big|_{\rho_{\mathrm{opt}}} \\ &= 1 - \frac{[E(\overline{y}_n^2)]^2}{E(\overline{y}_n^4)} \\ &= 1 - \frac{1}{3 + J_{\mathrm{SW}}(\overline{y}_n)}. \end{aligned}$$

The key point is that the reduced Godard criterion (which is radially optimized using ρ_{opt}) is a monotonic deformation of the Shalvi–Weinstein criterion, since one verifies readily that $d\overline{J}_{\mathrm{CMA}}/dJ_{\mathrm{SW}} > 0$. This then implies that the stationary points and minima of the two functions occur for the same angular orientation of the equalizer vector, and therefore that the two criteria are equivalent. All results below, although phrased for the Shalvi–Weinstein criterion with unit-norm scaling on the combined response $\mathbf{c}$, will thus apply to the Godard algorithm as well, upon scaling the combined response so that the second- and fourth-order moments of the equalizer output coincide.

4 Stationary points and extrema

To characterize stationary points and extrema, let $\mathbf{D}$ be a $Q \times Q$ diagonal matrix containing the fourth-order cumulants of the Q sources. (Gaussian innovation sources will contribute zero-valued cumulants to this matrix). Likewise, let $\mathcal{D}$ be a diagonal matrix containing copies of $\mathbf{D}$:

$$\mathcal{D} = \begin{bmatrix} \mathbf{D} & & \\ & \mathbf{D} & \\ & & \ddots \end{bmatrix}.$$

We then have [11] the following result:

Result 1 *The combined response* $\mathbf{c}$ *(scaled to unit* ℓ_2 *norm) is a stationary point of* J_{SW} *over the set of attainable combined responses if and only if*

$$\mathcal{P}\mathcal{D}\mathbf{c}^{\odot 3} = J_{\mathrm{SW}}(\mathbf{c}) \cdot \mathbf{c},$$

where $\mathbf{c}^{\odot 3}$ *is the Hadamard (or componentwise) cube of the vector* $\mathbf{c}$*, that is,* $[\mathbf{c}^{\odot 3}]_k = \mathbf{c}_k^3$.

To characterize which stationary points are extrema of J_{SW}, let $\mathrm{diag}(\mathbf{x})$ denote a diagonal matrix containing the elements of a vector $\mathbf{x}$ along its main diagonal, and introduce the symmetric matrix

$$\mathcal{T} \triangleq \mathcal{P}\mathcal{D}\,\mathrm{diag}(\mathbf{c}^{\odot 2})\,\mathcal{P}.$$

We can then observe that, at any stationary point, the pair $(J_{\mathrm{SW}}(\mathbf{c}), \mathbf{c})$ form an eigenpair of $\mathcal{T}$, since

$$\begin{aligned} \mathcal{T}\mathbf{c} &= \mathcal{P}\mathcal{D}\,\mathrm{diag}(\mathbf{c}^{\odot 2})\,\mathcal{P}\mathbf{c} && \text{with } \mathcal{P}\mathbf{c} = \mathbf{c} \\ &= \mathcal{P}\mathcal{D}\mathbf{c}^{\odot 3} && \text{since } \mathrm{diag}(\mathbf{c}^{\odot 2})\,\mathbf{c} = \mathbf{c}^{\odot 3} \\ &= J_{\mathrm{SW}}(\mathbf{c})\,\mathbf{c} && \text{by Result 1.} \end{aligned}$$

Let now λ_+ (resp., λ_-) denote the most positive (resp., most negative) eigenvalue of $\mathcal{T}$. Extrema may be characterized as follows [11]:

Result 2 *The combined response* $\mathbf{c}$ *(scaled to unit* ℓ_2 *norm) lies at a local maximum (resp., local minimum) of* J_{SW} *if and only if the following two conditions hold:*

(1) $J_{\mathrm{SW}} = \lambda_+$ *(resp.,* $J_{\mathrm{SW}} = \lambda_-$*) with this eigenvalue simple;*

(2) $\lambda_+ \geq 3\lambda_k$ *(resp.,* $\lambda_- \leq 3\lambda_k$*) for all other eigenvalues* $\lambda_k \neq \lambda_+$ *(resp.* $\lambda_k \neq \lambda_-$*).*

In the sufficient order case ($\mathcal{P} = \mathbf{I}$) solutions verifying these results are explicit. Result 1 simplifies to $\mathcal{D}\mathbf{c}^{\odot 3} = J_{\mathrm{SW}}\,\mathbf{c}$; since $\mathcal{D}$ is diagonal, this reads componentwise as

$$c_k\,(d_k\,c_k^2 - J_{\mathrm{SW}}) = 0, \qquad \text{for all } k.$$

This says that all nonzero terms, once squared and scaled by the source cumulant values d_k, share a common amplitude J_{SW}. The matrix $\mathcal{T}$ simplifies to the diagonal matrix

$$\mathcal{T} = \mathcal{D}\,\mathrm{diag}(\mathbf{c}^{\odot 2}) = \begin{bmatrix} d_0 c_0^2 & & & \\ & d_1 c_2^2 & & \\ & & d_2 c_2^2 & \\ & & & \ddots \end{bmatrix}$$

whose nonzero terms now all share the same value J_{SW}. Therefore, $\mathcal{T}$ has J_{SW} as its sole nonzero eigenvalue, with multiplicity given by the number of nonzero terms from the combined response. By Result 2, an extremum occurs iff a sole term from the combined response is nonzero. This, of course, represents an ideal combined response.

For the undermodelled case ($\mathcal{P} \neq \mathbf{I}$), by contrast, analytic solutions satisfying Results 1 and 2 are less explicit. For the special case in which the source cumulants all coincide, that is, $d_k \equiv d$ for all k, the Shalvi–Weinstein cost function assumes the simple form

$$J_{\mathrm{SW}} = d\left(\sum_{k=0}^{\infty} c_k^4\right) \Big/ \left(\sum_{k=0}^{\infty} c_k^2\right)^2 = d\left(\frac{\|\mathbf{c}\|_4}{\|\mathbf{c}\|_2}\right)^4, \tag{4.1}$$

involving the ℓ_2 and ℓ_4 norms of the combined response $\mathbf{c}$; the extrema correspond to maxima of $\|\mathbf{c}\|_4/\|\mathbf{c}\|_2$. If we instead consider the ratio

$$J_\infty = \frac{\|\mathbf{c}\|_\infty}{\|\mathbf{c}\|_2},$$

then each maximum yields, to within a scale factor, a Wiener response [9], that is, a best least-squares approximation to a pure delay from the set of attainable combined responses. Now, by the consistency of norms, there exist constants β_1 and β_2 such that

$$\beta_1\|\mathbf{c}\|_\infty \leq \|\mathbf{c}\|_4 \leq \beta_2\|\mathbf{c}\|_\infty, \qquad \text{for all attainable } \mathbf{c}.$$

By dividing these equalities by $\|\mathbf{c}\|_2$, we see that $\|\mathbf{c}\|_4/\|\mathbf{c}\|_2$ is upper and lower bounded by scalar multiples of $\|\mathbf{c}\|_\infty/\|\mathbf{c}\|_2$. To the extent that the functionals $\|\mathbf{c}\|_4/\|\mathbf{c}\|_2$ and $\|\mathbf{c}\|_\infty/\|\mathbf{c}\|_2$ then exhibit similar (but not identical) behaviour, it is reasonable to expect that their respective maximizing arguments be not too disparate.

Let us now return to the general case of differing source cumulant values, so that the expression (4.1) becomes

$$J_{\mathrm{SW}} = \left(\sum_{k=0}^{\infty} d_k\, c_k^4\right) \Big/ \left(\sum_{k=0}^{\infty} c_k^2\right)^2$$

The case where the source cumulants are all nonnegative or nonpositive is called the *sign-definite cumulant case*, whereas the case where some source cumulants are positive and others negative is called the *mixed cumulant case*. The former case still affords writing $|J_{\mathrm{SW}}|$ as the ratio of weighted (semi-) norms, and similarity of extrema with Wiener equalizers follows as above.

5 Gradient search procedure

Having characterized stationary points and extrema, we consider now a gradient search procedure for J_{SW}:

$$\begin{aligned}
\mathbf{c}^+ &= \mathbf{c}_n \pm \frac{\mu_n}{4}\nabla J_{\mathrm{SW}}(\mathbf{c}_n), \qquad \mu_n > 0 \\
&= [1 \mp \mu_n\, J(\mathbf{c}_n)]\,\mathbf{c}_n \pm \mu_n\, \mathcal{P}\mathcal{D}\mathbf{c}^{\odot 3} \\
\mathbf{c}_{n+1} &= \mathbf{c}^+/\|\mathbf{c}^+\|.
\end{aligned}$$

We choose $\pm\mu_n = +\mu_n$ (resp., $-\mu_n$) if the algorithm is to ascend (resp., descend), as determined by the sign of the fourth-order cumulant of the source to be restored. For the particular stepsize choice $\pm\mu_n = 1/J_{\text{SW}}(\mathbf{c}_n)$, the algorithm simplifies to

$$\begin{aligned} \mathbf{c}^+ &= \mathcal{P}\mathcal{D}\mathbf{c}^{\odot 3}/|J_{\text{SW}}(\mathbf{c}_n)| \\ \mathbf{c}_{n+1} &= \mathbf{c}^+/\|\mathbf{c}^+\| \end{aligned}$$

which may be recognized as the super-exponential algorithm [13]. Practical implementations involve estimated cumulants of the received signal $\mathbf{u}_n$; see [13].

Since $J_{\text{SW}}(\mathbf{c})$ is a nonquadratic function, a bound on the stepsize sequence $\{\mu_n\}$ which ensures convergence is not elementary. Nonetheless, in the sign definite cumulant case, the cost function J_{SW} is likewise sign definite, and its magnitude $|J_{\text{SW}}|$ can be written as the ratio of two convex functions. This is exploited in [6, 10] to obtain the following stepsize bound.

Result 3 *Suppose the cumulants are sign definite, so that* $\pm\mu_n = [\text{sgn}\, J_{\text{SW}}]\mu_n$. *If* $\mathbf{c}_n$ *is not a stationary point, the inequality* $|J_{\text{SW}}(n{+}1)| > |J_{\text{SW}}(n)|$ *results if*

$$0 < \mu_n < \frac{2\,|J_{\text{SW}}(n)|}{2\,|J_{\text{SW}}(n)|^2 - \|\mathcal{P}\mathcal{D}\mathbf{c}_n^{\odot 3}\|_2^2}\,.$$

Observe that the upper bound varies with $\mathbf{c}$, as is typical of nonquadratic optimization problems. We can check, however, that $|J_{\text{SW}}| \leq \|\mathcal{P}\mathcal{D}\mathbf{c}^{\odot 3}\|_2$, since from the Cauchy–Schwarz inequality we have (with $\|\mathbf{c}\|_2 = 1$)

$$\begin{aligned} |J_{\text{SW}}| &= |\mathbf{c}^t\mathcal{P}\mathcal{D}\mathbf{c}^{\odot 3}| \\ &\leq \|\mathbf{c}\|_2\,\|\mathcal{P}\mathcal{D}\mathbf{c}^{\odot 3}\|_2 = \|\mathcal{P}\mathcal{D}\mathbf{c}^{\odot 3}\|_2\,, \end{aligned}$$

with equality iff $\mathbf{c}$ and $\mathcal{P}\mathcal{D}\mathbf{c}^{\odot 3}$ are colinear (corresponding to a stationary point by Result 1). The upper bound on μ_n from Result 3 may therefore be lower bounded as

$$\frac{2\,|J_{\text{SW}}(n)|}{2\,|J_{\text{SW}}(n)|^2 - \|\mathcal{P}\mathcal{D}\mathbf{c}_n^{\odot 3}\|_2^2} \geq \frac{2}{|J_{\text{SW}}(n)|} \geq \frac{2}{\max_k |d_k|}\,,$$

since $|J_{\text{SW}}(n)| \leq \max_k |d_k|$ at each iteration. This shows that any fixed stepsize in the range $0 < \mu < 2/(\max_k |d_k|)$—requiring knowledge only of an extremal source cumulant value—will ensure convergence of the gradient search algorithm in the sign definite cumulant case.

Remark Since the sources are scaled to unit variance, it is easy to show that their fourth-order cumulants are minorized as $d_k \geq -2$ for all k, where equality holds for constant modulus sources. Thus if all source cumulants are negative, then the stepsize range $0 < \mu < 1$ will do. Whether a comparable bound can be developed for the mixed cumulant case remains open.

6 Low-rank tensor approximation

We consider next the case where a non-i.i.d. sources and/or nonlinear channels are present. Suppose the received signal $\mathbf{u}_n$ is (pre-)whitened to second order, and for any n, re-index its components as

$$[\mathbf{u}_n^t \ \mathbf{u}_{n-1}^t \ \cdots \ \mathbf{u}_{n-M}^t] = [u_1 \ u_2 \ \cdots \ u_{(M+1)P}],$$

and apply a similar reindexation to $[\mathbf{g}_0^t \ \cdots \ \mathbf{g}_M^t]$. Introduce the fourth-order cumulant tensor as a four-way array:

$$\mathcal{E}_{i,j,k,l} = \mathrm{cum}(u_i, u_j, u_k, u_l).$$

Then the cost function J_{SW} appears as the ratio

$$J_{\mathrm{SW}}(\mathbf{g}) = \Bigg(\sum_{i,j,k,l=1}^{(M+1)P} \mathcal{E}_{i,j,k,l}\, g_i\, g_j\, g_k\, g_l \Bigg) \Bigg/ \Bigg(\sum_{i=1}^{(M+1)P} g_i^2 \Bigg)^2, \tag{6.1}$$

in which we may assume that $\mathbf{g}$ is scaled to unit ℓ_2 norm to simplify things. The four-way outer product $\mathbf{g} \star \mathbf{g} \star \mathbf{g} \star \mathbf{g}$ will denote the rank one tensor [2, 3]

$$[\mathbf{g} \star \mathbf{g} \star \mathbf{g} \star \mathbf{g}]_{i,j,k,l} = g_i\, g_j\, g_k\, g_l \,.$$

The rank-one tensor approximation problem [2] is to find a unit-norm vector $\mathbf{g}$ and a scalar λ which minimize

$$\|\mathcal{E} - \lambda\, \mathbf{g} \star \mathbf{g} \star \mathbf{g} \star \mathbf{g}\|_F^2 \,,$$

where $\|\cdot\|_F^2$ is the Frobenius norm squared, that is, the sum of squares of all elements of the array.

Result 4. (2, 5) . *The unit-norm vector* $\mathbf{g}$ *and the scalar* λ *yield a local minimum of*

$$\|\mathcal{E} - \lambda\, \mathbf{g} \star \mathbf{g} \star \mathbf{g} \star \mathbf{g}\|_F^2$$

iff $\mathbf{g}$ *is a local maximum of* $|J_{\mathrm{SW}}|$ *from (6.1), at which point* $\lambda = J_{\mathrm{SW}}(\mathbf{g})$.

As such, blind deconvolution in the nonlinear channel and/or non-i.i.d. source case is equivalent to rank-one tensor approximation. Iterative algorithms for constructing rank-one approximations have been developed in [2, 1], variants of which yield the earlier super-exponential algorithm [5].

7 Neural optimization algorithms

We return to the case in which the received vector $\mathbf{u}$ has been prewhitened so that $E(\mathbf{u}\mathbf{u}^t) = \mathbf{I}$. A generalized deconvolution criterion can be written as

$$J(\mathbf{g}) = \Big(E[F(\mathbf{g}^t\mathbf{u})] - E[F(\nu)] \Big)^2, \tag{7.1}$$

where ν is a Gaussian random variable of unit variance, and where $F(x)$ is a user-chosen function. The special choice $F(x) = x^4$, combined with a unit-norm constraint on $\mathbf{g}$, gives $J(\mathbf{g})$ as the square of the fourth-order cumulant of $y = \mathbf{g}^t\mathbf{u}$; maximizing $J(\mathbf{g})$ subject to $\|\mathbf{g}\|_2 = 1$ can therefore be accomplished using the algorithms referenced earlier.

Hyvärinen [4] provides ample motivation for considering other choices of $F(x)$, particularly for 'super-Gaussian' sources which are more commonly encountered in speech and image deconvolution, in contrast to 'sub-Gaussian' sources which dominate in digital communications. Candidate choices for $F(x)$ include [4]

$$\begin{aligned} F_1(x) &= \frac{1}{a_1}\log\cosh(a_1 x); \\ F_2(x) &= -\frac{1}{a_2}\exp(-a_2 x^2/2); \\ F_3(x) &= x^4; \end{aligned}$$

with $1 \le a_1 \le 2$ and $a_2 \approx 1$. Observe that $F_1(x)$ and $F_3(x)$ are convex, although $F_2(x)$ is not, where we recall that a convex function satisfies, for all $0 \le \alpha \le 1$,

$$F\big(\alpha x_1 + (1-\alpha)x_2\big) \le \alpha\, F(x_1) + (1-\alpha)\, F(x_2), \qquad \text{for all } x_1 \text{ and } x_2.$$

For super-Gaussian sources, maximizing (7.1) reduces to maximizing the functional $E[F(\mathbf{g}^t\mathbf{u})]$ itself, subject to the constraint $\|\mathbf{g}\|_2 = 1$ [4]. If we relax momentarily the unit-norm constraint on $\mathbf{g}$, we may exploit the following elementary result.

Result 5 *If $F(x)$ is a convex function of x, then $H(\mathbf{g}) \triangleq E[F(\mathbf{u}^t\mathbf{g})]$ is a convex function of $\mathbf{g}$ over $\mathrm{I\!R}^{(M+1)P}$.*

Proof If $p(\mathbf{u}) \ge 0$ denotes the joint probability density function of the components of $\mathbf{u}$, then by direct calculation we have

$$\begin{aligned} H\big(\alpha\mathbf{g}_1 + (1-\alpha)\mathbf{g}_2\big) &= \int_{\mathbf{u}} F\big(\alpha\mathbf{u}^t\mathbf{g}_1 + (1-\alpha)\mathbf{u}^t\mathbf{g}_2\big)\, p(\mathbf{u})\, d\mathbf{u} \\ &\le \int_{\mathbf{u}} \Big(\alpha\, F(\mathbf{u}^t\mathbf{g}_1) + (1-\alpha)\, F(\mathbf{u}^t\mathbf{g}_2)\Big)\, p(\mathbf{u})\, d\mathbf{u} \\ &= \alpha\, H(\mathbf{g}_1) + (1-\alpha)\, H(\mathbf{g}_2), \end{aligned}$$

in which the second line follows from convexity of $F(x)$ combined with nonnegativity of $p(\mathbf{u})$. □

Let now $h(\mathbf{g})$ denote the gradient of $H(\mathbf{g})$ over $\mathrm{I\!R}^{(M+1)P}$:

$$h(\mathbf{g}) \triangleq \frac{dH(\mathbf{g})}{d\mathbf{g}} = E[\mathbf{u}\, f(\mathbf{u}^t\mathbf{g})],$$

where $f(x) = dF(x)/dx$. The gradient inequality for convex functions [12] asserts that

$$H(\mathbf{g}_2) - H(\mathbf{g}_1) \geq h^t(\mathbf{g}_1)\,(\mathbf{g}_2 - \mathbf{g}_1),$$

for all $\mathbf{g}_1$ and $\mathbf{g}_2$ in $\mathbb{R}^{(M+1)P}$. The super-exponential algorithm presented in section 5 then admits a natural generalization as

$$\begin{aligned} \mathbf{g}^+ &= h(\mathbf{g}_i) = E[\mathbf{u}\, f(\mathbf{u}^t \mathbf{g}_i)], \qquad \|\mathbf{g}_0\|_2 = 1 \\ \mathbf{g}_{i+1} &= \mathbf{g}^+/\|\mathbf{g}^+\|_2 \end{aligned}$$

which maps the unit sphere $\|\mathbf{g}\|_2 = 1$ to itself. This differs slightly from the fixed-point algorithm developed in eq. (19) of Hyvärinen [4], for which only local convergence was shown. If $F(x)$ is chosen convex, in fact a stronger result of global convergence can be developed.

Result 6 *If* $\mathbf{g}_{i+1} \neq \mathbf{g}_i$, *then* $H(\mathbf{g}_{i+1}) > H(\mathbf{g}_i)$.

Remark This then implies that the sequence of iterates $\{\mathbf{g}_i\}$ converges to a local maximum of the restriction of $H(\mathbf{g})$ to the unit sphere $\|\mathbf{g}\|_2 = 1$.

Proof Since the gradient inequality for $H(\mathbf{g})$ applies to any two vectors in $\mathbb{R}^{(M+1)P}$, it applies in particular to successive iterates $\mathbf{g}_i$ and $\mathbf{g}_{i+1}$, to give

$$H(\mathbf{g}_{i+1}) - H(\mathbf{g}_i) \geq h^t(\mathbf{g}_i)\,(\mathbf{g}_{i+1} - \mathbf{g}_i). \tag{7.2}$$

We need only show that the right-hand side is positive. To this end, the Cauchy–Schwarz inequality implies that, for any unit-norm vector $\mathbf{g}$,

$$h^t(\mathbf{g}_i)\,\mathbf{g} \leq \|h(\mathbf{g}_i)\|_2$$

where the upper bound is attained iff $\mathbf{g} = h(\mathbf{g}_i)/\|h(\mathbf{g}_i)\|_2$. Since the unit-norm vector $\mathbf{g}_{i+1} = h(\mathbf{g}_i)/\|h(\mathbf{g}_i)\|_2$ attains the upper bound while $\mathbf{g}_i \neq \mathbf{g}_{i+1}$ does not, we must have

$$h^t(\mathbf{g}_i)\,\mathbf{g}_{i+1} > h^t(\mathbf{g}_i)\,\mathbf{g}_i\,.$$

This gives the positivity of the right-hand side of (7.2), as desired. □

8 Concluding remarks

We have presented a semi-tutorial overview of recent results in blind signal restoration, emphasizing an equivalence between the constant modulus algorithm and the Shalvi–Weinstein (or kurtosis) criterion. Results concerning stationary points, extrema, and convergence of iterative algorithms have been reviewed for the general 'undermodelled' case, in which the set of attainable combined responses does not encompass the entire combined response space.

Bounds for a stepsize parameter which ensure convergence of a gradient descent procedure have also been presented. It is noteworthy that the super-exponential algorithm arises from an optimal choice of the stepsize parameter

[11]. Extensions of the super-exponential algorithm to handle possibly nonlinear channels leads naturally to low-rank tensor approximation and the corresponding higher-order power method [2], or generalized constrast optimization and the corresponding fixed-point algorithm [4], of which an improved version has been developed here. That these may all be viewed as variants on a common algorithm may hopefully serve to unify seemingly disparate approaches to an ultimately common signal restoration objective.

Bibliography

[1] Comon, P. (1998). Blind channel identification and extraction of more sources than sensors, *SPIE Conf. Adv. Sig. Process. VIII*, San Diego, CA, 2–13.

[2] De Lathauwer, L, De Moor, B., Comon, P., and Vandewalle, J. (1995). The higher-order power method, *Proc. NOLTA'95*, (Las Vegas, NV).

[3] Grigorascu, V.S. and Regalia, P.A. (1999). Tensor displacement structures and polyspectral matching. In: *Fast Reliable Algorithms for Matrices with Structure*, (eds, T. Kailath and A. H. Sayed), SIAM, Philadelphia.

[4] Hyvärinen, A. (1999). Fast and robust fixed-point algorithms for independent component analysis, *IEEE Trans. Neural Networks*, **10**, 626–634.

[5] Kofidis, E. and Regalia, P.A. (2001). Tensor approximation and signal processing applications. In: *Structured Matrices in Numerical Analysis, Control, Signal and Image Processing*, (ed. Olshevsky, V.), AMS., Providence, RI.

[6] Mboup, M. and Regalia, P.A. (2000). A gradient search interpretation of the super-exponential algorithm, *IEEE Trans. Information Theory*, **46**, 2731–2734.

[7] Li, Y. and Ding, Z. (1996). Global convergence of fractionally spaced Godard (CMA) blind equalizers, *IEEE Trans. Signal Processing*, **44**, 818–826.

[8] Regalia, P.A. (1999). On the equivalence between the Godard and Shalvi–Weinstein schemes of blind equalization, *Signal Processing*, **73**, 185–190.

[9] Regalia, P.A. and Mboup, M. (1999a). Undermodeled equalization: A characterization of stationary points for a family of blind criteria, *IEEE Trans. Signal Processing*, **47**, 760–770.

[10] Regalia, P.A. and Mboup, M. (1999b). Equalization in noisy multi-user channels, *IEEE/EURASIP Int. Workshop on Nonlinear Signal and Image Processing*, (Antalya, Turkey).

[11] Regalia, P.A. and Mboup, M. (2001). Properties of some blind equalization criteria in noisy multi-user environments, *IEEE Trans. Signal Processing*, **49**, 2001, to appear.

[12] Rockafellar, R.T. (1970). *Convex Analysis*, Princeton University Press, Princeton, NJ.

[13] Shalvi, O. and Weinstein, E. (1993). Super-exponential methods for blind deconvolution, *IEEE Trans. Information Theory*, **39**, 504–519.

An Algebraic Algorithm for Independent Component Analysis With More Sources Than Sensors

L. De Lathauwer

Equipe Traitement des Images et du Signal, Ecole Nationale Supérieure de l'Electronique et de ses Applications and Université de Cergy-Pontoise, France

B. De Moor and J. Vandewalle

Department of Electrical Engineering, Katholieke Universiteit Leuven, Belgium

Abstract

In this paper we derive an algorithm to identify the mixing matrix in the context of an independent component analysis with more sources than sensors. The technique can cope with $I(I+1)/2$ sources for only I sensors. It exploits the fact that for non-circular complex-valued data, depending on the type of complex symmetry, two different fourth-order cumulants are available.

1 Introduction

Assume the following basic statistical model:

$$Y = \mathbf{M}X + N, \tag{1.1}$$

in which $Y \in \mathbb{C}^I$ is referred to as the *observation vector*, $X \in \mathbb{C}^R$ is called the *source vector* and $N \in \mathbb{C}^I$ represents additive *noise*. $\mathbf{M} \in \mathbb{C}^{I \times R}$ is the *mixing matrix*. The goal of independent component analysis now consists of the estimation of the mixing matrix and/or the corresponding realizations of the source vector X, given only realizations of the observation vector Y. The key assumption is that the components of X are mutually statistically independent, as well as statistically independent from the noise components. This is a very strong hypothesis, but also quite natural in lots of applications. It means that the aim can often be rephrased as splitting the dataset into components 'of a different nature', which contributed to the data in a linear way.

The solution of an independent component analysis is subject to two basic indeterminacies. First, it is impossible to determine the norm of the columns of $\mathbf{M}$ in (1.1), since a rescaling of these vectors can be compensated by the inverse scaling of the source signal values. Similarly, the ordering of the source signals, having no physical meaning, cannot be identified.

Uniqueness is usually guaranteed by claiming that the columns of $\mathbf{M}$ are linearly independent and that the sources are non-Gaussian; it can be shown that, under these conditions, the two basic indeterminacies are indeed the only way in which the solution is not unique [1, 2]. However, in this paper we focus on the situation in which there are more sources than sensors, that is, the mixing matrix has more columns than rows. Hence, in Section 4, we have to follow another approach. For non-circular complex-valued data (like for instance in telecom applications) the condition of linear independence will be replaced by a weaker condition.

Independent component analysis is related to the classical problem of principal component analysis. Here, by exploiting that the source signals are uncorrelated, one is able to find the sources as well as the mixing matrix up to a unitary transformation. Let us define the covariance matrix $\mathbf{C}_Z^{(1,1)}$ of a random vector Z by the element-wise equation

$$(\mathbf{C}_Z^{(1,1)})_{i_1 i_2} = \mathtt{E}\{(z_{i_1} - \mathtt{E}\{z_{i_1}\})(z_{i_2} - \mathtt{E}\{z_{i_2}\})^*\}, \tag{1.2}$$

in which $\cdot^*$ denotes a complex conjugation and $\mathtt{E}$ the expectation. The covariance matrices of Y and X are related by

$$\mathbf{C}_Y^{(1,1)} = \mathbf{M} \cdot \mathbf{C}_X^{(1,1)} \cdot \mathbf{M}^H + \mathbf{C}_N^{(1,1)}, \tag{1.3}$$

in which $\mathbf{C}_X^{(1,1)}$ is positive diagonal ($\cdot^H$ denotes the complex conjugated transpose). Since the sources may be arbitrarily rescaled, we can assume that they have unit variance. Then we have (we omit the noise term at this point, for clarity):

$$\mathbf{C}_Y^{(1,1)} = \mathbf{M} \cdot \mathbf{M}^H. \tag{1.4}$$

Hence the mixing matrix can be estimated, up to a unitary transformation, as a square root of the observed covariance. In principal component analysis the solution is made essentially unique by selecting a matrix of which the columns are orthonormal. (This factor may of course also be found, in a numerically more reliable way, from the singular value decomposition of the observed dataset [3].)

The solution to the independent component analysis problem lies in the fact that the assumption of *statistical independence* is stronger than the notion of *uncorrelated* signals. Statistical independence is not only a claim on the second-order statistics of the signals, but also on their higher-order statistics [4]. More precisely, it is not sufficient that the source covariance $\mathbf{C}_X^{(1,1)}$ is a diagonal matrix—in addition, the higher-order cumulants of the source vector should be diagonal higher-order tensors. (A higher-order tensor can intuitively be imagined as a multi-way matrix, of which the entries are characterized by more than two indices; its diagonal is defined as the entries for which all the indices are equal; the definition of a higher-order cumulant will be given in Section 2.) This leads to a sufficient amount of constraints to solve the problem. Actually, it is not necessary to resort to the second-order information to be able to find the solution: the higher-order statistics alone contain enough information. The advantage of the latter approach is that one is able to work in a way that is conceptually blind to the presence of additive Gaussian noise, as we will explain in Section 2; the technique derived in this paper belongs to this class.

After a short introduction to higher-order cumulants in Section 2, we will show in Section 3 that our problem can be formulated as the simultaneous decomposition of two higher-order cumulant tensors in a minimal number of rank-1 terms, in which each rank-1 term consists of the contribution of one particular source signal to the overall linear combination. In Section 4 we will consider the two cumulant tensors as the tensor equivalent of a matrix pencil; by associating an eigenvalue decomposition with it, a mixing matrix with $I(I+1)/2$ columns can be identified. For moderate values of I, this number is higher than for any other method published so far. Nevertheless, our contribution should be considered as a new line of approach that can be further improved (as we will explain in Section 4), rather than as an end result. Therefore we found it unnecessary to add a section on numerical simulations at this stage. In Section 5 we go into some more detail with respect to a numerical aspect of the technique.

Once the mixing matrix has been estimated, it is common to estimate the source values by premultiplying the dataset with its pseudo-inverse. However, the rectangular size of the mixing matrix in our problem setting makes this approach impossible. Nevertheless, the source values may be reconstructed by exploiting prior knowledge on the source distributions, like for example the finite alphabet property [5, 6].

Let us end this introduction by referring to some related papers. In [7] the structure of a complex fourth-order cumulant is exploited to extract, for a generic mixing matrix, R columns, bounded by $2R(R-1) \leqslant (I(I-1))^2$. The special case of (2×3) independent component analysis was first addressed in [5]. Further results can be found in [8]. The idea of the current paper was first presented in [9]; this paper also explains how [7] can be adapted to take two fourth-order cumulant tensors into account.

2 Cumulants

The basic quantities of higher-order statistics are higher-order moments and higher-order cumulants. Cumulants have a number of important properties, that are not shared with higher-order moments, such that in practice cumulants are more frequently used. In this section we will give a definition of cumulants up to order 4, and list some of the important properties. For a more comprehensive introduction we refer to [4, 10].

Definition 2.1. (Cumulant) *For zero-mean random variables* x_1, x_2, x_3, x_4

the cumulants up to order 4 are defined by

$$\mathtt{Cum}(x_1) \stackrel{\text{def}}{=} \mathrm{E}\{x_1\}, \tag{2.1}$$

$$\mathtt{Cum}(x_1, x_2) \stackrel{\text{def}}{=} \mathrm{E}\{x_1 x_2\}, \tag{2.2}$$

$$\mathtt{Cum}(x_1, x_2, x_3) \stackrel{\text{def}}{=} \mathrm{E}\{x_1 x_2 x_3\}, \tag{2.3}$$

$$\begin{aligned}\mathtt{Cum}(x_1, x_2, x_3, x_4) \stackrel{\text{def}}{=} & \ \mathrm{E}\{x_1 x_2 x_3 x_4\} \\ & -\mathrm{E}\{x_1 x_2\}\mathrm{E}\{x_3 x_4\} \\ & -\mathrm{E}\{x_1 x_3\}\mathrm{E}\{x_2 x_4\} \\ & -\mathrm{E}\{x_1 x_4\}\mathrm{E}\{x_2 x_3\}.\end{aligned} \tag{2.4}$$

For every variable x_i $(1 \leqslant i \leqslant 4)$ that has a non-zero mean, x_i has to be replaced in these formulas, except (2.1), by $x_i - \mathrm{E}\{x_i\}$.
The cumulants of stochastic vectors $X^{(1)}, \ldots, X^{(K)}$ $(K \leqslant 4)$ are defined by the element-wise equation

$$(\mathtt{Cum}(X^{(1)}, \ldots, X^{(K)}))_{i_1, \ldots, i_K} \stackrel{\text{def}}{=} \mathtt{Cum}(x^{(1)}_{i_1}, \ldots, x^{(K)}_{i_K}). \tag{2.5}$$

We have the following properties:

(1) *Multilinearity:* if the stochastic vectors $X^{(1)}, \ldots, X^{(K)}$ are transformed into the stochastic vectors $\tilde{X}^{(1)}, \ldots, \tilde{X}^{(K)}$ by the matrix multiplications $\tilde{X}^{(k)} = \mathbf{A}^{(k)} \cdot X^{(k)}$, with $\mathbf{A}^{(k)} \in \mathbb{C}^{J_k \times I_k}$ $(1 \leqslant k \leqslant K)$, then we have

$$\begin{aligned}&(\mathtt{Cum}(\tilde{X}^{(1)}, \ldots, \tilde{X}^{(K)}))_{j_1, \ldots, j_K} \\ &= \textstyle\sum_{i_1, \ldots, i_K} a^{(1)}_{j_1 i_1} \ldots a^{(K)}_{j_K i_K} (\mathtt{Cum}(X^{(1)}, \ldots, X^{(K)}))_{i_1, \ldots, i_K}.\end{aligned} \tag{2.6}$$

(2) *Partitioning of independent variables:* if a subset of K stochastic variables $x_1, x_2, \ldots, x_K$ is independent of the other variables, then we have:

$$\mathtt{Cum}(x_1, x_2, \ldots, x_K) = 0. \tag{2.7}$$

A consequence of this property is that, for a stochastic vector X having mutually independent components, the cumulant $\mathtt{Cum}(X, \ldots, X)$ is a diagonal tensor. (The same holds if in the definition of the cumulant X is replaced a couple of times by X^*, as in Section 3.) This very strong algebraic condition is the basis of all algebraic techniques for independent component analysis.

(3) *Sum of independent variables:* if the stochastic variables $x_1, x_2, \ldots, x_K$ are mutually independent from the stochastic variables $y_1, y_2, \ldots, y_K$, then we have

$$\begin{aligned}&\mathtt{Cum}(x_1 + y_1, x_2 + y_2, \ldots, x_K + y_K) \\ &= \mathtt{Cum}(x_1, x_2, \ldots, x_K) + \mathtt{Cum}(y_1, y_2, \ldots, y_K).\end{aligned} \tag{2.8}$$

The cumulant tensor of a sum of independent random vectors is the sum of the individual cumulants. This property explains the term 'cumulant'.

(4) *Non-Gaussianity:* higher-order cumulants of a Gaussian variable are 0. In combination with the previous property, we conclude that higher-order cumulants have the interesting property to be blind for additive Gaussian noise. Cumulants of a random variable are some measure of its non-Gaussianity.

Generally speaking, it becomes harder to estimate higher-order statistics from sample data as the order increases, that is, longer datasets are required to obtain the same accuracy [11, 12]. Hence in practice the use of higher-order statistics is usually restricted to third- and fourth-order cumulants. In what follows, we will only consider fourth-order cumulants because in most applications the probability density functions of the random variables exhibit symmetries that cause the third-order cumulants to vanish.

3 Two expansions in rank-1 terms

We start from the observation that for non-circular complex-valued sources two types of fourth-order cumulants are available, even when the probability density is symmetric about the origin:

$$\begin{aligned} \mathcal{C}_Y^{(4,0)} &\stackrel{\text{def}}{=} \texttt{Cum}(Y,Y,Y,Y), \\ \mathcal{C}_Y^{(2,2)} &\stackrel{\text{def}}{=} \texttt{Cum}(Y,Y,Y^*,Y^*). \end{aligned}$$

Due to the multilinearity property, the observation cumulants are related to the source cumulants $\kappa_r^{(4,0)}$ and $\kappa_r^{(2,2)}$ $(1 \leqslant r \leqslant R)$ in the following way:

$$(\mathcal{C}_Y^{(4,0)})_{ijkl} = \sum_{r=1}^{R} \kappa_r^{(4,0)} m_{ir} m_{jr} m_{kr} m_{lr}, \tag{3.1}$$

$$(\mathcal{C}_Y^{(2,2)})_{ijkl} = \sum_{r=1}^{R} \kappa_r^{(2,2)} m_{ir} m_{jr} m_{kr}^* m_{lr}^*. \tag{3.2}$$

for all index values. Note that (3.1) and (3.2) form a clear fourth-order counterpart of (1.3). Moreover, there is no explicit noise contribution if the noise is assumed to be Gaussian. Of course, when dealing with sample cumulants, (3.1) and (3.2) are only valid up to some perturbation.

Each of the terms in (3.1) and (3.2) corresponds to a generalization of an outer product of two vectors; actually we are dealing with outer products of four vectors; in multilinear algebraic terminology we say that $\mathcal{C}_Y^{(4,0)}$ and $\mathcal{C}_Y^{(2,2)}$ are decomposed in a sum of fourth-order rank-1 terms. Higher-order tensors have the advantage, compared to matrices, that for the decomposition to be unique, the number of rank-1 terms is not bounded by the dimension of the column nor the row space, which makes it *a priori* possible to address the problem of independent component analysis in a context where the number of sources exceeds the number of sensors.

4 A generalized eigenvalue problem

Consider $\mathcal{C}^{(4,0)}$ and $\mathcal{C}^{(2,2)}$ as linear matrix-to-matrix mappings, according to

$$(\mathcal{C}(\mathbf{A}))_{ij} = \sum_{p,q=1}^{N} c_{ijpq} a_{pq},$$

and let $\mathbf{C}^{(4,0)}, \mathbf{C}^{(2,2)} \in \mathbb{C}^{I^2 \times I^2}$ be the corresponding matrix representations. In this format, (3.1) and (3.2) can be expressed as follows:

$$\mathbf{C}_Y^{(4,0)} = \tilde{\mathbf{M}} \cdot \mathbf{C}_X^{(4,0)} \cdot \tilde{\mathbf{M}}^T, \quad (4.1)$$

$$\mathbf{C}_Y^{(2,2)} = \tilde{\mathbf{M}} \cdot \mathbf{C}_X^{(2,2)} \cdot \tilde{\mathbf{M}}^H, \quad (4.2)$$

in which $\mathbf{C}_X^{(4,0)}, \mathbf{C}_X^{(2,2)} \in \mathbb{C}^{I^2 \times I^2}$ are diagonal matrices, containing the source autocumulants, and in which $\tilde{\mathbf{M}} \stackrel{\text{def}}{=} \mathbf{M} \odot \mathbf{M}$, with $\odot$ the Kathri–Rao product (if $\mathbf{M} = [M_1 M_2 \ldots M_R]$ then $\mathbf{M} \odot \mathbf{M} \stackrel{\text{def}}{=} [M_1 \otimes M_1 \; M_2 \otimes M_2 \; \ldots \; M_R \otimes M_R]$, with $\otimes$ the Kronecker product).

One can associate a generalized eigenvalue problem with the 'tensor pencil' $(\mathcal{C}_Y^{(4,0)}, \mathcal{C}_Y^{(2,2)})$ as follows:

$$\mathbf{C}_Y^{(4,0)} (\mathbf{C}_Y^{(2,2)})^{\dagger *} = \tilde{\mathbf{M}} \cdot \left(\mathbf{C}_X^{(4,0)} (\mathbf{C}_X^{(2,2)})^{\dagger *} \right) \cdot \tilde{\mathbf{M}}^{\dagger *}, \quad (4.3)$$

in which $\cdot^{\dagger}$ denotes the pseudo-inverse. This is the unsymmetric version of the Tagaki eigenvalue decomposition [13]; as far as we know, it has not been studied in the literature yet.

The decomposition can be reduced to an ordinary eigenvalue decomposition by multiplying each side of the equation with its complex conjugate:

$$\mathbf{C}_Y^{(4,0)} \cdot (\mathbf{C}_Y^{(2,2)})^{\dagger *} \cdot (\mathbf{C}_Y^{(4,0)})^{*} \cdot (\mathbf{C}_Y^{(2,2)})^{\dagger} = \tilde{\mathbf{M}} \cdot \mathbf{D} \cdot \tilde{\mathbf{M}}^{\dagger}, \quad (4.4)$$

in which

$$\mathbf{D} \stackrel{\text{def}}{=} \mathbf{C}_X^{(4,0)} \cdot (\mathbf{C}_X^{(2,2)})^{\dagger *} \cdot (\mathbf{C}_X^{(4,0)})^{*} \cdot (\mathbf{C}_X^{(2,2)})^{\dagger} \quad (4.5)$$

is diagonal. If there are no mutually collinear pairs in the set $\{(|\kappa_r^{(4,0)}|, |\kappa_r^{(2,2)}|)\}$ and if the matrices $\{M_r M_r^T\}_{1 \leqslant r \leqslant R}$ are linearly independent, then the eigenvalue decomposition (4.4) is unique. The eigenvectors correspond to $\{M_r \odot M_r\}_{1 \leqslant r \leqslant R}$, from which the columns of the mixing matrix may be derived. If the sample cumulants are perturbed by noise then the eigenvectors may not exactly correspond to (vectorized representations of) rank-1 matrices. Best rank-1 approximations can then be used to initialize an optimization routine in which both sides of (3.1) and (3.2) are matched.

Taking into account that the range of the mapping in (4.4) only contains symmetric matrices, we conclude that $R = I(I+1)/2$ factors can be identified in this way.

For generic mixing matrices the multiplicity condition on the eigenvalues of (4.4) is not crucial. The problem then becomes the determination of a set of rank-1 matrices, given the vector space they span. This problem is addressed in [7, 14]. One can allow a multiplicity S, bounded by $2S(S-1) \leqslant (I(I-1))^2$; then the rank-1 matrices in the eigenspace can still be identified in an essentially unique way. One can use the same technique to deal with the possible ill-conditioning of the eigenvectors when eigenvalues are close.

Note that, apart from the remark in the preceding paragraph, we did not exploit the fact that the eigenvectors correspond to rank-1 matrices. Incorporating this knowledge will lead to improved techniques. This forms a topic of future research.

5 Complex eigenvalues

Equation (4.5) shows that the eigenvalues in decomposition (4.4) should be real. On the other hand, not every matrix can be decomposed as in (4.3), which may lead to the presence of complex eigenvalues in (4.4).

If the noise level is so high that complex eigenvalues do occur in (4.4), then one can still proceed in an approximate sense as follows. Instead of the eigenvalue decomposition (4.4), let us consider the Schur decomposition

$$\mathbf{B} \cdot \mathbf{B}^* = \mathbf{Q} \cdot \mathbf{R} \cdot \mathbf{Q}^H, \tag{5.1}$$

in which $\mathbf{R}$ is upper triangular and $\mathbf{B}$ is a short-hand notation for $\mathbf{C}_Y^{(4,0)} \cdot (\mathbf{C}_Y^{(2,2)})^{\dagger *}$. From this unitary decomposition one can find a least-squares approximation of $\mathbf{BB}^*$ that has real eigenvalues, by simply setting the diagonal of $\mathbf{R}$ equal to its real part; we denote the result by $\tilde{\mathbf{R}}$.

In terms of the QR-decomposition $\tilde{\mathbf{M}} = \mathbf{QU}$, the remaining task in the estimation of $\tilde{\mathbf{M}}$ is then the computation of the upper triangular eigenmatrix $\mathbf{U}$ of $\tilde{\mathbf{R}}$:

$$\tilde{\mathbf{R}} = \mathbf{U} \cdot \tilde{\Lambda} \cdot \mathbf{U}^{\dagger}, \tag{5.2}$$

in which $\tilde{\Lambda}$ is the diagonal part of $\tilde{\mathbf{R}}$. Complex conjugate eigenvalues of $\mathbf{BB}^*$ yield eigenvalues of multiplicity 2 in (5.2). In the application under consideration, the indeterminacy in the eigenspace can be met by claiming that the corresponding columns of $\tilde{\mathbf{M}}$ represent rank-1 matrices, as in the previous section.

6 Conclusion

In this paper a multilinear algebraic technique for more-sources-than-sensors independent component analysis was developed. The technique was based on a tensor generalization of an unsymmetric variant of the Tagaki eigenvalue decomposition.

Acknowledgements

Lieven De Lathauwer holds a permanent research position with the French Centre National de la Recherche Scientifique (CNRS); he is also a post-doctoral researcher with the Fund for Scientific Research-Flanders (FWO), affiliated with the K.U.Leuven. Bart De Moor and Joos Vandewalle are Full Professor with the K.U.Leuven. Part of this work was supported by the Flemish Government (Research Council K.U.Leuven (GOA-Mefisto-666, IDO), FWO (G.0240.99, G.0256.97, Research Communities ICCoS and ANMMM), BIL 98 with South Africa) and by the Federal State (IUAP P4-02, IUAP P4-24). The scientific responsibility is assumed by the authors.

Bibliography

[1] Comon, P. (1994). Independent component analysis, a new concept? *Signal Processing*, **36**, 287–314.

[2] Tong, L., *et al.* (1991). Indeterminacy and identifiability of blind identification. *IEEE Trans. Circuits and Systems*, **38**, 499–509.

[3] Golub, G. and Van Loan, C. (1996). *Matrix computations* (3rd edn). Johns Hopkins University Press, Baltimore, MD.

[4] Nikias, C. and Mendel, J. (1993). Signal processing with higher-order spectra. *IEEE Signal Process. Mag.*, 10–37.

[5] Comon, P. (1998). Blind channel identification and extraction of more sources than sensors. *Proc. SPIE Conf.*, (San Diego, CA, USA, July 19–24), 2–13.

[6] Comon, P. and Grellier, O. (1999). Non-linear inversion of underdetermined mixtures. *Proc. ICA'99*, (Aussois, France, Jan. 11–15), 461–465.

[7] Cardoso, J.-F. (1991). Super-symmetric decomposition of the fourth-order cumulant tensor. Blind identification of more sources than sensors. *Proc. ICASSP-91*, Vol. 5, 3109–3112.

[8] De Lathauwer, L., *et al.* (1999). ICA algorithms for 3 sources and 2 sensors. *Proc. IEEE Signal Processing Workshop on Higher-Order Statistics* (*HOS'99*, Caesarea, Israel, June 14–16), 116–120.

[9] De Lathauwer, L., *et al.* (1999). ICA techniques for more sources than sensors. *Proc. IEEE Signal Processing Workshop on Higher-Order Statistics* (*HOS'99*, Caesarea, Israel, June 14–16), 121–124.

[10] De Lathauwer, L., *et al.* (2000). An introduction to independent component analysis. *J. Chemom.*, **14**, 123–149.

[11] Kendall, M. and Stuart, A. (1977). *The Advanced Theory of Statistics*, Vol. 1. Griffin, London.

[12] Bourin, C. and Bondon, P. (1995). Efficiency of high-order moment estimates. *Proc. IEEE Signal Processing / (ATHOS Workshop on Higher-Order Statistics*, Girona, Spain), 186–190.

[13] Horn, R. and Johnson, C. (1991). *Topics in matrix analysis.* Cambridge University Press, New York.

[14] van der Veen, A.-J. and Paulraj, A. (1996). An analytical constant modulus algorithm. *IEEE Trans. Signal Processing*, **44**, 1136–1155.

Quasi-Newton Cross-Correlation and Constant Modulus Adaptive Algorithm for Space-Time Blind Equalization

Yuhui Luo and Jonathon Chambers

Signal and Image Processing Group, Department of Electronic and Electrical Engineering, University of Bath, Bath BA2 7AY, UK

Abstract

The problem of blind source separation and equalization in a generic multiuser wireless communication system is considered. A new quasi-Newton cross-correlation and constant modulus algorithm (QN-CCCMA) is proposed to overcome the slow convergence of the conventional stochastic gradient descent cross-correlation and constant modulus algorithm (LMS-CCCMA) by exploiting the Hessian information within the update equation. The rapid convergence property of the proposed algorithm is confirmed by simulations.

1 Introduction

In a multiuser wireless communication system, it may be assumed that the baseband digital signals originating from d spatially separated sources are transmitted through multiple linear channels and captured by an array of M antennas, representative of a multiple input multiple output (MIMO) model. Due to the potentional large delay spread introduced by multipath propagation and the presence of co-channel system users, the received signals are corrupted not only by intersymbol interference (ISI) but also by interuser interference (IUI). In this context, the objective for a blind space-time equalizer is to jointly combat ISI and IUI without recourse to exploit training sequences. A schematic illustration is shown in Fig. 1.

Among various blind adaptive algorithms proposed, the constant modulus algorithm (CMA) [5] exploits the underlying constant modulus property of the signals and has the advantage of computational simplicity. For its application in a multiuser environment, to prevent retrieval of the same source, a mixed cross-correlation and constant modulus (CC-CM) cost function has been proposed based on the fact that the sources are mutually uncorrelated, [4, 3]. In minimizing the CC-CM cost function, the conventional approach ultilizes the stochastic gradient descent technique and yields the LMS-CCCMA algorithm. As this algorithm exhibits slow rate of convergence, its application in many practical situations will be limited, therefore algorithms with rapid convergence property are required. In this paper, based on the CC-CM cost function, we em-

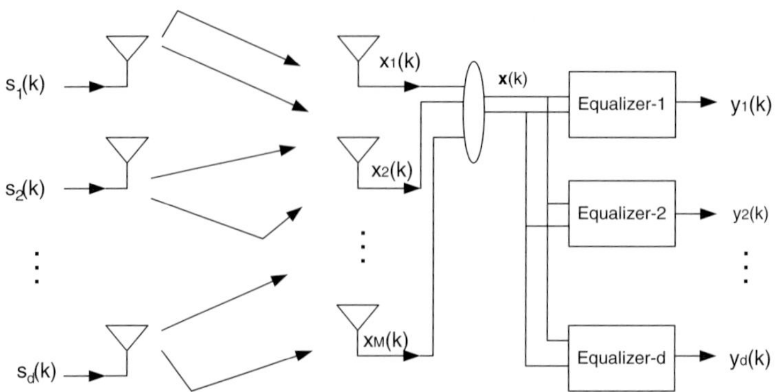

FIG. 1. A generic multiple input multiple output wirelss communication system model together with space-time equalizers.

ploy the Hessian matrix to obtain a better approximation of the curvature of the cost function and propose the new quasi-Newton cross-correlation and constant modulus algorithm (QN-CCCMA).

2 The cross-correlation and constant modulus cost function

The antenna outputs are modelled as convolution mixtures of the independent and indentical distributed complex source signals. They are processed by a bank of d parallel space-time equalizers to mitigate the channel distortion and therefore retrieve all the source, as shown in Fig. 1. The following notations are used in this paper: $(\cdot)^H$ denotes Hermitian transpose, $(\cdot)^T$ transpose, and $(\cdot)^*$ conjugate. The channel from the ith source to the jth antenna is modelled as a FIR filter of order L. Let the order of the sub-equalizers be N. We denote $\triangle_{ij}$ as the corresponding channel convolution matrix of dimension $(N+L+1)\times(N+1)$,

$$\triangle_{ij} = \begin{bmatrix} c_{ij}(0) & c_{ij}(1) & \dots & c_{ij}(L) & 0 & \dots & 0 \\ 0 & c_{ij}(0) & c_{ij}(1) & \dots & c_{ij}(L) & 0 & \dots \\ \vdots & \dots & \ddots & & \dots & & \vdots \\ 0 & \dots & 0 & c_{ij}(0) & c_{ij}(1) & \dots & c_{ij}(L) \end{bmatrix}^T \tag{2.1}$$

Denote the ith source signal at time k as $\mathbf{s}_i(k) = [s_i(k)\dots s_i(k-N-L)]^T$. The jth antenna output is written as

$$\begin{aligned} \mathbf{x}_j(k) &= [x_j(k)\dots x_j(k-N)]^T \\ &= \triangle_{ij}^T \mathbf{s}_i(k). \end{aligned} \tag{2.2}$$

For all d sources, the composite source vector is represented by $\mathbf{s}(k) = [\mathbf{s}_1^T(k)\dots\mathbf{s}_d^T(k)]^T$. By defining an overall channel convolution matrix as

$$\Delta = \begin{bmatrix} \Delta_{11} & \dots & \Delta_{1M} \\ \vdots & \ddots & \vdots \\ \Delta_{d1} & \dots & \Delta_{dM} \end{bmatrix} \tag{2.3}$$

the space-time equalizer regressor is given by

$$\begin{aligned} \mathbf{x}(k) &= [\mathbf{x}_1^T(k)\dots\mathbf{x}_M^T(k)]^T \\ &= \triangle^T \mathbf{s}(k) \end{aligned} \tag{2.4}$$

The lth space-time equalizer tap vector is denoted as $\mathbf{w}_l(k) = [w_l^1(k)\ w_l^2(k)\dots w_l^{M(N+1)}(k)]^T$, its output is

$$y_l(k) = \mathbf{w}_l^T(k)\mathbf{x}(k) \tag{2.5}$$

$$= \mathbf{h}_l^T(k)\mathbf{s}(k) \tag{2.6}$$

where $\mathbf{h}_l(k) = \triangle \mathbf{w}_l(k)$ is the combined channel + lth equalizer impulse response. The CC-CM cost function for the lth equalizer is

$$J_l(k) = E\left\{\left(|y_l(k)|^2 - R_2\right)^2\right\} + \frac{\gamma}{2}\sum_{m=1}^{l-1}\sum_{\delta=-(N+L)}^{N+L} |E\{y_l(k)y_m^*(k-\delta)\}|^2 \tag{2.7}$$

where $E\left\{\left(|y_l(k)|^2 - R_2\right)^2\right\}$ is the constant modulus cost, the term $\sum_{m=1}^{l-1}\sum_{\delta=-(N+L)}^{N+L} |E\{y_l(k)y_m^*(k-\delta)\}|^2$ penalizes the cross-correlation between the lth equalizer and the other $l-1$ retrieved sources, $R_2 = \frac{E\{|s|^4\}}{E\{|s|^2\}}$ is the dispersion constant, and $\gamma \in R^+$ is the mixing parameter.

3 Quasi-Newton cross-correlation and constant modulus algorithm

In the classical Newton method, the update equation is given by

$$\mathbf{w}(k+1) = \mathbf{w}(k) - \mu H^{-1}(k)\mathbf{g}(k) \tag{3.1}$$

where $H^{-1}(k)$ and $\mathbf{g}(k)$ are respectively the Hessian matrix and the gradient vector of the objective function and $\mu \in R^+$ is the step size parameter. To ensure that the update is a descent direction of the cost surface, the inverse Hessian matrix $H^{-1}(k)$ must be positive definite.

For the CC-CM cost function, the gradient with respect to the tap weight vector of the lth equalizer $\mathbf{w}_l(k)$, at the kth iteration, is described by

$$\begin{aligned} \mathbf{g}_l(k) &= 2E\{\left(|y_l(k)|^2 - R_2\right) y_l^*(k)\mathbf{x}(k)\} \\ &+\gamma \sum_{m=1}^{l-1} \sum_{\delta=-(N+L)}^{N+L} E\{y_l^*(k)y_m(k-\delta)\}E\{y_m^*(k-\delta)\mathbf{x}(k)\}. \end{aligned} \tag{3.2}$$

During implementation, the gradient in (3.2) is usually approximated by its stochastic estimator.

The Hessian of the CC-CM objective function is written as

$$\begin{aligned} H_l(k) &= 2E\{\left(2|y_l(k)|^2 - R_2\right) \mathbf{x}(k)\mathbf{x}^H(k)\} \\ &+\gamma \sum_{m=1}^{l-1} \sum_{\delta=-(N+L)}^{N+L} E\{y_m^*(k-\delta)\mathbf{x}(k)\}E\{y_m(k-\delta)\mathbf{x}^H(k)\}. \end{aligned} \tag{3.3}$$

As in [6], the statistical expectation is estimated by a time average. To handle the situation of time-varying channels, by using a forgetting factor $\lambda \in (0,1)$, data from distant past is less weighted than the current inputs. Hence, we obtain

$$\begin{aligned} \widehat{H}_l(k) &= 2\frac{1-\lambda}{1-\lambda^k} \sum_{i=1}^{k} \lambda^{k-i} \left(2|y_l(i)|^2 - R_2\right) \mathbf{x}(i)\mathbf{x}^H(i) \\ &+\gamma \left(\frac{1-\lambda}{1-\lambda^k}\right)^2 \sum_{m=1}^{l-1} \sum_{\delta=-(N+L)}^{N+L} \left(\sum_{i=1}^{k} \lambda^{k-i} y_m^*(i-\delta)\mathbf{x}(i)\right) \\ &\times \left(\sum_{i=1}^{k} \lambda^{k-i} y_m(i-\delta)\mathbf{x}^H(i)\right). \end{aligned} \tag{3.4}$$

For notational convenience, we denote

$$\Psi(k) = \sum_{i=1}^{k} \lambda^{k-i} \left(2|y_l(i)|^2 - R_2\right) \mathbf{x}(i)\mathbf{x}^H(i). \tag{3.5}$$

It can be recursively updated as

$$\Psi(k) = \lambda\Psi(k-1) + \left(2|y_l(k)|^2 - R_2\right) \mathbf{x}(k)\mathbf{x}^H(k). \tag{3.6}$$

In order to reduce the computational complexity of the proposed QN-CCCMA algorithm, we use the diagonal terms of the Hermitian matrix $\sum_{m=1}^{l-1} \sum_{\delta=-(N+L)}^{N+L} \left(\sum_{i=1}^{k} \lambda^{k-i} y_m^*(i-\delta)\mathbf{x}(i)\right) \left(\sum_{i=1}^{k} \lambda^{k-i} y_m(i-\delta)\mathbf{x}^H(i)\right)$. Therefore, representing the square root of its diagonal as $\mathbf{b}(k)$, we use the diagonal matrix $\left(\mathbf{b}(k)\mathbf{b}^H(k)\right).*\mathbf{I}$ to account for the second derivative of the cross-correlation cost, that is,

$$\sum_{m=1}^{l-1}\sum_{\delta=-(N+L)}^{N+L}\left(\sum_{i=1}^{k}\lambda^{k-i}y_m^*(i-\delta)\mathbf{x}(i)\right)\left(\sum_{i=1}^{k}\lambda^{k-i}y_m(i-\delta)\mathbf{x}^H(i)\right)$$
$$\approx \left(\mathbf{b}(k)\mathbf{b}^H(k)\right).*\mathbf{I} \tag{3.7}$$

where $.*$ is the operation of point-by-point multiplication and $\mathbf{I}$ is the identity matrix. By so doing, only the diagonal elements of the Hermitian matrix in (3.7) are required in the calculation of $\mathbf{b}(k)$, which are equal to the modulus of the vector $\rho(k) = \sum_{m=1}^{l-1}\sum_{\delta=-(N+L)}^{N+L}\left(\sum_{i=1}^{k}\lambda^{k-i}y_m^*(i-\delta)\mathbf{x}(i)\right)$ and can be recursively computed as

$$\rho(k) = \lambda\rho(k-1) + \sum_{m=1}^{l-1}\sum_{\delta=-(N+L)}^{N+L} y_m^*(i-\delta)\mathbf{x}(i). \tag{3.8}$$

With respect to eqn (3.5) and eqn (3.7), the Hessian matrix in eqn (3.4) is rewritten as

$$\widehat{H}_l(k) = 2\frac{1-\lambda}{1-\lambda^k}\Psi(k) + \gamma\left(\frac{1-\lambda}{1-\lambda^k}\right)^2\left(\mathbf{b}(k)\mathbf{b}^H(k)\right).*\mathbf{I}. \tag{3.9}$$

Applying Woodbury's identity [2] to eqn (3.9), the approximated inverse Hessian matrix is given by

$$\widehat{H}_l^{-1}(k) = \frac{1-\lambda^k}{2(1-\lambda)}\Psi^{-1}(k) - \frac{\frac{1-\lambda^k}{2(1-\lambda)}\Psi^{-1}(k)\ \mathrm{diag}\ \left(\mathbf{b}(k)\mathbf{b}^H(k)\right)\frac{1-\lambda^k}{2(1-\lambda)}\Psi^{-1}(k)}{\gamma^{-1}\left(\frac{1-\lambda^k}{1-\lambda}\right)^2 + \mathbf{b}^H(k)\frac{1-\lambda^k}{2(1-\lambda)}\Psi^{-1}(k)\mathbf{b}(k)}. \tag{3.10}$$

A problem in the operation of eqn (3.10) is the matrix $\Psi^{-1}(k)$ involved. We may apply Woodbury's identity again and effectively compute $\Psi^{-1}(k)$ by

$$\Psi^{-1}(k) = \frac{1}{\lambda}\Psi^{-1}(k-1) - \frac{\lambda^{-2}\Psi^{-1}(k-1)\mathbf{x}(k)\mathbf{x}^H(k)\Psi^{-1}(k-1)}{\left(2\left|y_l(k)\right|^2 - R_2\right)^{-1} + \lambda^{-1}\mathbf{x}^H(k)\Psi^{-1}(k-1)\mathbf{x}(k)}. \tag{3.11}$$

It is feasible for the matrix in eqn (3.11) to lose its positive definite nature and monitoring may therefore be necessary in practice. However, no problem was observed in the simulation studies, as in [6].

Combining the expression for $\widehat{H}_l^{-1}(k)$ and the stochastic gradient, the QN-CCCMA update equation for the lth equalizer at the kth iteration is written as

$$\mathbf{w}(k+1) = \mathbf{w}(\mathrm{k}) - 2\mu \left(\begin{array}{c} (\,|y_l(k)|^2 - R_2) y_l^*(k) + \\ \gamma \sum_{m=1}^{l-1} \sum_{\delta=-(N+L)}^{N+L} \widehat{E}\{y_l^*(k) y_m(k-\delta)\} y_m^*(k-\delta) \end{array} \right) \times \widehat{H}_l^{-1}(k)\mathbf{x}(k) \tag{3.12}$$

where the step size μ can be adaptively chosen with respect to the inverse Hessian matrix $\widehat{H}_l^{-1}(k)$. That is, as in [7],

$$\mu = \frac{\alpha}{\varepsilon + \mathbf{x}^H(k)\widehat{H}_l^{-1}(k)\mathbf{x}(k)}. \tag{3.13}$$

Here the parameters ε and α are small constants within the range of $(0, 1]$.

4 Remarks

(1) Convergence. By inspection of eqn (3.9), provided the positive definite property of the approximated Hessian matrix is retained, the update direction remains a descent direction in the adaptation [1]. When the parameters α and ε are appropriately chosen, which means that step size μ is neither too large nor too small, the QN-CCMA algorithm is able to converge to a local minimum in the cost surface.

In terms of the convergence speed, the Newton method is able to provide quadratic convergence rate compared with the linear convergence rate of the stochastic gradient descent type algorithms. As the matrix $\widehat{H}_l^{-1}(k)$ is only an approximation to the true inverse Hessian matrix, quadratic convergence is unlikely in the QN-CCCMA algorithm. However, since more accurate curvature information is built up in the adaptation of QN-CCCMA than that in LMS-CCCMA, faster convergence speed is achieved by the QN-CCCMA algorithm, as confirmed by the simulation results.

(2) Computational Complexity. With a p tap space-time equalizer, $6p^2 + 2p + (l-1)((N+L)(2p+4) + 2p + 5)) + 5$ complex multiplications, 4 complex divisions and p square root operations are required at each iteration of the QN-CCCMA algorithm. Hence, the price for fast convergence is also in the increase of computational complexity when compared with the conventional LMS-CCCMA algorithm. But if the inputs of the equalizer are statistically stationary or slowly varying compared with the convergence of the filter tap weights, the update of the inverse Hessian matrices can be performed every d samples once steady state is reached. Furthermore, it is possible to reduce the complexity if the Levinson–Durbin recursion is used to calculate efficiently the quantity $\widehat{H}_l^{-1}(k)\mathbf{x}(k)$.

5 Simulation

A QPSK system with source alphabet $\left\{\pm\frac{1}{\sqrt{2}} \pm \frac{1}{\sqrt{2}}j\right\}$ is assumed. In the first simulation, we assume $d = 2$ users, $M = 3$ sensors and six random complex

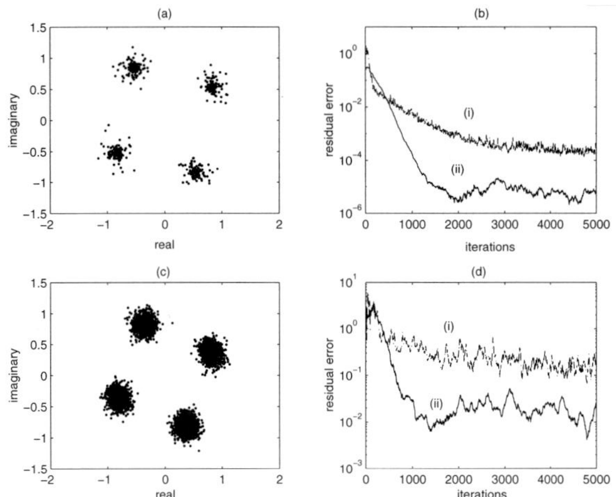

FIG. 2. Comparison between QN-CCCMA and CCCMA: i—CC-CMA algorithm ii—QN-CCCMA algorithm. (a) Eye diagram of EQ-1 after 300 samples; (b) residual error of EQ-1; (c) eye diagram of EQ-2 after 700 samples; (d) residual error of EQ-2.

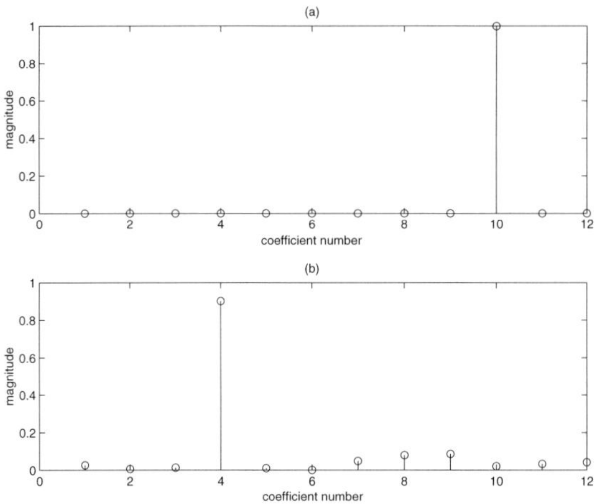

FIG. 3. Retrieval of all sources with the QN-CCCMA algorithm. (a) Recovery of source-2: combined channel + EQ-1 impulse response; (b) recovery of source-1: combined channel + EQ-2 impulse response.

channels of order 2 ($L = 2$). The order of the sub-equalizer is $N = 3$. Thus the length of the space-time equalizer is 12 ($thatis$, $M(N + 1) = 12$). The channel convolution matrix $\triangle^T$ is a full rank square matrix of dimension 12. White Gaussian noise of $SNR = 30$ dB is assumed to be present at the channel output.

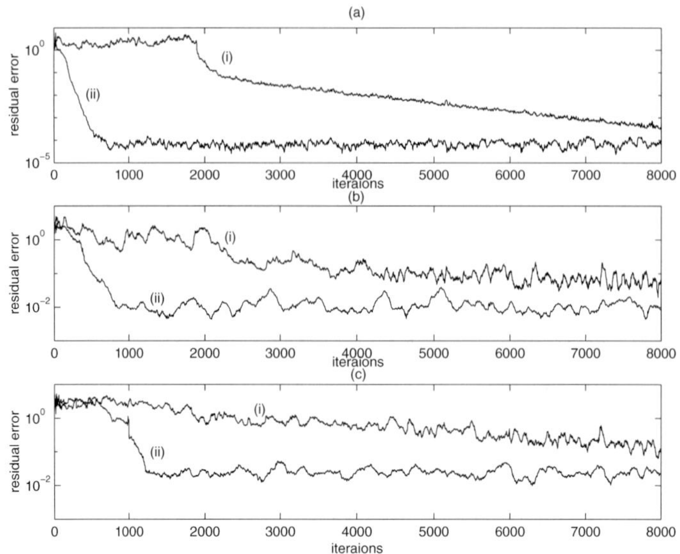

FIG. 4. Comparison between QN-CCCMA and CCCMA: i—CC-CMA algorithm ii—QN-CCCMA algorithm. (a) Residual error of EQ-1; (b) residual error of EQ-2; (c) residual error of EQ-3.

The performance comparison between QN-CCCMA and LMS-CCCMA is shown in Fig. 2, where the forgetting factor λ and the constant ε are respectively set to 0.99 and 0.5. Comparing the residual error, which is defined as:

$$\frac{\|\mathbf{h}_l(k)\|_2^2 - \max(|\mathbf{h}_l(k)|)^2}{\max(|\mathbf{h}_l(k)|)^2} \tag{5.1}$$

the proposed QN-CCCMA algorithm shows much faster convergence speed than the conventional LMS-CCCMA algorithm, (see Figs 2(b) and (d)). The signal constellations of the equalizer outputs when adapted by QN-CCCMA are given in Figs 2(a) and (c). After approximately 700 data samples, both equalizer-1 and equalizer-2 give open-eye patterns. In comparison, with LMS-CCCMA, clear open-eye constellation cannot be achieved even with 4000 samples for equalizer-2. In Fig. 3, the combined channel + equalizer impulse responses of the two equalizers are described. The retrieval of all three sources is confirmed by the different position sections of the largest impulse. That is, equalizer-1 and 2 respectively retrieve source-2 with delay of 3 and source-1 with delay of 3.

In the second simulation, the system has $d = 3$ users and $M = 4$ sensors. The order of the channel is $L = 1$. The order of the sub-equalizer is $N = 2$, which implies that the space-time equalizers have a length of $M(N + 1) = 12$ and the channel convolution matrix is full rank. Additive white Gaussian noise of $SNR = 30$ dB is also assumed. The simulation results shown in Fig. 4 also confirm the desirable fast convergence property of the QN-CCCMA algorithm

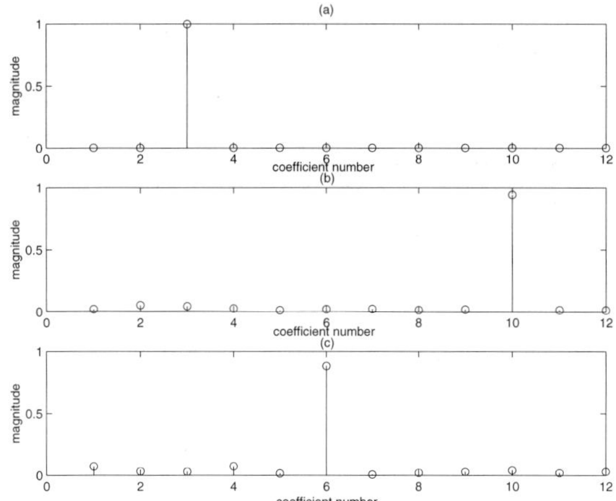

FIG. 5. Retrieval of all sources with the QN-CCCMA algorithm. (a) Recovery of source-1: combined channel + EQ-1 impulse response; (b) recovery of source-3: combined channel + EQ-2 impulse response; (c) recovery of source-2: combined channel + EQ-3 impulse response.

over the LMS-CCCMA algorithm. For equalizer-1, only 200 data samples are required for the open-eye constellation diagram to be achieved, compared with 1500 samples in LMS-CCCMA. For equalizer-2, for the eye to be open, 450 samples are required by the QN-CCCMA. But when adapted with the LMS-CCCMA algorithm, approximately 2000 samples are respectively needed. For equalizer-3, 800 data samples are required for QN-CCCMA and at least 5000 samples for LMS-CCCMA. All the different sources are reconstructed, as shown in Figs 5(a–c). Equalizer-1 retrieves source-1 with two delays. Equalizer-2 and equalizer-3 respectively retrieve source-3 and source-2, both with one delay.

6 Conclusion

In order to overcome the slow convergence of the conventional LMS-CCCMA algorithm, we proposed the quasi-Newton cross-correlation and constant modulus algorithm (QN-CCCMA) by using the Hessian information of the cost function and repeated application of the matrix inversion lemma to invert the Hessian matrix. Simulation results show that an order of magnitude improvement in convergence rate is achieved with the proposed algorithm, at the expense of an order of magnitude increase in computational complexity.

Bibliography

[1] Gill, P.E. (1981). *Practical Optimization*, Academic Press, New York.

[2] Haykin, S. (1983). *Communication Systems*. Wiley, New York.

[3] Papadias, C.B. and Paulraj, A (1997). A constant modulus algorithm for

multiuser signal separation in the presence of delay spread using antenna arrays *IEEE Signal Process. Lett.*, **4**, 178–181

[4] Haykin, S (2000). Blind separation of independent sources based on multiuser kurtosis optimization criteria. In *Unsupervised Adaptive Filtering* (ed. Papadias, C.B.). Vol. 2, 147–179, Wiley, New York

[5] Godard, D.N. (1980). Self-recovering equalization and carrier tracking in two dimensional data communication system *IEEE Trans. on Communications*, **28**, 1867–1875

[6] Yan, G. and Fan, H. (2000). A Newton-like algorithm for complex variables with applications in blind equalization *IEEE Trans. on Signal Processing*, **48**, 553–556

[7] Schirtzinger, T.R. and Li, X. and Jenkins, W.K. (1995). A comparison of three algorithms for blind equalization based on the constant modulus error criterion *Proc. IEEE Int. Conf. Acoust. Speech Signal Process.* Detroit, 1049–1052.

Audio Source Separation

Mike Davies
Signal Processing Laboratory, King's College London, The Strand, London WC2R 2LS, UK

Abstract

We examine the problem of blind audio source separation. This is an extension of independent component analsysis (ICA) to the case where the mixing process is convolutional. As such we approach the problem in the frequency domain in a similar manner to [15]. However, this introduces a permutation problem. Here we suggest solutions based upon appropriate time–frequency audio models (priors) using the Short Time Fourier Transform (STFT) which provides a discriminator that we can use to identify the correct permutations. Unfortunately, this does not guarantee that the incorrect permutations do not persist as local minima. We show that empirically this does happen. We also present two solutions to overcome this problem.

1 Introduction

In 1991 Jutten [10] proposed a Neural Network for separating out independent signals from a set of mixtures. This has led to the development of Independent Component Analysis (ICA) or Blind Signal Separation (BSS). The problem is deceivingly simple. Let

$$x_n = As_n \tag{1.1}$$

where x_n is a vector of observed data, A is an unknown mixing matrix and s_n is a vector of independent components. Then, given only x_n, the problem is to try to identify the sources s_n. When A is invertible this is equivalent to estimating $W = A^{-1}$ and $y = Wx$. When the sources are independent and non-Gaussian no additional information is necessary to identify W (up to trivial ambiguities) and hence y. For some excellent reviews of the subject and its relationship to Higher Order Statistics and Maximum Likelihood methods see [6, 4, 7].

Currently a number of extensions to ICA are being explored. Of particular interest in audio is the extension to convolutional mixing since an acoustic environment typically includes delays, echoes and reverberation. However, we are by no means the first to tackle the convolutional mixing problem and a number of algorithms have been proposed [5, 16, 17, 11, 13, 15]. These approaches either exploit a signal's nonGaussian structure or its 2nd order nonstationary structure. Both techniques are in fact strongly related since nonstationary in this context is being used to describe signals with slowly varying variance profiles [14]. The

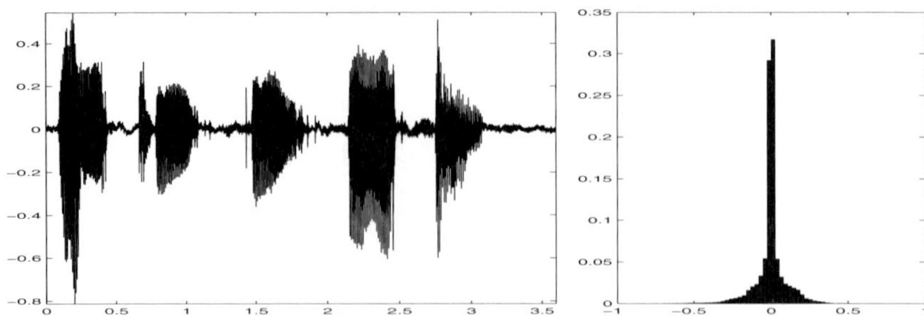

FIG. 1. Time history of the speech 'one, two, three, four, five' along with a histogram of the signal values.

sample distribution of such signals is typically highly non-Gaussian. Figure 1 shows a short time history of some speech. Such signals can be modelled equally as nonstationary or nonGaussian.

Another issue is whether to work in the time or frequency domain. Working in the time domain has the disadvantage that repeated calculations of convolutions are required and this is computationally expensive. Furthermore, signals such as audio are typically much more nonGaussian in the frequency domain [2, 18] and the convolution becomes multiplication. Unfortunately, this last point introduces a permutation problem: how do we choose which source in one frequency bin to match to one at another frequency?

In this paper we review the permutation problem and argue that there must be some coupling of the frequencies in a separation algorithm. If, as we do here, we build a probablistic model this requires us to have frequency coupling in either the source priors or the unmixing priors. Here we consider priors on the sources and introduce a simple time–frequency model based on the STFT. This provides us with an efficient means to discriminate between the different permutations. Unfortunately, this does not guarantee that the incorrect permutations do not persist as local minima. Indeed, as we will see, this does seem to happen. We propose two different approaches to overcome this problem.

2 Previous work

Initial solutions based on nonGaussian statistics were developed in the time domain [5, 16]. However, a new approach was suggested in [11] that calculated updates for an unmixing process in the frequency domain. Here the convolutions become multiplications. Therefore, at each frequency bin we must estimate an unmixing matrix $W(\omega)$. However, really this is still a time domain method, since the nonGaussian model is being applied in that domain. This requires the algorithm to transform between time and frequency at each iteration. The resulting algorithm is given by

$$\Delta W(\omega) \propto (I - \mathrm{fft}(\phi(y(t)), \omega) y(\omega)^H) W(\omega) \tag{2.1}$$

where by $\text{fft}(\phi(y(t)), \omega)$ we mean the ω frequency components of the signals $\phi(y)$: see [11] for details.

Smaragdis [15] proposed working directly in the frequency domain applying a nonlinear activation function to coefficients from the STFT. Since the activation function can be viewed as representing the derivative of the log prior of the sources this is essentially applying a probability model in the frequency domain. The advantage of this method is computational simplicity since the procedure works completely in the frequency domain. The update rule is

$$\Delta W(\omega) \propto (I - \phi(y(\omega))y(\omega)^H)W(\omega) \tag{2.2}$$

where ϕ is an appropriate nonlinear activation function (see Section 5.1 below). The price we pay for this though is a new problem: permutation ambiguity. Since we are treating each frequency bin as independent there is a permutation ambiguity for each bin. We have no means by which we can associate a permutation choice at one frequency with that at another.

Another avenue of research has been to use second-order statistics (known to be inadequate for normal ICA) along with a number of additional constraints. In Weinstein *et al.* [17] they forced the unmixing model to consist of fixed-length FIR filters. The authors also noted that a nonstationarity property could be used. This latter idea has been pursued in [13] where the authors worked in the frequency domain and came across the same permutation problem as [15]. Their solution was to apply a fixed-length FIR filter constraint as in [17]. This is acheived by applying a projection operator P to the filter estimates at each iteration, where $P = FZF^{-1}$, F is the Fourier transform, and Z is a diagonal operator that projects onto the first q terms. Like the algorithm of [11], this requires transforming between time and frequency at each iteration. There is also some evidence that the algorithm can get trapped in local minima [9].

3 Model formulation

Let $x(t) = [x_1(t), x_2(t), \ldots, x_d(t)]^T$ be d observed signals that are convolutive mixtures of d statistically independent sources $s(t) = [s_1(t), s_2(t), \ldots, s_d(t)]^T$ such that

$$x(t) = \sum_{\tau} A(\tau)s(t - \tau). \tag{3.1}$$

We wish to identify the unknown source signals given $x(t)$. For simplicity we have ignored any additive noise, noting that Cardoso [4] observed that it is still an open question to determine which application domains would benefit from explicit noise modelling. Although there are a number of indeterminacies in this model, such as permutation, sign, amplitude, and spectral shape, these can be effectively ignored if we aim to reconstruct the source signals *observed by the sensors*, as proposed in [3]. This leaves only an arbitrary ordering of the sources.

Transforming to the frequency domain gives

$$X(\omega) = A(\omega)S(\omega) \tag{3.2}$$

where $A(\omega)$ is a $d \times d$ complex matrix. As in the standard ICA problem we can define an unmixing matrix $W(\omega) \simeq A(\omega)^{-1}$. The kth *observed source* can then be estimated as

$$X_k(\omega) \simeq W(\omega)^{-1} P_k W(\omega) X(\omega) \tag{3.3}$$

where P_k projects out all but the kth source.

Putting this into a probablistic framework we can estimate $W(\omega)$ by maximizing its probability given the observations X (MAP estimate):

$$\log P(W|X) \propto \log P(W) + \log P(Y) + \sum_{\omega} \log \det W(\omega). \tag{3.4}$$

3.1 Frequency domain permutation symmetries

Constructing frequency independent priors for W and Y allows the MAP solution to be derived separately for each frequency bin:

$$\log P(W(\omega)|X(\omega)) \propto \log P(W(\omega)) + \log P(Y(\omega)) + \log \det W(\omega). \tag{3.5}$$

Indeed this is the function that the Smaragdis algorithm maximises. The problem now comes when we try to reconstruct the sources since there is the inherent permutation ambiguity in the rows of $W(\omega)$ [13, 15]. This is more than just an ordering ambiguity because we have broken one BSS problem into N BSS problems which manifests itself as a highly symmetric posterior distribution with an identical minimum for each permutation (correct or otherwise). To avoid this the model *must* have some frequency coupling.

3.2 Frequency coupling priors

It is interesting to examine how previous solutions have effectively introduced frequency coupling. In [11] the frequency coupling comes from applying the non-Gaussian prior to y in the time domain (generating inter-frequency dependencies through the higher-order spectra). We can also interpret the constraints of [13] as very strong priors on W. Frequency coupling is evident by the fact that the linear projection matrix FZF^{-1} is not diagonal. Finally, we can interpret the heuristic solution in [15] as placing weakly coupled priors on $W(\omega)$ of the form

$$P(W(\omega)|W(\omega-1)) \propto \exp(-\frac{1}{2\sigma^2}\|W(\omega) - W(\omega-1)\|_F). \tag{3.6}$$

This imposes some weak smoothness constraint across frequency which should have a similar effect to the constraints in [13].

Below we construct a prior on y that allows us to work in the frequency domain while at the same time providing the necessary frequency coupling.

4 Source modelling

In [11] the application of a tanh nonlinear activation function in the time domain imposes the assumption of super-Gaussianity upon the time domain signals.

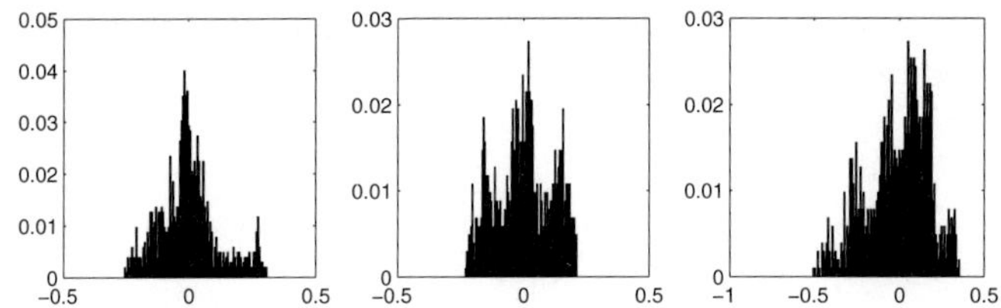

FIG. 2. Three histograms of sections from the signal in Fig. 1

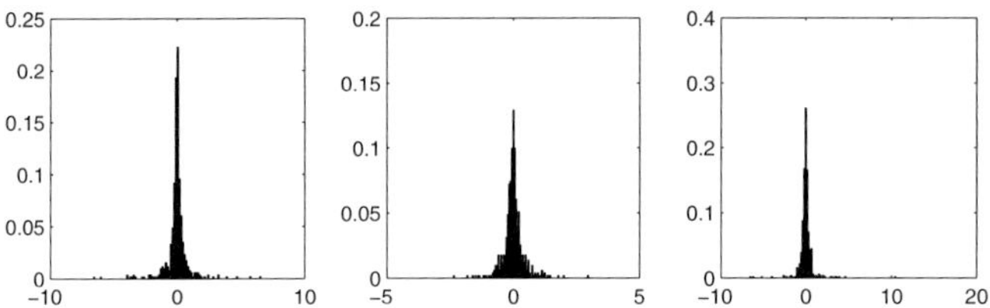

FIG. 3. Three histograms of the pre-whitened STFT coefficients for the same sections of signal as used in Fig. 2.

Figure 1 showed a time history of a speech signal along with a histogram of the signal values. The sample distribution is super-Gaussian. However, if we examine the speech over shorter quasi-stationary periods the data are not well modelled as super-Gaussian (Fig. 2). Indeed the kurtosis of the three sections shown are all about −3. The strong super-Gaussianity comes mainly from the modulated amplitude of the signal.

In contrast to this, if we model in the frequency domain then even the quasi-stationary sections have very heavy-tailed distributions [2, 18]. Figure 3 shows the distributions of the real part of the frequency coefficents for the same sections as above after normalizing.

This suggests that on short time scales the signal is best modelled as non-Gaussian in the frequency domain. However the slowly varying amplitude profile also gives us valuable information that can be exploited for source separation and does not seem to be affected by the permutation problem. This motivates us to introduce the following time–frequency model.

5 A time–frequency model

Let us assume the frequency coefficients from an STFT are distributed by some nonGaussian distribution. However, we want our model to explicitly take into account the scaling of the signal (that is, the signal envelope) which we will assume is approximately constant over the analysis window. This is modelled by a nonstationary time varying scale parameter β_k. Thus we have

$$P(y_k(t, \omega)) \sim \beta_k(t)^{-1} \exp(-h(y_k(t, \omega)/\beta_k(t))) \tag{5.1}$$

where $h(\cdot)$ defines the density's form, the index t represents the timeframe index and k is the source index. The key feature is that $\beta_k(t)$ is *not* a function of frequency. This restriction provides us with a sufficient coupling between the frequencies to break the permutation symmetries.

Let us now consider how this alters the simple gradient algorithm proposed in [15]. Incorporating $\beta_k(t)$ into (2.2) we get

$$\Delta W(\omega) = \delta(I - \beta(t)^{-1}\phi(y(t,\omega))y(t,\omega)^H)W \tag{5.2}$$

where $\beta(t) = \text{diag}(\beta_1(t), \ldots, \beta_{d_s}(t))$ is an estimated scaling matrix. The additional computation here is a single multiplication by a diagonal $d \times d$ matrix. However, we still have to estimate $\beta_k(t)$. Although we could place a prior on the value of $\beta_k(t)$, mimicking the effect of the nonlinearity in [11] we choose to simply estimate its value from the N frequency samples in the STFT. Its exact form will depend on our choice of activation function ϕ.

5.1 Complex activation functions

One complication with defining our distribution in the frequency domain is that we must come up with a complex the activation function. Smaragdis [15] shows that the standard Bell and Sejnowski nonlinearity, tanh, cannot be used since it introduces singularities in the complex plane. Instead he proposes the use of $\phi_k(y_k(\omega)) = \tanh(\text{Re}\, y_k(\omega)) + \mathrm{i}\tanh(\text{Im}\, y_k(\omega))$ which is bounded in the complex plane and provides a legitimate activation function. Here, however we prefer to use a slightly different activation function, based upon the expected behaviour of $y_k(\omega)$. A probablistic interpretation of the Smaragdis activation function is a prior proportional to $(\cosh(\text{Re}\, s_k(\omega)) \times \cosh(\text{Im}\, s_k(\omega)))^{-1}$. This prior is not rotationally symmetric and imposes preferences upon the phase angles expected, treating the real and imaginary components as independent.

It seems more intuitive to impose no preference on the phase angle, that is to introduce circularly symmetric priors on complex variables. This is essentially the same as the priors on subspaces used by Hyvärinen [8] in his Independent Subspace Analysis (ISA). An activation function for such a prior takes the form of $(s/|s_i|)f(|s_i|)$. If we assume a complex Laplacian then $f(|s_i|)$ is 1. Similarly we could use $f(|s_i|) = \tanh(|s_i|)$, etc. We note that [1] have also recently proposed these models for complex activation fuinctions and noted the relationship between ISA and ICA with complex variables.

Assuming a Laplacian distribution for the coefficients provides us with the following estimate for $\beta_k(t)$:

$$\beta_k(t) = \frac{1}{N}\sum_{\omega} |y_k(t,\omega)|. \tag{5.3}$$

Note that here we are explicitly coupling the behavior in the individual frequency bins. We will use this Laplacian model in our subsequent experiments.

5.2 Permutation problem revisited

Let us now look at what effect such a model has upon the permutation symmetries. Recall that without the β term the log likelihood function has an identical minimum for every permutation of the sources at each frequency.

Introducing β weights the likelihood of an unmixing matrix at a given frequency with the time envelope induced by the components at other frequencies. Thus β allows the matching of time envelopes. This provides us with a discriminator for the different permutations.

Unfortunately, a direct application of (5.2) and (5.3) does not guarantee that the correct permutation will be found. The introduction of β will of course break the symmetry; however, it will not necessarily change the cost function enough to completely remove spurious minima. Thus using a gradient based optimization scheme is likely to get stuck in a local minimum. This explains the behaviour noted by [15] and is probably related to the poor performance of the Parra and Spence algorithm [13] observed by [9].

To overcome this problem we propose two possible modifications.

5.3 A nonstationary second-order solution

In the analysis above the persistance of the local minima can be attributed to the nonlinearity ϕ. Let us consider what happens if, instead, we choose ϕ to be linear. This implies that our prior is now Gaussian. Applying this niavely to the Smaragdis algorithm will clearly fail since Gaussian statistics introduce a rotational ambiguity. That is there will be a single degenerate minimum containing all the different permutation solutions as a subset. However, the inclusion of the β term in (5.2) allows us to exploit the nonstationarity which permits a valid ICA solution [14].

In this case we have $\phi_k(y_k) = y_k/\beta_k$ where β_k is the standard deviation of y_k and our updates become

$$\Delta W(\omega) = \delta(I - \beta(t)^{-2}y(t,\omega)y(t,\omega)^H)W \tag{5.4}$$

$$\beta_k(t)^2 = \frac{1}{N}\sum_{\omega} |y_k(t,\omega)|^2. \tag{5.5}$$

Such an algorithm does not suffer from the permutation problem and can succesfully separate out convolutionally mixed sources.

In fact it is possible to view this algorithm as modelling the sources in the time domain since a Gaussian i.i.d. process in time transforms to an i.i.d. Gaussian process in frequency. We can therefore regard this algorithm as being very closely related to the algorithm of Lee *et al.* [11], the difference being that we are modelling the fluctuation of the amplitude in the time domain as a slowly varying i.i.d. process whereas in [11] it is modelled as a fixed nonGaussian i.i.d. process. The advantage of our algorithm is that it works purely on the STFT and does not require the expensive transformation from time to frequency at each iteration.

This solution is not ideal, however, since it does not exploit the superGausianity present in the frequency domain and experimental results suggest that this reduces the quality of the separation and slows down the algorithms' convergence.

5.4 A likelihood ratio jump solution

An alternative solution is to live with the multiple minima but introduce a mechanism by which the correct one is obtained. Fortunately, due to the symmetry of the problem, if we know where one minimum is we know where they all are. It is therefore possible to introduce a jump step into the update that chooses the permutation that is most likely.

Here we describe a solution for $d = 2$ using the Laplacian prior model developed in Section 5.1 (see the appendix for full derivation). At each step of the algorithm we can compare the likelihood of the unmixing matrix $W(\omega)$ with that of $\left[\begin{smallmatrix} 0 & 1 \\ 1 & 0 \end{smallmatrix}\right] W(\omega)$ (taking into account possible incorrect scaling). Assuming Laplacian priors, we can define the quantities:

$$\gamma_{ij} = \sum_{t=1}^{T} \frac{|y_i(t)|}{\beta_j(t)}. \tag{5.6}$$

where we are summing over T STFT frames. Then we can define the test in terms of

$$\mathrm{LR} = \left(\frac{\gamma_{12}\gamma_{21}}{\gamma_{11}\gamma_{22}}\right) \tag{5.7}$$

If $\mathrm{LR} < 1$ we permute $W(\omega)$ before proceeding.

There are two drawbacks of this approach. First, the likelihood test becomes more complicated for more than two sources (although one possible solution would be to consider the sources in a pairwise fashion as in [7]). Secondly, a simple adaptive version similar to [11, 15] is not possible using a one-sample likelihood. However, the algorithm does perform well in a batch mode which could therefore be used in a similar manner to that of [13].

6 Numerical experiments

In this section we compare the Smaragdis algorithm with the two solutions described above. Initially we applied it to some real data available from [12] of two people speaking simultaneously in a room. However, it appears that the room acoustics are relatively simple and a direct application of Smaragdis' algorithm (2.2) performs reasonably well when the unmixing matrices are all initialized to the identity. In this experiment the difference between the two algorithms is inaudible. However, if the unmixing matrices are randomized the Smaragdis algorithm completely fails to separate the signals even with the heuristic coupling (3.6). Similarly, the time–frequency algorithm (5.2) using only gradient descent failed to separate the sources with random initialization.

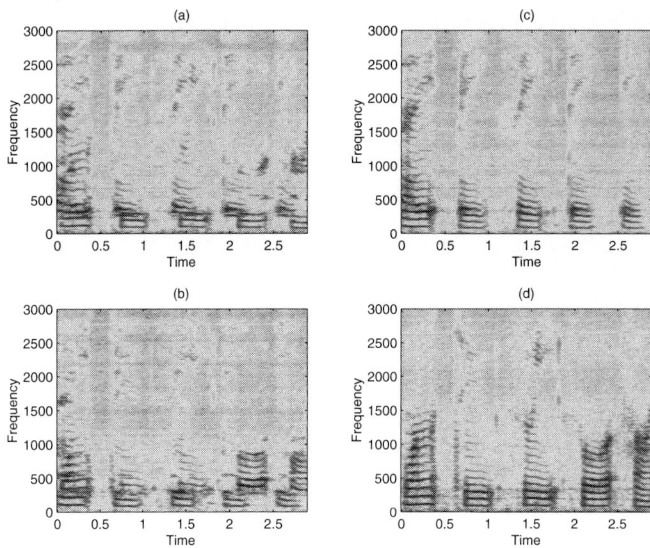

FIG. 4. Spectrograms for the sources estimates using the Smaragdis algorithm, (a) and (b), and using the Likelihood Ratio Jump algorithm, (c) and (d).

In contrast, both the nonstationary second-order algorithm and the likelihood ratio jump algorithm separated the signals successfully with random initialization. Comparing the two, the second-order algorithm took about twice as long to converge and did not produce as high a quality of separation (the interfering source was still audible).

We then constructed a synthetic mixture of two sources. This time the room acoustics had non-trivial echo components of around 5–10 ms. In this situation even with sensible initialization the Smaragdis algorithm failed to separate the sources. Figure 4 shows the spectrograms for the estimated sources using the Smaragdis algorithm and the Likelhood Ratio Jump test. The separation in the latter is nearly perfect (little audible difference from the original sources). These plots illustrate the permutation problem very well. Whereas both algorithms appear to have performed reasonable separation on a frequency-by-frequency comparison, it is clear that there is a permutation change in the Smaragdis estimates at about 400 Hz and another one (less obvious) at around 900 Hz. The result is that each source estimate contains large proportions of both sources which are both audible.

7 Conclusions

In this work we have explored the permutation problem that arises when implementing source separation algorithms in the frequency domain. We argued that it is necessary to include priors either on the source signals y or the unmixing process W that contain frequency coupling. This led us to construct a time–frequency source prior that could be implemented efficiently. However, a direct application of the gradient rule for this model is not guaranteed to work. The robustness of the ICA update to incorrect priors means that the incorrect permutations can still persist as local maxima, albeit with lower likelihood levels than a correct permutation. Two solutions have been presented. The first abandons the super-Gaussian prior to remove the incorrect minima. The second solution uses a likelihood ratio test to jump between permutations to identify the most likely one. In preliminary tests this second method appears to be superior. We also note that this latter solution could potentially be applied to the prior on W suggested in Section 3.6.

In future we hope to examine the performance of these algorithms further and to explore the use of more sophisticated time–frequency models.

Bibliography

[1] Bingham, E. and Hyvärinen, A. (2000). A fast fixed-point algorithm for independent component analysis of complex-valued signals. *Int. J. of Neural Systems*, **10**, pp. 1–8.

[2] Bofill, P. and Zibulevsky, M. (2000). Blind Separation of more source than mixtures using sparsity of their short-time Fourier transform. *Proc. ICA2000*, pp., 87–92.

[3] Cardoso, J.-F. (1998). Multidimensional independent component analysis, *Proc. ICASSP '98*, Seattle.

[4] Cardoso, J.-F. (1998). Blind signal separation: statistical principles, *Proc. IEEE,* **90**, 2009–2026.

[5] Cichocki, A., Amari, S.-I. and Cao, J. (1996). Blind separation of delayed and convolved signals with self-adaptive learning rate, *Proc. Int Symp. on Nonlinear Theory and Applications.*

[6] Comon, P. (1994). Independent compoinent analysis—a new concept? *Signal Process.*, **36**, 287–314.

[7] Hyvärinen, A. (1999). Survey on independent component analysis. *Neural Comput. Surveys* **2**, 94–128.

[8] Hyvärinen, A. and Hoyer, P. (1999), Independent subspace analysis shows emergence of phase and shift invariant features from natural images. *Proc. Int. Joint Conf. on Neural Networks.*

[9] Ikram, M. and Morgan, D. (2000). Exploring permutation inconsistency in blind separation of speech signals in a reverberant environment. *Proc. ICASSP 2000.*

[10] Jutten, C. and Herault, J. (1991). Blind separation of sources, part I: An adaptive algorithm based on the neuromimetric architecture. *Signal Process.* **24**, 1–10.

[11] Lee, T-W., Bell, A. J. and Lambert, R. (1997). Blind separation of delayed and convolved sources. *Proc. NIPS 97.*

[12] http://www.cnl.salk.edu/~tewon/ica_cnl.html

[13] Parra, L. and Spence, C. (1998). Convolutive Blind Signal Separation based on Multiple Decorrelation. *Proc. IEEE Workshop on Neural Networks for Signal Processing.*

[14] Pham, D.-T. and Cardoso, J.-F. (2000). Blind separation of instantaneous mixtures of non-stationary sources. *Proc. ICA2000*, 187–192.

[15] Smaragdis, P. (1997). Information Theoretic Approaches to Source Separation, *Masters Thesis*, MAS Department, Massachusetts Institute of Technology.

[16] Torkkola, K. (1996). Blind separation of convolved sources based on information maximisation, *Proc. IEEE Workshop on Neural Networks and Signal Processing*, Kyota, Japan.

[17] Weinstein, E., Feder, M. and Oppenheim, A.V. (1993). Multi-channel signal separation by decorrelation. *IEEE Trans. on Speech and Audio Processing*, **1**, 405–413.

[18] Zibulevsky, M. and Pearlmutter, B. (2000). Blind source separation by sparse decomposition. *Proc. ICA2000.*

Appendix

Here we derive the likelihood ratio test used in Section 5.4 for two sources with the assumption of Laplacian priors. For a given set of known $W(\omega)$, $\beta_k(t)$ and $y(\omega) = W(\omega)x(\omega)$ we wish to compare two possible choices for source estimates $\tilde{y}(\omega)$:

$$(1)\ \begin{bmatrix} \gamma_{11} & 0 \\ 0 & \gamma_{22} \end{bmatrix} \tilde{y}(\omega) = y(\omega)$$

$$(2)\ \begin{bmatrix} 0 & \gamma_{12} \\ \gamma_{21} & 0 \end{bmatrix} \tilde{y}(\omega) = y(\omega).$$

γ_{ij} are rescaling parameters that account of incorrect scaling (note that $W(\omega)$ is likely to have been adapted to best suit the current permutation). To compare these two possibilities we will evaluate their likelihood over T time frames. This gives

Case 1:

$$\log P(y|\gamma_{11}, \gamma_{22}) = -T \log(\gamma_{11}\gamma_{22}) + \log P(\tilde{y}(\omega))$$

Case 2:

$$\log P(y|\gamma_{12}, \gamma_{21}) = -T \log(\gamma_{12}\gamma_{21}) + \log P(\tilde{y}(\omega))$$

with the values of γ_{ij} chosen to maximize the likelihood. For the Laplacian model these are

$$\gamma_{ij} = \frac{1}{T} \sum_t \frac{|y_i(t, \omega)|}{\beta_j(t)}$$

which gives (5.6). We can now evaluate the likelihood of $\tilde{y}(\omega)$ in terms of the known quantities $y(\omega)$ and γ. For case 1 we have

$$\log P(\tilde{y}(\omega)) \propto -\gamma_{11}^{-1} \sum_t \frac{|y_1(t,\omega)|}{\beta_1(t)} - \gamma_{21}^{-1} \sum_t \frac{|y_2(t,\omega)|}{\beta_2(t)}$$

which reduces to $\log P(\tilde{y}(\omega)) \propto -2T$. The analysis for case 2 is identical. Thus, overall we get

$$\log P(\text{"case 1"}) - \log P(\text{"case 2"}) = -T \log(\gamma_{11}\gamma_{22}) + T \log(\gamma_{12}\gamma_{21})$$

giving

$$\frac{P(\text{"case 1"})}{P(\text{"case 2"})} = \left(\frac{\gamma_{12}\gamma_{21}}{\gamma_{11}\gamma_{22}}\right)^T.$$

Ignoring the power term gives the test defined in (5.7).

A Novel Approximate Joint Diagonalization Algorithm

M. Klajman and J. A. Chambers
Communications and Signal Processing Research Group, Imperial College of Science, Technology and Medicine, London SW7 2BT, UK

Abstract

Most existing blind source separation algorithms unmix the sources in two stages. First the data are decorrelated. Then, the transformation that relates the independent sources to the decorrelated mixture is found as a pure rotation since these two, the sources and the decorrelated data, are spatially white vectors. This rotation is found by jointly diagonalizing a set of covariance matrices, or alternatively a set of cumulant slices, by a unitary matrix. The fundamental problem is to retain the unitary property of the operators, and this is achieved in this paper via the Cayley transformation. Based on this approach, we present two novel joint unitary diagonalization algorithms. Simulation studies are undertaken to evaluate the proposed schemes.

1 Introduction

Blind source separation (BSS) addresses the separation of mixed spatially independent sources. BSS algorithms operate blindly in the sense that, except for the statistical independence assumption, no *a priori* knowledge about the sources or mixing matrix is available. If the source signals have non-overlapping spectra, separation on the basis of only their second-order statistics is possible. The second-order blind identification (SOBI) algorithm [1] unmixes the sources in two phases. First, a whitening matrix is applied to the observed data. At this stage, the data are decorrelated, but not yet independent. Next, a unitary diagonalizer is found by performing successive Jacobi rotations on a set of whitened covariance matrices in order to minimize their off-diagonal elements [2]. Knowledge of the whitener together with the unitary matrix is sufficient to estimate the mixing matrix. Although separation could be achieved by using a single extra covariance matrix, approximate joint diagonalization of several covariance matrices offers several advantages. The statistical efficiency of the procedure is increased by inferring the mixing matrix from a larger set of statistics. More importantly, however, it reduces the possibility of choosing an unfortunate lag that would prevent separation completely. Thus, after prewhitening, the BSS problem is essentially reduced to searching for a unitary matrix that diagonalizes the different whitened covariance matrices.

In this paper, we present two novel diagonalization algorithms based on the

Cayley transformation. The Linear Combination Diagonalization (LCD) algorithm finds some kind of 'average' unitary matrix by forming a linear combination of the individual unitary matrices in the Cayley domain. The Cayley-based Joint Unitary Diagonaliser (CJUD) simply searches for the optimal diagonalizer of all the covariance matrices.

Section 2 discusses briefly the fundamentals of second-order BSS. In Section 3 the Cayley transform is explained. The next section describes the cost function of the joint unitary diagonalizer and in Section 5 the analytical gradient of the cost function is derived. Section 6 describes the CJUD algorithm and in Section 7 we present some simulations. Finally, Section 8 concludes this paper.

2 Preliminaries

The noiseless BSS problem, with the assumption of an equal number of sources and sensors, can mathematically be described as follows:

$$\mathbf{x}(t) = A\mathbf{s}(t). \tag{2.1}$$

In this context, vector $\mathbf{s}(t) = [s_1(t), \ldots, s_n(t)]^T$ contains the original sources, vector $\mathbf{x}(t) = [x_1(t), \ldots, x_n(t)]^T$ contains the output sampled at time t and the matrix A is the $n \times n$ mixing matrix or transfer function between the sources and sensors. An implicit assumption is that the sources have unity variance.

First, we obtain a whitening matrix W that satisfies the following equation:

$$WR_x(0)W^H = I \tag{2.2}$$

where $R_x(0)$ is the covariance matrix of the mixture at zero lag, I is the $n \times n$ identity matrix, and $(\cdot)^H$ denotes the Hermitian transpose operator. Next, the whitener is applied to all the non-zero lag mixture covariance matrices:

$$\underline{R}_x(\tau) = WR_x(\tau)W^H. \tag{2.3}$$

Define $\Re : \{\underline{R}_x(\tau_k), k = 1, \ldots, K\}$ and rewriting (2.3) in terms of the sources yields

$$\underline{R}_x(\tau) = WAR_s(\tau)A^HW^H. \tag{2.4}$$

Equation (2.4) shows that the matrix $U = WA$ is a unitary matrix. Hence,

$$\underline{R}_x(\tau) = UR_s(\tau)U^H \tag{2.5}$$

Thus, by diagonalizing the matrices of the set $\Re$ by a unitary matrix, we obtain a matrix $\widehat{U}$ that is essentially equal, that is, up to a scalar factor and a permutation, to U. From this, the mixing matrix is derived as $\widehat{A} = W^{-1}U$. It is important to clarify our notation. With U we denote a single unitary matrix that diagonalizes *all* the matrices of the set $\Re$, while the matrix U_k is a matrix that only diagonalizes *a single* matrix $\underline{R}_x(\tau_k)$ of the set $\Re$. This matrix is not necessarily a unitary matrix. For the sake of brevity, the whitened covariance matrices $\underline{R}_x(\tau_k)$ will be denoted as R_k in the sequel.

In theory, the matrices of the set $\Re$ are a family of normal commuting matrices and will therefore have a single unique joint unitary diagonalizer [3]. But due to estimation errors and noise, the whitened covariance matrices will not be perfectly Hermitian. Consequently, U_k, the diagonalizer of the individual matrices of the set $\Re$ will almost certainly not be unitary. But, more importantly, there will not exist a single unitary matrix that will diagonalize all matrices of the set $\Re$. The former problem can be solved by forcing the whitened covariance matrices to be Hermitian. This can easily be achieved by substituting the off diagonal elements of R_k with their respective arithmetic average. The latter problem is a bit more complicated. At most, we can look for a kind of average unitary matrix that will *approximately* diagonalize the different covariance matrices.

3 The Cayley transform

In essence we are looking for some way to average the individual unitary matrices U_k. But a linear combination of unitary matrices does not remain unitary. The mathematical tool we propose to employ is the generalised Cayley transform. (I am grateful to Dr. G. Weiss of Imperial College for bringing the Cayley transform to my attention.) It transforms a unitary matrix into a skew Hermitian one and vice versa. The generalized Cayley transform is given by [7]:

$$C(S) : U = (I - S)(I + S)^{-1} \tag{3.1}$$

and the generalised inverse Cayley transform by

$$C^{-1}(U) : S = (I - U)(I + U)^{-1} \tag{3.2}$$

where U and S are respectively a unitary and a skew-Hermitian matrix. Provided that neither $(I + S)$ nor $(I + U)$ are singular, (3.1) transforms a unitary matrix into a skew-Hermitian one and (3.2) reverses the transformation.

A linear combination of skew-Hermitian matrices is still a skew-Hermitian matrix. Thus we can transform the unitary matrices into the Cayley domain, average them out and then transform them back into the normal domain. The resulting unitary matrix will in some sense be an average of the matrices U_k.

4 The cost function

We want to find a unitary matrix U that diagonalizes all the matrices of the set $\Re$, that is, that minimises the off-diagonal elements of the whitened covariance matrices R_k with $k = 1, \ldots, K$. Thus we define the following cost function:

$$\psi(U, \Re) = \sum_{k=1}^{K} \text{off}\left(U^H R_k U\right) \tag{4.1}$$

where off$(\cdot)$ is defined as

$$\text{off}\,(A) = \sum_{i \neq j} |\, a_{ij} \,|^2. \tag{4.2}$$

Hence the matrix U that minimises (4.1) is the 'approximate joint unitary diagonalizer' of the set $\Re$. As a unitary transformation preserves the energy of a matrix, minimizing (4.1) is equivalent to maximizing the energy of the diagonal elements of $U^H R_k U$ for $k = 1, \ldots, K$. In matrix notation, the modified cost function can be written as:

$$\phi\,(U, \Re) = \sum_{k=1}^{K} \left\{ \text{diag}^H \left[U^H R_k U \right] \ \text{diag} \left[U^H R_k U \right] \right\} \tag{4.3}$$

where diag$\,[\cdot]$ denotes a vector formed from the diagonal elements of a matrix.

Let the skew-Hermitian matrices C_k with $k = 1, ..., K$ be the respective Cayley transformations of U_k, and

$$\overline{C} = \sum_{k=1}^{K} w_k C_k \tag{4.4}$$

where the weights w_k, $k = 1, \ldots, K$ have to be determined. Also

$$\overline{U} = \left(I - \overline{C}\right) \left(I + \overline{C}\right)^{-1} \tag{4.5}$$

that is, $\overline{U}$ is the inverse Cayley transform of a weighted average of the individual unitary matrices in the Cayley domain. Using (4.4), (4.5) and (4.3), the optimization problem can be summarized as

$$\min_{\mathbf{w}} \ \phi\left(\overline{U}, \Re\right) \tag{4.6}$$

where $\mathbf{w}$ is the $K \times 1$ vector containing the different weights of C_k. Hence by setting the gradient of (4.3) to zero, we can find the weights.

5 The analytical gradient

We first introduce a useful identity. Let the non-singular matrix A be a function of the scalar x. Then

$$\frac{\partial}{\partial x}\left(A^{-1}\right) = -A^{-1}\frac{\partial A}{\partial x}A^{-1} \tag{5.1}$$

which can be easily shown as follows. If A is non-singular, then $AA^{-1} = I$, and taking the derivatives of both sides yields:

$$\frac{\partial A}{\partial x}A^{-1} + A\frac{\partial}{\partial x}\left(A^{-1}\right) = 0. \tag{5.2}$$

Rearranging the terms gives (5.1). Differentiation of the cost function given in (4.3) with respect to the weights yields

$$\frac{\partial\phi}{\partial w_p} = 2\sum_{k=1}^{K} \operatorname{diag}^H\left[\overline{U}^H R_k \overline{U}\right] \operatorname{diag}\left[\overline{U}^H R_k \frac{\partial\overline{U}}{\partial w_p} + \left(\frac{\partial\overline{U}}{\partial w_p}\right)^H R_k \overline{U}\right] \tag{5.3}$$

where w_p denotes the weight of R_p. Note that

$$\operatorname{diag}\left[\left(\frac{\partial\overline{U}}{\partial w_p}\right)^H R_k \overline{U}\right] = \operatorname{diag}\left[\overline{U}^H R_k^H \frac{\partial\overline{U}}{\partial w_p}\right] \tag{5.4}$$

so that (5.3) can be simplified to

$$\frac{\partial\phi}{\partial w_p} = 2\sum_{k=1}^{K} \operatorname{diag}^H\left[\overline{U}^H R_k \overline{U}\right] \operatorname{diag}\left[\overline{U}^H \left(R_k + R_k^H\right) \frac{\partial\overline{U}}{\partial w_p}\right]. \tag{5.5}$$

Due to noise and estimation errors, R_k will generally not be Hermitian so that no further simplification is possible. The next step is to find the derivative of the averaged unitary diagonalizer $\overline{U}$ with respect to the individual weights:

$$\begin{aligned}
\frac{\partial\overline{U}}{\partial w_p} &= -\frac{\partial\overline{C}}{\partial w_p}\left(I+\overline{C}\right)^{-1} - \left(I-\overline{C}\right)\left(I+\overline{C}\right)^{-1}\frac{\partial\overline{C}}{\partial w_p}\left(I+\overline{C}\right)^{-1} && (5.6)\\
&= -\left\{I + \left(I-\overline{C}\right)\left(I+\overline{C}\right)^{-1}\right\}\frac{\partial\overline{C}}{\partial w_p}\left(I+\overline{C}\right)^{-1} && (5.7)\\
&= -\left(I+\overline{U}\right)\frac{\partial\overline{C}}{\partial w_p}\left(I+\overline{C}\right)^{-1} && (5.8)\\
&= -\left(I+\overline{U}\right)C_p\left(I+\overline{C}\right)^{-1} && (5.9)
\end{aligned}$$

where C_p is the Cayley transform of the unitary diagonalizer of R_p. Using (5.5) and (5.9), and writing $\overline{U}$ in terms of the different matrices C_k and their respective weights w_k, it is possible to construct a set of K equations containing K unknowns, namely the weights. These equations will be highly nonlinear and their solution is by no means trivial. We have therefore chosen the steepest-ascent algorithm in order to find the weights that maximize (4.3):

$$\mathbf{w}(k+1) = \mathbf{w}(k) + \mu\nabla|_{\mathbf{w}(k)} \tag{5.10}$$

where $\mathbf{w}(k)$ is the vector containing the weights at iteration k and $\nabla|_{\mathbf{w}(k)}$ denotes the gradient evaluated at $\mathbf{w}(k)$. These weights will also be one of the solutions of (5.5). We will call this algorithm the LCD as it uses a linear combination of the individual unitary matrices in the Cayley domain to find an average unitary diagonalizer.

The performance of the LCD algorithm is very similar to that of the SOBI algorithm. If the number of matrices in the set $\Re$, and hence the number of weights, is equal to the number of unknown elements of U, that is, $n(n-1)/2$ where n is the dimension of the matrix, then it is always possible to determine the weights that SOBI assigns to each of the individual unitary matrices. This is interesting, considering that SOBI does not make any explicit or even implicit use of the different individual unitary matrices.

Despite its conceptual attractiveness, the LCD algorithm turns out to have several disadvantages. As the LCD algorithm forms a linear combination of the individual unitary diagonalizers in the Cayley domain, it is necessary to calculate them, which might be computationally expensive. Moreover, the whitened covariance matrices will most probably not even have a unitary diagonalizer as they are not Hermitian. Another problem is that in order to use the Cayley transform, the eigenvalues of U_k must be significantly different from 1. We next develop an algorithm that finds the average unitary diagonalizer independently from the individual diagonalizers.

6 The modified diagonalization algorithm

Let H be a real $n \times n$ skew-Hermitian matrix with elements h_{ij}, where $h_{ij} = 0$ if $i = j$ and $h_{ij} = -h_{ji}$ if $i \neq j$. Define ∂h_{ij} as a small change δ in element h_{ij} and $-\delta$ in h_{ji}. Then

$$\frac{\partial H}{\partial h_{ij}} = T_{ij} \tag{6.1}$$

where all the elements of T_{ij} are zero, except $t_{ij} = 1$ and $t_{ji} = -1$. Replacing $\overline{C}$ by H and w_p by h_{ij}, we can rewrite (5.8) as

$$\frac{\partial \overline{U}}{\partial h_{ij}} = -\left(I + \overline{U}\right) \frac{\partial H}{\partial h_{ij}} \left(I + H\right)^{-1}. \tag{6.2}$$

Using (6.1), this simply becomes

$$\frac{\partial \overline{U}}{\partial h_{ij}} = -\left(I + \overline{U}\right) T_{ij} \left(I + H\right)^{-1}. \tag{6.3}$$

Similar to (5.5), the new gradient can be written as:

$$\frac{\partial \phi}{\partial h_{ij}} = 2 \sum_{k=1}^{K} \operatorname{diag}^{H} \left[\overline{U}^{H} R_k \overline{U}\right] \operatorname{diag} \left[\overline{U}^{H} \left(R_k + R_k^{H}\right) \frac{\partial \overline{U}}{\partial h_{ij}}\right] \tag{6.4}$$

where $\frac{\partial \overline{U}}{\partial h_{ij}}$ is given by (6.3). Here too, we can use (5.10) and (6.4) in order to find the maximum of (4.3). This algorithm, which we call the CJUD algorithm, is much more powerful than the LCD as it does not suffer from any of the latter's disadvantages. Note that whereas the LCD algorithm finds a *vector* with weights, the CJUD searches for the optimal *matrix*.

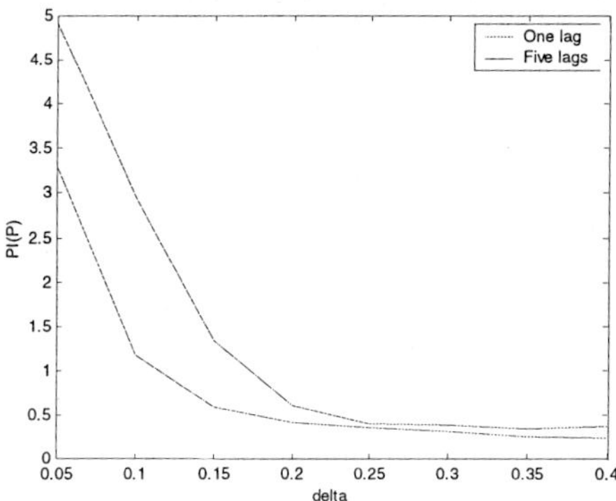

FIG. 1. Performance of the CJUD for the first lag (solid) and the first five lags (dotted). The spectral difference is varied but the SNR is fixed to 25 dB.

7 Simulations

We present some simulations to show the performance of the CJUD algorithm. In the noiseless case, we have used three sources and three sensors. The sources are derived from second-order autoregressive filters with complex conjugate poles and driven by white Gaussian noise. A data length of a thousand samples was chosen. The performance is measured by the *Performance Index (PI)* given by

$$PI\,(P) = \sum_{i=1}^{N}\left(\sum_{j=1}^{n}\frac{\mid p_{ij}\mid}{\max_k\left(\mid p_{ik}\mid\right)} - 1\right) + \sum_{j=1}^{N}\left(\sum_{i=1}^{n}\frac{\mid p_{ij}\mid}{\max_k\left(\mid p_{kj}\mid\right)} - 1\right) \tag{7.1}$$

where $\max_k\left(\mid p_{ik}\mid\right)$ and $\max_k\left(\mid p_{kj}\mid\right)$ are the maxima per row and per column respectively and $P = A^{-1}\widehat{A}$. The $PI\,(P)$ index measures the distance of the matrix P to a scaled permutation matrix.

In the first experiment, the signal-to-noise-ratio (SNR) was kept fixed at 25 dB, while the positions of the poles of the filters are changed. We have used a 3×3 real mixing matrix with elements picked from a Gaussian distribution. The results were averaged over 100 trials with repect to the sources. The poles of the filters where placed at $\exp\left\{\pm\mathrm{j}\left(0.5-\delta\right)\right\}$, $\exp\left\{\pm\mathrm{j}0.5\right\}$ and $\exp\left\{\pm\mathrm{j}\left(0.5+\delta\right)\right\}$ where δ is the spectral difference. In Fig. 1, performance based on diagonalization of the first five matrices is compared with the performance obtained from diagonalizing only one matrix. We can see that as the spectral difference increases, separation becomes better. This is due to the fact that the spectral content of the signals become more distinct. Using five lags gives better performance than using only one lag as the CJUD searches for a kind of average matrix that will approximatly diagonalize all five whitened covariance matrices and hence the found diagonalizer will be based on more information.

In the next simulation, shown in Fig. 2, we have fixed the poles to $\exp\left(\pm\mathrm{j}0.45\right)$, $\exp\left(\pm\mathrm{j}0.5\right)$ and $\exp\left(\pm\mathrm{j}0.55\right)$ and have varied the SNR. The sources

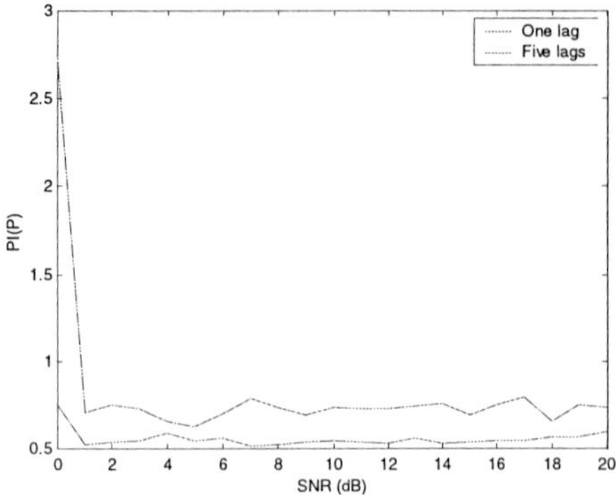

FIG. 2. Performance of the CJUD for the first lag (solid) and the first five lags (dotted). The spectral difference is fixed but the SNR is varied.

are again real, but this time we use a 4×3 real mixing matrix to account for the noise. Here too, we compare the performance of the first five lags with the performance of the first lag. We have averaged the results over 100 trials with respect to the sources.

The same argument applies here. Separation based on a set of lags means that we use more information, and therefore the estimation of the rotation matrix will be statistically more efficient.

8 Conclusion

We have introduced two novel methods for the joint unitary diagonalization of a set of matrices. The LCD algorithm is conceptually easy, but might occasionally be difficult to implement. The CJUD algorithm is essentially an algebraic method, but we have to use some kind of iterative method, in our case the steepest0ascent method, in order to find the maximum of the cost function. This is the major disadvantage of the CJUD algorithm when compared to the Joint Diagonalisation method [2]: it is computational more expensive and its convergence is much slower. However, the CJUD algorithm has some significant advantages. In some cases, it might be useful to make the cost function dependent on the sample length, the noise or some *a priori* information. Whereas the cost function of SOBI is fixed to the energy of the diagonal elements, the cost function of the CJUD algorithm can be easily changed [4]. Another difference between SOBI and CJUD which might turn out to be useful, is that in the SOBI algorithm, the elements of the unitary matrix change in discrete steps, two elements at a time, towards the optimal matrix. In the CJUD algorithm all elements of the matrix adjust continuously to the optimal unitary matrix.

Acknowledgements

The author would like to thank Professor Constantinides, Professor McWhirter and Paul Baxter for their valuable comments. This work was carried out as part of Technology Group TG10 of the MOD Corporate Research Programme, supported by the Defence Evaluation and Research Agency.

Bibliography

[1] Belouchrani, A. Abed-Meraim, K. Cardoso, J. and Moulines, E. A blind source separation technique using second-order statistics, *IEEE Trans. on Signal Process.* **45**, 434–444, 1997.

[2] Cardoso, J.F. Jacobi angles for simultaneous diagonalization, *SIAM J. Matrix Anal. Appl.* **17**, 161–164, 1996.

[3] Horn, R.A. and Johnson, C.A. *Matrix Analysis.* Cambridge: Cambridge University Press, 1985.

[4] Horn, R. and Johnson, C.R. *Topics in Matrix Analysis.* Cambridge: Cambridge University Press, 1999.

[5] Klajman, M. and Constantinides, A.G. A variable cost function for second-order blind source separation, in preparation.

An Adaptive Blind CMOE-CMA Receiver for DS-CDMA Systems

Peerapol Yuvapoositanon

Department of Electronic Engineering, Mahanakorn University of Technology, Bangkok 10530, Thailand

Jonathon Chambers

Department of Electronic and Electrical Engineering, University of Bath, Bath BA2 7AY, UK

Abstract

A new initialization scheme for a constant modulus algorithm (CMA) equalizer for asynchronous direct-sequence code-division multiple access (DS-CDMA) systems is proposed. The solution from minimizing the constrained minimum output energy (CMOE) criterion is adopted as an initialization for the successive CMA equalizer. Simulations confirm the near-far resistance of the proposed receiver over a wide range of near-far situations.

1 Introduction

The use of the constant modulus algorithm (CMA) for blind interference cancellation in multiuser applications, as in DS-CDMA systems, has been widely studied. However, without proper control of its convergence behaviour, the CMA receiver may lock onto an interfering user rather than the desired user. This misconvergence problem is aggravated when the interference is stronger than the desired user. Such a power disparity situation is widely known as the *near–far problem.* Lambotharan *et al.* [4] propose the idea of including a cross-correlation term with the CMA cost function so as to prevent different interference cancellers (equalizers) converging to the same solution. Doroslovački and Vojčić [1] impose a constraint on the adaptation of the equalizer weight vector in the form of a cone around a vector given by the desired user spreading code. Locking onto the desired user is guaranteed if the cone angle is chosen appropriately. Treichler and Larimore [9] show, however, that initialization is the most important factor for CMA equalizers to converge to global solutions and this evidence is supported by Li and Ding [7].

Hence the problem is reduced to acquisition of a 'good' initialization. It is suggested by Lee *et al.* [5] that a strategy consisting of initialising the CMA equalizer with the desired user spreading sequence and pre-amplifying the amplitudes of transmitted symbols of the desired user results in the correctly detected symbols. However, the pre-amplification of the transmitted symbols of the desired user invalidates the near–far resistance of the receiver. Schniter and Johnson [8] introduced the kurtosis-based technique called the minimum-entropy (ME) criterion as an initialization strategy for the CMA equalizer. The ME-CMA receiver is proven to be near–far resistant in frequency-selective fading asynchronous DS-CDMA channels provided that the received signal is prewhitened. An additional constraint for the ME-CMA receiver, called the non-constrained for interchip interference cancellation (NCICI) constraint, is proposed in [6]. The NCICI-CMA receiver outperforms the ME-CMA receiver in terms of signal to interference plus noise ratio (SINR).

In this study, we propose a new initialization strategy for a CMA equalizer for frequency-selective fading asynchronous DS-CDMA channels. The minimum output energy (MOE) criterion is appropriate in this case since it possesses a unimodal cost surface and offers scaled mean-square error (MSE) solutions. We adopt the adaptive constrained MOE (CMOE) criterion proposed by Tsatsanis [10]. The constraint neglects the effect of multipath distortion which is not fundamentally solved by the original CMOE detector by Honig *et al.* [3]. From a bank of delayed CMOE detectors, the maximum output energy detector is selected as the detector with correct time alignment and is used to initialise the successive CMA equalizer. By combining the CMOE and the CM criteria in the form of the CMOE-CMA receiver, we obtain a new receiver structure in which CMOE performs the interference suppression and CMA mitigates the distortion caused by the channel and improves the MSE performance of the receiver. Simulations show the near–far resistance of the proposed CMOE-CMA receiver compared with those of existing receivers.

2 System model

For the real system model from Honig *et al.* [3], the baseband received signal for a K-user asynchronous CDMA channel is defined as

$$r(t) = \sum_{i=-\infty}^{\infty} \sum_{k=1}^{K} A_k b_k[i] c_k(t - iT - \tau_k) + v(t), \tag{2.1}$$

where A_k represents the received amplitude of the kth user. The data bits $b_k[i]$ are independent identically distributed (i.i.d.) and assumed to be drawn from the finite alphabet $\{-1, +1\}$. The symbol period is denoted by T. The spreading (or signature) waveform of the kth user $c_k(t)$ is L_c-dimensional and has unit energy property, that is, $||c_k||^2 = 1$ and $\tau_k \in [0, T)$ are the relative offsets of the asynchronous signals at the receiver. The zero-mean additive white Gaussian channel noise $v(t)$ has power spectral density σ^2. If we incorporate the amplitude A_k and delay τ_k in the channel response $h_k(t)$, we can replace the spreading code sequence $c(t)$ with the discrete-time combined channel-code response

$$g_k[n] = \sum_{i=0}^{L_c-1} c_k[i] h_k[n - i], \tag{2.2}$$

where $c_k[i]$ is the ith element of the code vector for the kth user $\mathbf{c}_k = (c_k[0], \ldots, c_k[L_c - 1])^T$ and the received signal $r(t)$ in (2.1) can be written in the discrete-time vector form

$$\mathbf{r}[n] = \sum_{k=1}^{K} \mathbf{r}_k[n] + \mathbf{v}[n] = \sum_{k=1}^{K} \mathbf{G}_k \mathbf{b}_k[n] + \mathbf{v}[n], \tag{2.3}$$

where $\mathbf{G}_k$ is the combined code-channel response matrix of the kth user and $\mathbf{b}_k[n] = (b_k[n], b_k[n-1], \ldots, b_k[n - L_b + 1])^T$ with $L_b = \lceil \frac{L_f + L_h - 1}{L_c} \rceil$ and $\mathbf{v}[n] = (v[nN - L_f + 1], \ldots, v[nN])^T$. Note that

$$\mathbf{G}_k = \mathbf{C}_k \mathbf{H}_k, \tag{2.4}$$

where the code matrix $\mathbf{C}_k$ and the channel matrix $\mathbf{H}_k$ are defined as

$$\mathbf{C}_k = \begin{pmatrix} c_k[L_c - 1] & & \\ \vdots & \ddots & \\ c_k[0] & & c_k[L_c - 1] \\ & \ddots & \vdots \\ & & c_k[0] \end{pmatrix}, \quad \mathbf{H}_k = \begin{pmatrix} \mathbf{h}_k & & & \\ & \mathbf{h}_k & & \\ & & \ddots & \\ & & & \mathbf{h}_k \end{pmatrix},$$

where the channel response vector for the kth user has length L_h, that is, $\mathbf{h}_k = (h[L_h - 1], \ldots, h[0])^T$, and all other entries of these matrices are zero.

3 CMOE-CMA receiver

We adopt the constraint proposed by Tsatsanis [10] which is shown to be

$$\mathbf{f}_{k,l}^T \mathbf{C}_k = (0, \ldots, 0, \underset{\substack{\uparrow \\ l}}{1}, 0, \ldots, 0) \triangleq \mathbf{1}_l^T \tag{3.1}$$

where $\mathbf{f}_{k,l}$ is the weight vector of the lth CMOE detector for the kth user. The nonzero term corresponds to the lth position of the multipath rays. Denote the output energy of the lth detector as

$$J_{\mathrm{MOE}_k}^{(l)} = \mathbf{f}_{k,l}^T \mathbf{R} \mathbf{f}_{k,l}, \tag{3.2}$$

where $\mathbf{R} = E\{\mathbf{r}[n]\mathbf{r}^T[n]\}$. By invoking the instantaneous value of the autocorrelation matrix $\mathbf{R}$, the adaptation rule for the lth CMOE detector of the kth user that minimizes $J_{\mathrm{MOE}_k}^{(l)}$ becomes [10]

$$\mathbf{f}_{k,l}[n+1] = \mathbf{f}_{k,l}[n] - \mu_{\mathrm{CMOE,l}} \mathbf{\Pi}_{\mathbf{C}_k}^{\perp} \mathbf{r}[n]\mathbf{r}^T[n]\mathbf{f}_{k,l}[n], \tag{3.3}$$

where $\mu_{\mathrm{CMOE,l}}$ is the stepsize of the lth detector and $\mathbf{\Pi}_{\mathbf{C}_k}^{\perp} = \mathbf{I} - \mathbf{C}_k(\mathbf{C}_k^T \mathbf{C}_k)^{-1} \mathbf{C}_k^T$ denotes the projection matrix onto the nullspace of $\mathbf{C}_k$. As shown in [10], the lth detector whose delay corresponds to that of the dominant path can be identified by using the maximum MOE criterion after the convergence of all L_c detectors. (If the approximate delay information of the desired user is known, less than L_c detectors can be set around the correct delay and the complexity of the algorithm can be reduced significantly.) The convergence of the lth detector is achieved at time n_s when the slope (difference) of the norm of the weight vector $\|\mathbf{f}_{k,l}[n]\|$ is less than a pre-assigned small constant

$$\text{slope}[n_s] := \frac{\|\mathbf{f}_{k,l}[n_s]\| - \|\mathbf{f}_{k,l}[n_s - \triangle]\|}{\triangle} < \epsilon, \tag{3.4}$$

where $\triangle$ is the interval for slope calculation and ϵ is the pre-assigned threshold. The estimated MOE of the lth detector is derived by

$$\hat{J}^{(l)}_{\mathrm{MOE_k}}[n] = \frac{1}{N_t} \sum_{j=n-N_t+1}^{n} y^2_{k,l}[j], \qquad \forall l \in \{0, \ldots, L_c - 1\}, \tag{3.5}$$

where $y_{k,l}[n] = \mathbf{f}^T_{k,l}[n]\mathbf{r}[n]$ denotes the output signal of the lth finger at time n and N_t is the sample size of the data symbols. We determine the maximum output energy detector

$$\mathbf{f}_{k,l_{\max}}[n_s] = \arg\{\max_l \hat{J}^{(l)}_{\mathrm{MOE_k}}[n_s]\}. \tag{3.6}$$

Specializing the CMA algorithm for real signals, the following cost is minimized [2]:

$$J^{(k)}_{rmCMA} = E\{(z_k^2 - \gamma_k)^2\}, \tag{3.7}$$

where $z_k = \mathbf{w}_k^T\mathbf{r}$ is the output symbol of the equalizer for the kth user and the dispersion constant γ_k defines the gain of the CMA equalizer. The adaptation rule for the CMA equalizer is

$$\mathbf{w}_k[n+1] = \mathbf{w}_k[n] - \mu_{\mathrm{cma}}\mathbf{r}[n]z_k[n]e_k[n], \tag{3.8}$$

where μ_{cma} is the stepsize and $e_k[n] = z_k^2[n] - \gamma_k$ is the error signal at time n. We summarize the steps for the CMOE-CMA acquisition scheme for the kth user as follows:

(1) From a bank of CMOE detectors for the kth user, compute the outputs of the lth detector $y_{k,l}[n] = \mathbf{f}^T_{k,l}[n]\mathbf{r}[n]$. The tap weight vector $\mathbf{f}_{k,l}$ is initialised by $\mathbf{f}^{(init)}_{k,l} = \mathbf{C}_k^\dagger \mathbf{1}_l$ where $\mathbf{C}_k^\dagger$ is the pseudo-inverse of $\mathbf{C}_k^T$.
(2) Operate the CMOE algorithm (3.3) for each lth detector and compute the $\hat{J}^{(l)}_{\mathrm{MOE_k}}[n]$ from (3.5).
(3) Calculate the slope of $\|\mathbf{f}_{k,l}[n]\|$ by (3.4). After all the CMOE detectors have converged at time n_s, select the lth weight vector that generates the maximum MOE to initialise the CMA equalizer: $\mathbf{w}_k[0] = \mathbf{f}_{k,l_{\max}}[n_s]$.
(4) Compute the estimate of the symbols of the kth user $z_k[n] = \mathbf{w}_k^T[n]\mathbf{r}[n]$ and update by (3.8).

The diagram of the CMOE-CMA receiver is shown in Fig. 1.

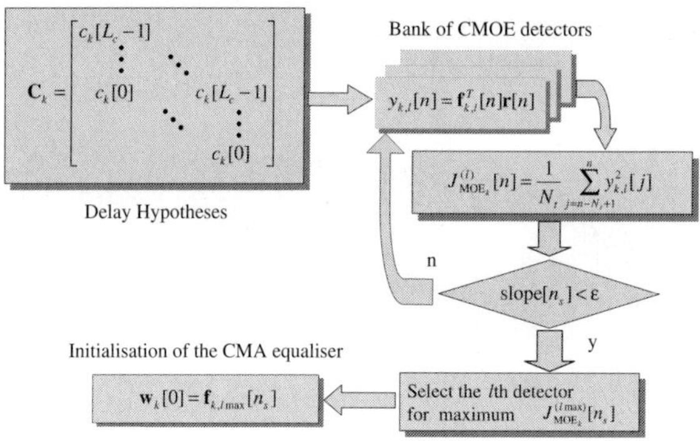

FIG. 1. The diagram of an adaptive blind CMOE-CMA receiver for the kth user in asynchronous DS-CDMA systems.

4 Simulations

We considered a symbol asynchronous system with processing gain $L_c = 31$ and number of users $K = 10$. User delays τ_k were uniformly distributed over $[0, 10T_c)$ and then kept fixed. The number of multipath rays was five, where the last four rays were uniformly distributed in delay over $[0, 5T_c)$. The channel length for all users was $5T_c$. We assumed without loss of generality that the first user is the user of interest with unity power. The background noise was zero-mean AWGN with SNR = 20 dB (referenced to the desired user). The receiver length was chosen to be twice that of the spreading gain ($2L_c$). The performance measure was the averaged SINR in dB and all SINR plots were averaged over 100 Monte Carlo runs. We considered various near–far situations and compared the performances of the ME-CMA receiver [8], the NCICI-CMA receiver [6], the CMOE detector [10], and the proposed CMOE-CMA receiver. To demonstrate the performance of the proposed algorithm in a near–far situation, all interfering users were set to be stronger than the desired user by 10 dB, that is, $\frac{A_1^2}{A_k^2} = -10$ dB where $k = \{2, \ldots, 10\}$, and the averaged SINR plots are shown in Fig. 2. For such near–far level, the performances of the ME-CMA and NCICI-CMA receivers are inferior to that of the CMOE-CMA receiver. The CMOE detector performs worst in this situation. Figure 3 compares the steady-state averaged SINR performances of the four receivers at different initial near–far situations. It is shown that the CMOE detector attains the SINR level at about 5 dB over the wide range of the initial near–far settings. The performance of the ME-CMA and NCICI-CMA receivers degrades gradually with the increase of near–far levels. For the CMOE-CMA receiver, it is shown that the SINR performance is comparable to the NCICI-CMA and ME-CMA receivers in mild near–far situations and is maintained over moderate to severe near–far level settings.

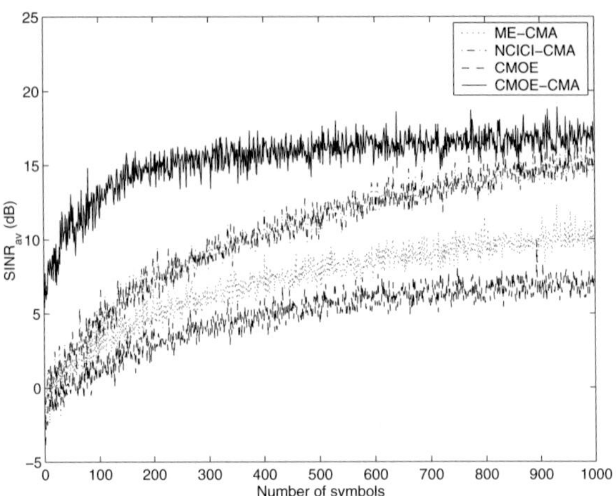

FIG. 2. Comparison of the ME-CMA receiver, the NCICI receivers, the CMOE detector, and the proposed CMOE-CMA receiver, at $\frac{A_1^2}{A_k^2} = -10$ dB.

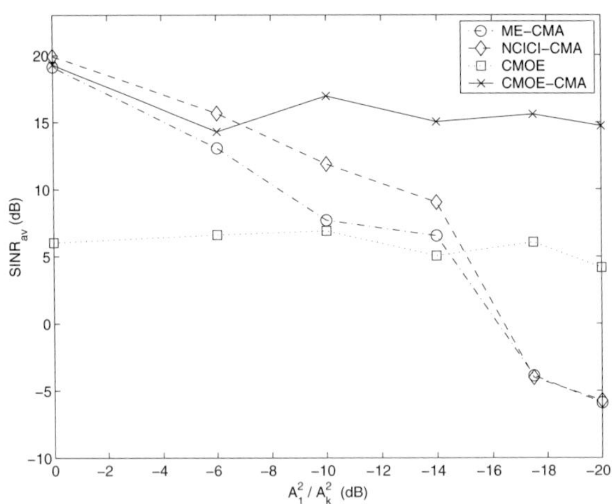

FIG. 3. The steady-state SINR performances of the CMOE detector, the ME-CMA, the NCICI-CMA and the proposed CMOE-CMA receivers at different initial interference powers.

5 Conclusions

Initialising the CMA equalizer by using the kurtosis criterion leads to global convergence to the desired user solution. On the other hand, initialising the CMA equalizer with taking into account the interference structure, i.e., using the CMOE criterion, not only leads to global convergence but also obtains a tap weight vector which is a solution from a decorrelating detector. Combining the CMOE and CM criteria is a new approach leading to a near–far resistant acquisition/equalization scheme for blind multiuser CDMA receivers.

Acknowledgement

This work is supported by the Mahanakorn University of Technology, Thailand.

Bibliography

[1] Doroslovački, M. I. and Vojčić, B. R. (1998) Cone constraint constant modulus algorithm for blind adaptive multiuser interference suppression. In *Proc. Asilomar Conf. on Signals, Systems and Computers* (Pacific Grove, CA), pp. 1500–1504.

[2] Godard, D.N. (1980) Self-recovering equalization and carrier tracking in two-dimensional data communication systems. *IEEE Trans. Commun.*, **28**, 1867–1875.

[3] Honig, M.L., Madhow, U. and Verdú, S. (1995) Blind adaptive multiuser detection. *IEEE. Trans. Inform. Theory*, **41**, 944–966.

[4] Lambotharan, S., Chambers, J.A. and Constantinides, A.G. (1999) Adaptive blind retrieval techniques for multiuser DS-CDMA signals. *Electron. Lett.*, **86**, 693–695.

[5] Lee, W., Vojic, B.R. and Pickholtz, R.L. (1996) Constant modulus algorithm for blind multiuser detection. In *Proc. IEEE Int. Symp. Spread Spectrum Techniques and Applications* (Mainz, Germany), Vol. 3, pp. 1262–1265.

[6] Li, L. and Fan, H.H. (2000) Blind CDMA detection and equalization using linearly constraint. In *Proc. Int. Conf. Acoustics, Speech and Signal Processing* (Istanbul, Turkey), Vol. 5, pp. 2905–2908.

[7] Li, Y. and Ding, Z. (1995) Convergence analysis of finite length blind adaptive equalizers. *IEEE Trans. Signal Processing*, **43**, 2120–2129.

[8] Schniter, P. and Johnson Jr., C.R. (1998) Minimum-entropy blind acquisition/equalization for uplink DS-CDMA. In *Proc. Allerton Conf. on Communications, Control, and Computing* (Monticello, IL), pp. 401–410.

[9] Treichler, J.R. and Larimore, M.G. (1985) The tone capture properties of CMA-based interference suppressors. *IEEE Trans. Acoust., Speech, Signal Processing*, **33**, 946–958.

[10] Tsatsanis, M.K. (1997) Inverse filtering criteria for CDMA systems. *IEEE Trans. Signal Processing*, **45**, 102–112.

Statistics of Impulse Noise in xDSL

Iain Mann and Stephen McLaughlin
Department of Electronics and Electrical Engineering, University of Edinburgh, Edinburgh EH9 3JL, UK

Abstract

A limiting factor in the design of any Digital Subscriber Loop system will be the amount of impulse noise on subscriber lines. In this paper a new method for detecting impulse noise events using a statistical measure is presented. This is described in comparison with an existing technique which uses a hard amplitude-time threshold to locate the impulses. Both techniques are then used to calculate impulse duration and amplitude PDFs from real subscriber line data, recorded in the UK. Next these PDFs are compared to an analytical model proposed by Deutsche Telekom (DT). It is found that the form of the model is correct, but that the parameters chosen by DT are not appropriate for the line data used in this study, which was obtained from the network of British Telecom. Finally, more suitable parameter values are suggested in order to arrive at a working model of impulse length and amplitudes.

1 Introduction

This paper is concerned with the statistical nature of impulsive noise on unshielded twisted copper pairs for use in a Digital Subscriber Loop (DSL) system. With increasing interest being shown in the various DSL technologies, the study of the limiting factors, such as noise, has become very important [1]. The contribution of this paper is twofold: firstly a novel method for detecting impulsive events in real subscriber line data is presented. This new technique is expected to be more robust than the amplitude-time thresholding model currently in use, hence allowing more accurate calculation of the impulse duration and amplitude probability density functions (PDFs). Secondly, the impulse duration and amplitude PDFs from real line data collected in the UK by British Telecommunications (BT), as calculated by both impulse detection techniques, are compared with statistical models proposed by Deutsche Telekom (DT).

The paper is structured as follows: initially the current method for locating impulses is reviewed and some of its drawbacks, in relation to BT line data, are highlighted. Then the new impulse location method is presented, which is based on a statistical measure rather than a set amplitude threshold. In Section 3.1 the parametric models for impulse length and amplitude PDFs, as proposed by DT [3], are presented. These are compared with the PDFs generated from the real line data by the two impulse event detection methods. The comparison is performed by both visual inspection of the overlaid pdf plots and by a χ^2 test, which provides a measure of the goodness of fit between the distributions. This analysis demonstrates that the general form of the DT models is reasonable, but that the actual parameters chosen by DT are not appropriate when applied to the BT line data. Thus, finally, some improved parameters for the DT model to fit it to the BT line data are suggested.

2 Locating impulses

In this section two methods to detect impulse events from real subscriber line data are described. These allow the impulse duration and energy statistics of the lines to be collected.

2.1 Amplitude test

The standard method for locating impulse events on subscriber lines up until this time is to examine the amplitudes of data samples and perform a straightforward amplitude thresholding, also taking the impulse length into account. This was proposed by DT [5] and operates as follows: the trigger amplitude level was set at 10 mV. A time resolution is then set (1 μs), and all samples exceeding the trigger amplitude level which are separated by a time not exceeding the time resolution, together with all intervening samples, are deemed to belong to the same impulse event. The next impulse will begin at the point where the trigger threshold is next exceeded, with the time between impulses separated by more than the time resolution.

The disadvantage of having a hard amplitude/time threshold can be seen in cases where the level of 'background' noise is high (where insignificant events may be incorrectly classified as impulses), and from events that (from visual inspection) should be classed as a single impulse of long duration but are incorrectly classed as a large number of very short impulses, due to their amplitudes. Other scenarios can be envisaged where a hard-limiting, data-independent, amplitude threshold is far from ideal. These problems may cause the impulse length and amplitude PDFs to be incorrectly calculated.

2.2 New statistical-based measure

A new test is proposed that uses the statistics of the signal in order to detect the presence of an impulse. By using the fact that the first 4 ms of a recorded event will not contain any significant impulse (due to the data recording technique used), it is possible to calculate a statistic noise of the *background.* (It is possible that the first 4 ms may contain impulsive events, but in the data analysed to date it has not been observed.) The basic idea is then to compare this with subsequent sections from the event file, with any significant deviations indicating an impulse. The clear advantage of this method is that it avoids the use of a hard amplitude threshold, and so will provide a more reliable way of assessing the length and energy of impulses. There are a number of different statistical measures that could be used to carry out this task. We have chosen to calculate amplitude histograms, and to use the χ^2 test in order to search for differences between histograms. This provides a robust and efficient algorithm, which can be described in more detail as follows:

- The initial 4 ms *run-in* of the event file is used to calculate a background, or expected, histogram. This is normalized in line with the window length subsequently used.
- The remainder of the event file is analysed on a frame-by-frame basis, using over-lapping frames.
- For each frame, an observed histogram is calculated. The difference between the expected and observed histograms is found from the χ^2 statistic, which is defined as

$$\chi^2 = \sum_{i=0}^{N-1} \frac{(X_i - x_i)^2}{x_i} \tag{2.1}$$

  where N is the number of histogram bins, X_i is the ith observed bin, and x_i is the ith expected bin. If the pre-set χ^2 threshold is exceeded, then the frame is marked as containing an impulse.
- The frame is then stepped forward through the data by some percentage of the frame length, called the frame step, and the process is repeated.

This basic algorithm will find those parts of the signal that deviate significantly from the background run-in period. In order to ensure that only impulses are marked, a variation of the previously proposed amplitude thresholding method is also used:

- When a frame is marked by the χ^2 test as containing an impulse, a check is made that there are significant amplitudes present in the frame.
- In order to overcome the problem caused by dc offsets in the data, the samples are zero-meaned, using the mean calculated over the 4 ms run-in.
- The zero-meaned frame is then checked to ensure that it contains amplitudes of over 10 mV (the previously defined hard amplitude threshold). Only if this is true is the frame declared to contain an impulse.

In our experiments, a frame size of 10 μs, with a frame step of 1 μs, was used. With a sampling rate of 30 MHz, this corresponds to 300 samples per frame (and hence per histogram). Shifting by 1 μs gives good time resolution, although obviously the use of a frame method does introduce some time resolution smearing. The impulse time is taken as the frame centre point. The χ^2 threshold was set manually after examination of a large number of records from five different lines.

3 Modelling impulse lengths and amplitude

We now consider how the impulse lengths and amplitudes as measured from actual line data using the two methods described above can be modelled in a compact manner. For the sake of clarity, the amplitude/time thresholding

Table 1 Parameters for DT amplitude and length models

	Central office	Customer premises
u_0	18 nV	3 nV
B	0.25	1
s_1	0.75	1.15
s_2	1.0	–
t_1	8 μs	18 μs
t_2	125 μs	–

method described in Section 2.1 will be referred to as method 1, and the new statistical based method from Section 2.2 will be method 2.

3.1 Existing DT models

The approach by DT for impulse modelling is described by Henkel and Kessler [3, 4]. The impulse voltage amplitude density is modelled by

$$f_i(u) = \frac{1}{140u_0}.e^{|-u/u_0|^{1/5}}. \tag{3.1}$$

The impulse length density is modelled by a log-normal form:

$$f_l(t) = B\frac{1}{\sqrt{2\pi}s_1 t}.e^{-\frac{1}{2s_1^2}\ln^2(t/t_1)} + (1-B)\frac{1}{\sqrt{2\pi}s_2 t}.e^{-\frac{1}{2s_2^2}\ln^2(t/t_2)}. \tag{3.2}$$

With $B = 1$, this reduces to a single log-normal function, which is used to represent measurements at customer premises. A value of $B = 0.25$ is used for the central office model. The parameters for the models are shown in Table 1. These models were obtained by analysis of real impulse data measurements carried out by DT on their network.

3.2 Comparison to real line data

3.2.1 *Impulse lengths*

Figure 1 shows the comparison between the DT model and the impulse length PDFs, with the impulse locations detected with method 1 (*that is*, amplitude thresholding). The corresponding plot using method 2 (χ^2 statistic test) is shown in Fig. 2. Average PDFs over the four lines for both methods were then calculated, and these are shown in comparison to the DT model in Fig. 3. The results from the χ^2 test are summarized in Table 2.

3.2.2 *Impulse amplitudes*

Following from Section 3.2.1, Fig. 4 shows the comparison between the DT model and the impulse amplitude PDFs, with the impulse locations detected with method 1 (amplitude thresholding). The corresponding plot using method 2 (χ^2 statistic test) is shown in Fig. 5. Average PDFs over the four lines for both methods are shown in comparison to the DT model in Fig. 6.

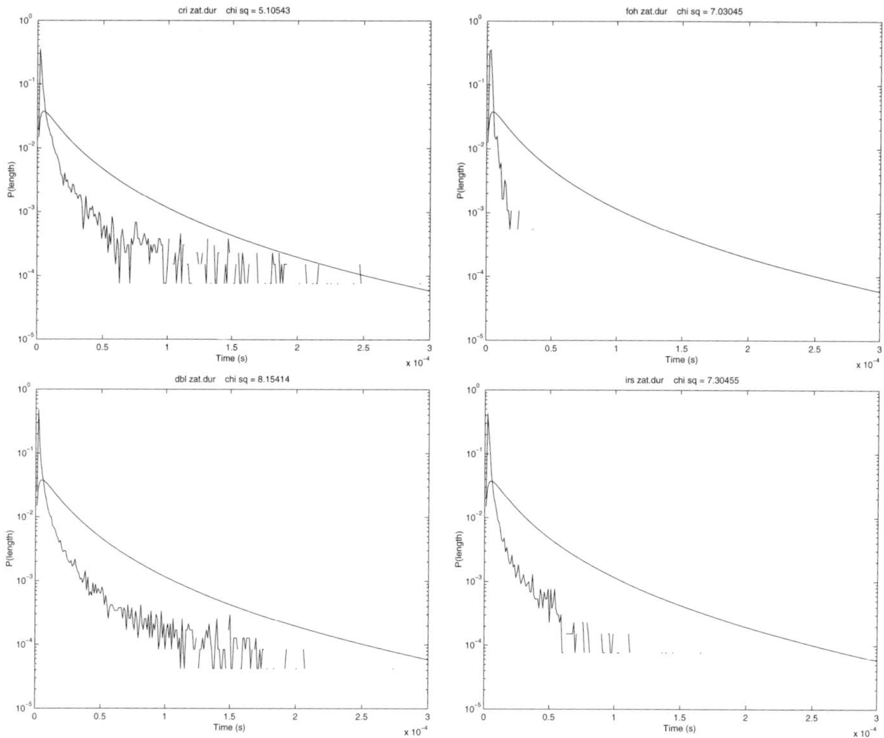

FIG. 1. Comparison of DT model with impulse length PDFs from the four BT subscriber lines, using method 1 for impulse location.

Table 2 Summary of χ^2 test between modelled length pdf and measured length pdf for both impulse detection strategies

Line	Amplitude threshold	χ^2 statistic
Cri	5.105	0.866
Dbl	8.154	1.456
Foh	7.030	2.890
Irs	7.305	1.589
Average	6.664	1.530

The results from the χ^2 test for the amplitude PDFs are summarised in Table 3.

3.3 Improved parameter selection

The previous results show that there is some disparity between the real line data and the models proposed by DT. However, the actual form of the models appears correct, and there is simply a requirement to modify the parameters used in the

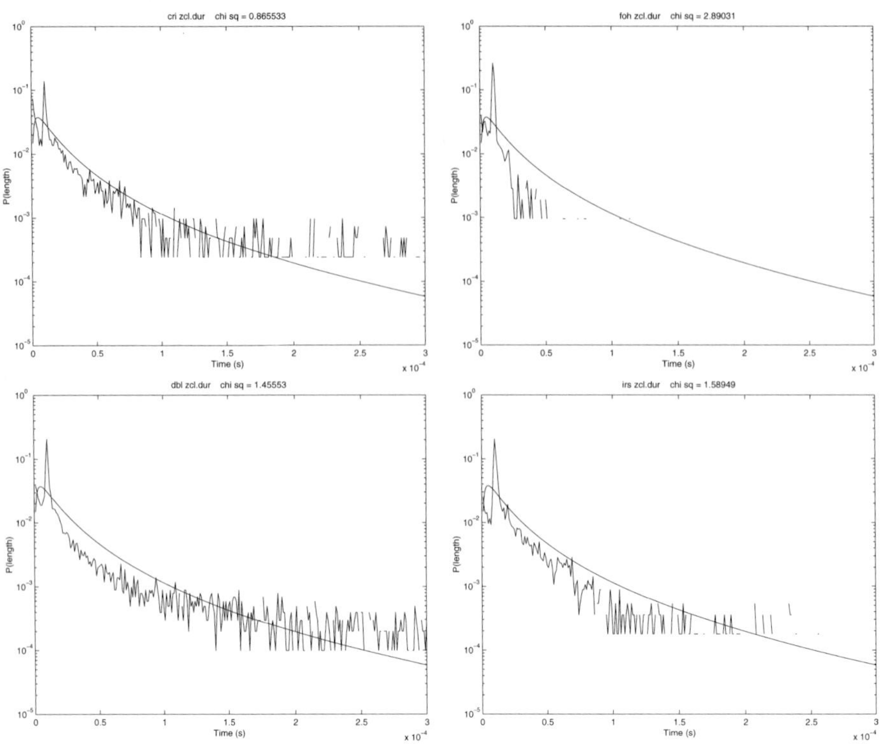

FIG. 2. Comparison of DT model with impulse length PDFs from the four BT subscriber lines, using method 2 for impulse location.

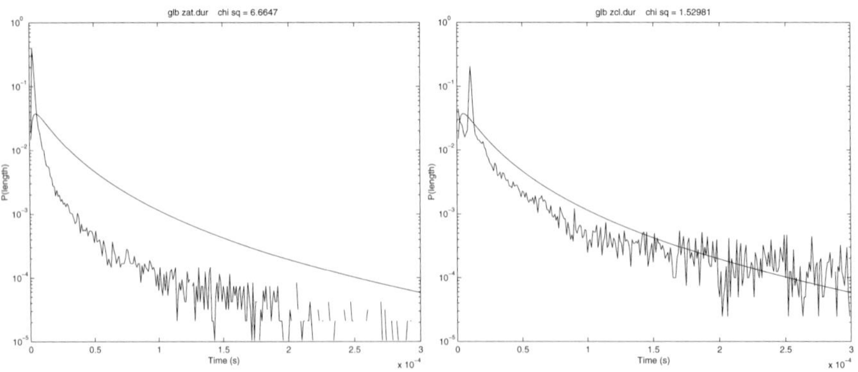

FIG. 3. Comparison of DT model with average impulse length PDFs using (left) the amplitude thresholding method and (right) the χ^2 statistic test.

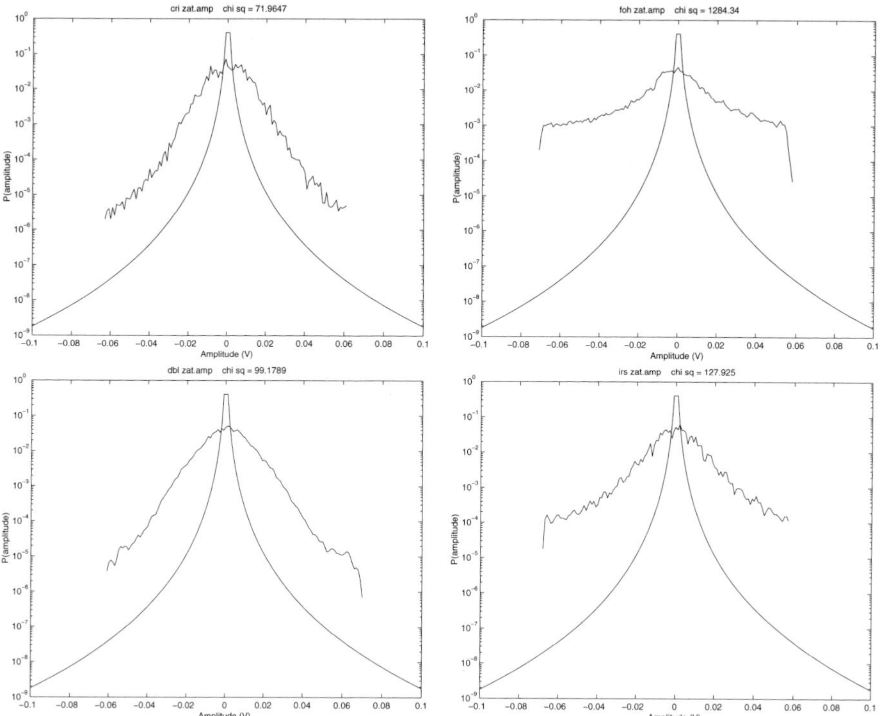

FIG. 4. Comparison of DT model with impulse amplitude PDFs from the four BT subscriber lines, using method 1 for impulse location.

Table 3 Summary of χ^2 test between modelled amplitude pdf and measured amplitude pdf for both impulse detection strategies

Line	Amplitude threshold	χ^2 statistic
Cri	71.96	41.30
Dbl	99.18	41.10
Foh	1284	155.2
Irs	127.9	38.07
Average	192.3	44.47

model to achieve a close fit with the actual data. This is a surprising but encouraging result since we are dealing with two completely separate measurement campaigns made on two entirely different networks.

The optimum parameter selection was made by calculating the parameters using the maximum likelihood method [2]. Considering (3.1), the likelihood

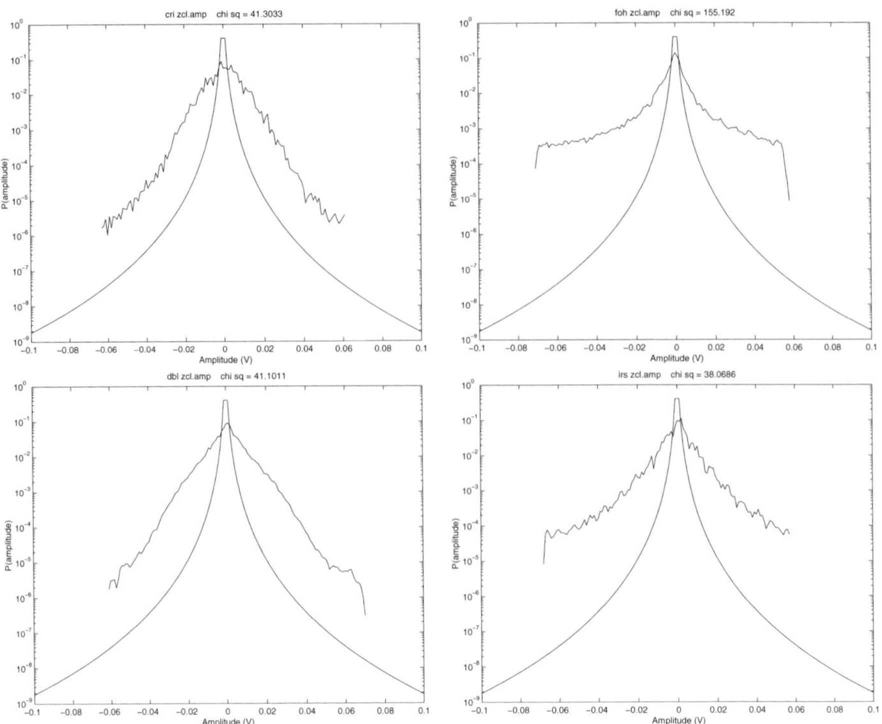

Fig. 5. Comparison of DT model with impulse amplitude PDFs from the four BT subscriber lines, using method 2 for impulse location.

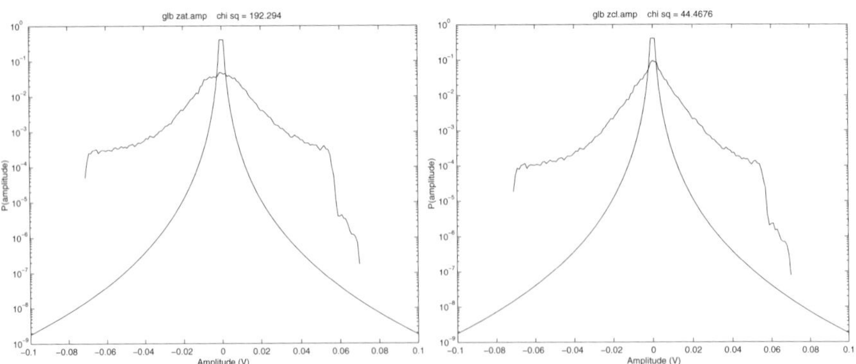

Fig. 6. Comparison of DT model with average impulse amplitude PDFs using (left) the amplitude thresholding method and (right) the χ^2 statistic test.

function in terms of u_0, $L(u_0)$, is

$$L(u_0) = \frac{1}{140^n {u_0}^n} e^{-\frac{1}{{u_0}^{1/5}} \sum_{i=1}^{n} u_i^{1/5}}. \tag{3.3}$$

Differentiating the log-likelihood function w.r.t. u_0, then solving to maximize $L(u_0)$ gives the maximum likelihood estimator of u_0:

$$u_0 = \left(\frac{1}{5n} \sum_{i=1}^{n} u_i^{1/5} \right)^5. \tag{3.4}$$

The duration function (3.2) requires the estimation of two parameters when considering the case of data recorded at the customer premises. The likelihood function in terms of s_1 and t_1 is

$$L(s_1, t_1) = \frac{1}{(\sqrt{2\pi} s_1)^n \sum_{i=1}^{n} t_i} e^{-\frac{1}{2s_1^2} \sum_{i=1}^{n} \ln^2(t_i/t_1)}. \tag{3.5}$$

Taking the partial differential of (3.5) w.r.t. t_1 and solving to maximise L gives the solution for t_1 as

$$t_1 = \left(\prod_{i=1}^{n} t_i \right)^{1/n}. \tag{3.6}$$

In the same way, s_1 is found by partially differentiating w.r.t. s_1 and maximising L:

$$s_1 = \frac{1}{\sqrt{n}} \left[\sum_{i=1}^{n} \ln^2 t_i/t_1 \right]^{1/2}. \tag{3.7}$$

Applying these equations to the averaged BT line data amplitude and duration PDFs, for both method 1 and 2, gives a revised set of parameters. These are shown in Table 4 along with the χ^2 statistic of goodness-of-fit. Figures 7 and 8 show the MLE approximated curves plotted in conjunction with the averaged data they are modelled on. It is immediately apparent that the fit for the amplitude data by both methods, as well as the fit for the duration distribution as calculated by Method 2 is much improved in comparison with the previous results. However, it is also clear that the general fit to the method 1 duration distribution is quite poor. Closer examination reveals that the fit is excellent for the initial part of the data, but that the function (3.2) cannot closely model the tails of the distribution.

4 Conclusion

It is clear from the previous results that the impulse amplitude model provides a good fit to the BT data, once the parameters have been modified, although

Table 4 Revised parameters for amplitude and length models at customer premises, with χ^2 statistic to indicate the improved goodness-of-fit

	Method 1	Method 2
u_0	1.67 μV	84.2 μV
χ^2 (amp)	0.345	0.170
s_1	0.81	1.06
t_1	3.54 μs	12.5 μs
χ^2 (dur)	21.15	1.281

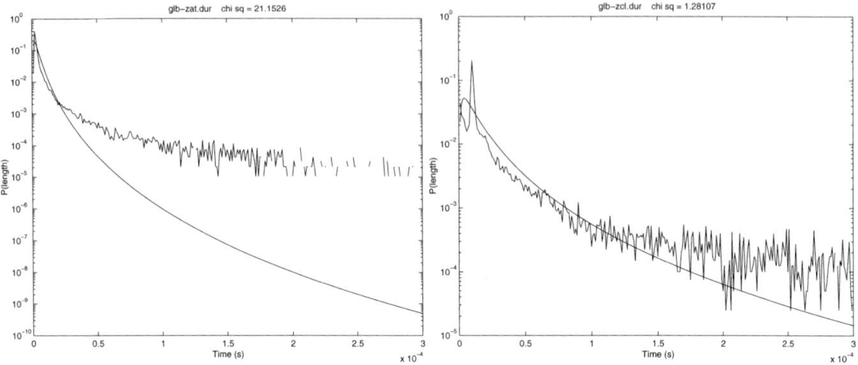

FIG. 7. Comparison of DT model calculated with revised parameters with average impulse duration PDFs using (left) the amplitude thresholding method and (right) the χ^2 statistic test.

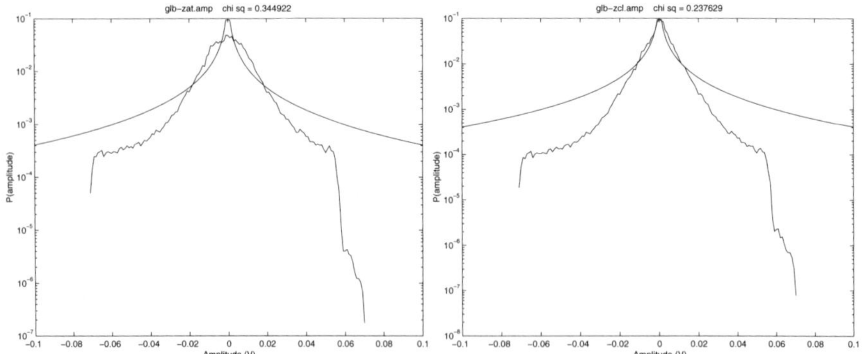

FIG. 8. Comparison of DT model calculated with revised parameters with average impulse amplitude PDFs using (left) the amplitude thresholding method and (right) the χ^2 statistic test.

it should be noted that the tails of the distributions are not modelled by (3.1) at all. The impulse length model does not provide quite such a convincing fit, since it does not seem possible to combine a sharp-enough peak with the correct roll-off in the latter part of the distribution. Further experimentation would be required to determine how significant an effect this would cause.

Acknoledgements

This research was supported by EPSRC, British Telecom and the Royal Society.

Bibliography

[1] Cook, J.W. et al. The noise and crosstalk environment for ADSL and VDSL systems. *IEEE Commun. Mag.*, **37**, 73–78, 1999.

[2] Clarke, G.M. and Cooke, D. *A Basic Course in Statistics.* London, Arnold, 1998.

[3] Henkel, W. and Kessler, T. Statistical description and modelling of impulsive noise on the German telephone network. *Electron. Lett.*, **30**, 935–936, 1994.

[4] Henkel, W. and Kessler, T. An impulse-noise model—a proposal for SDSL. *Submission to Standards Committe ETSI WG TM6, WD11*, May 1999.

[5] Komp, G. Impulse noise measurement in the VDSL environment. *Submission to Standards Committe ETSI WG TM6*, June 1997.

Nonlinear Thoughts about Linear Signal Processing

D. S. Broomhead, J. P. Huke, M. R. Muldoon
University of Manchester Institute of Science and Technology, Department of Mathematics, UMIST, PO Box 88, Manchester M60 1QD, UK

A. G. Brown
Defence Evaluation Research Agency, St Andrews Road, Great Malvern, Worcs WR14 3PS, UK

Abstract

Recent work on modelling digital channels using iterated function systems suggests a general approach to the theory of signal processing in digital communications which uses so-called *delay methods* developed for deterministic nonlinear timeseries analysis. Here we make the connection between this work and the more conventional approach to digital communications by casting linear channel models as iterated function systems and showing how the use of delay methods gives a nice connection with the theory of observability in the control of linear systems.

1 Introduction

The culture of signal processing is steeped in the mathematics of linear systems. Linear models are used across the range of signal processing applications and can be very effective. In contrast, the mathematics of nonlinear processes has had comparatively little impact. Nonlinearity, it is often suggested, leads to a confusion of special cases with no unifying theoretical picture. This is an unfortunate state of affairs since it is easy to anticipate circumstances in which nonlinearities unavoidably—by accident or design—play a significant role in the transmission of information. Indeed, we might imagine that a better developed theory of nonlinear signal processing might encourage the design of novel systems capable of exploiting nonlinear phenomena.

The assumption of linearity permits the invocation of a well-developed mathematical framework and, with it, the possibility of making generalizations about signals: that they can be added together, for example, and that filtering sums of signals is equivalent to summing the separately filtered signals. The disadvantage is, of course, that these generalities may not be relevant. We would like, therefore, to develop an approach to signal processing which has a general utility but which is based on less specialized assumptions [3, 4]. Of course, *something* should be assumed, and in our approach—which focuses on digital signal processing—it is the discreteness of the alphabet of input symbols which is exploited. Digital technology imposes discrete structures on continuous natural processes; by understanding better the implications of this we hope to make progress.

The starting point of this work is the inclusion of a model of the signal source within the model of the digital channel. As a result, a class of mathematical objects known as *iterated function systems* (IFSs) [7, 2] arises in a natural way. Moreover it is possible, for these systems, to adapt methods for the analysis of time series data which were first developed for deterministic, nonlinear dynamical systems [1, 13, 10, 9, 11]. This marriage of the theory of IFS with so-called *delay methods* for time series analysis is at the heart of our approach to digital signal processing.

In this paper we focus on establishing a connection with the usual theory of linear digital channels. It would make a pleasing picture if our results were to take a sensible form when restricted to the linear case. We shall show that this is indeed the case and that linear digital channels fall naturally within the ambit of the IFS-based theory. Actually, it is a peculiarity of the mathematical arguments employed here that we shall need a special theorem to deal with the particular case of linear channels. This will be the main result of the paper. During the discussion it will emerge that an IFS model of a digital linear channel is not actually linear. We shall show, however, that there is an elegant theoretical foundation on which we can build quite simple—nonlinear—equalization algorithms.

2 State space models of linear filters

In the following we make a strong appeal to geometry to provide a picture of the basic concepts. In this spirit it is helpful to cast linear channel models in the language of state space [6].

A linear, mth order, all-pole filter subject to a sequence of inputs $\{b_n\}$ can represented by the following system of non-homogeneous linear difference equations:

$$\mathbf{x}_{n+1} = \mathbf{A}\mathbf{x}_n + \mathbf{b}_{n+1} \tag{2.1}$$

where the states, $\mathbf{x}_n \in \mathbb{R}^m$, of the filter can be thought of as sequences of m numbers specifying the contents of a tapped delay line, and the input vectors, $\mathbf{b}_n \in \mathbb{R}^m$, have the form $\mathbf{b}_n = (b_n, 0, \ldots, 0)^T$. The matrix $\mathbf{A}$—which specifies the filter—has the form of a companion matrix

$$\mathbf{A} = \begin{pmatrix} a_0 & a_1 & \ldots & a_{m-2} & a_{m-1} \\ 1 & 0 & \ldots & 0 & 0 \\ 0 & 1 & \ldots & 0 & 0 \\ \vdots & \vdots & \ddots & \vdots & \vdots \\ 0 & 0 & \ldots & 1 & 0 \end{pmatrix}$$

where the usual filter coefficients are given by the top row, $\mathbf{a}^T$. The structure of $\mathbf{A}$ has the effect of shifting all the components of $\mathbf{x}_n$ down one place (the mth component is thereby lost) and replacing the first component with the linear combination $\mathbf{a}^T \cdot \mathbf{x}_n$.

The output of the filter is generated by making measurements corresponding to a linear function $v : \mathbb{R}^m \to \mathbb{R}$ of the state of the filter. This is equivalent to forming the scalar product of the state with a fixed vector $\mathbf{v}^T$, $v(\mathbf{x}) = \mathbf{v}^T \cdot \mathbf{x}$. The sequence of observations $\{v(\mathbf{x}_n)\}$ is then the output of a pole-zero filter given the input sequence $\{b_n\}$.

Let us now shift the viewpoint slightly. Our particular interest here is in digital channels and so we can assert that the possible values taken by the inputs, $\{b_n\}$, are drawn from a finite alphabet (containing, say, p symbols). A different interpretation of (2.1) is that the channel state evolves in one sampling interval under the action of one of p different maps $\mathbf{w}_b : \mathbb{R}^m \to \mathbb{R}^m$ defined as follows:

$$\mathbf{w}_b(\mathbf{x}) = \mathbf{A}\mathbf{x} + \mathbf{b}. \tag{2.2}$$

Assume, for simplicity, that the sequence of symbols input to the channel is an independent, identically distributed random process. (This is a reasonable initial assumption—efficiently coded data will appear random—but is not crucial to what follows.) The channel state $\mathbf{x}_n$ then evolves under a random iteration procedure according to which one of the p maps is selected at random at each time step and applied to the current state. Thus the nth state is obtained from the initial state by composition of a random sequence of maps:

$$\mathbf{x}_n = \mathbf{w}_{b_n} \circ \mathbf{w}_{b_{n-1}} \circ \cdots \circ \mathbf{w}_{b_1}(\mathbf{x}) \tag{2.3}$$

and the corresponding output is given by

$$v_n = \mathbf{v}^T \cdot \mathbf{w}_{b_n} \circ \mathbf{w}_{b_{n-1}} \circ \cdots \circ \mathbf{w}_{b_1}(\mathbf{x}). \tag{2.4}$$

From this point of view, the digital channel is seen as an IFS. There is now a considerable amount of interest among pure and applied mathematicians in this kind of dynamical system (see, for example, the recent review by Diaconis and Freedman [5]). A few basic results will suffice here. We shall assume that in some suitable norm the maps of the IFS are contractions, so that for each map $\mathbf{w}_b$

$$\|\mathbf{w}_b(\mathbf{x}) - \mathbf{w}_b(\mathbf{y})\| < \|\mathbf{x} - \mathbf{y}\|$$

for all $\mathbf{x}$ and $\mathbf{y}$ in some closed bounded subset of $\mathbb{R}^m$. This assumption is essentially one about the stability of the channel. It implies, for instance, that if the channel is repeatedly subjected to the same input symbol, the output will converge to a constant value which is independent of the initial state of the channel. (There has been a lot of recent interest in obtaining results for IFSs under weaker conditions than strict contractivity, and it is possible that these results could be of relevance to the modelling of digital channels. This, however, is work for the future.)

For the present purposes we note that with the above assumptions the IFS has a unique attractor, $\mathcal{A}$, which is a compact invariant subset of the region of $\mathbb{R}^m$. Supported on this set is a unique ergodic probability measure. We note also that $\mathcal{A}$ satisfies the following equation:

$$\mathcal{A} = \bigcup_b \mathbf{w}_b(\mathcal{A}). \tag{2.5}$$

That is to say, the attractor is the union of p sets, each of which is the image of the attractor itself under one of the mappings in the IFS. This result has (at least) two interesting consequences: the first is that $\mathcal{A}$ is often a fractal set since it is the union of contracted copies of itself, each of which is a union of contracted copies, and so on; the second—related—result is that every point in $\mathcal{A}$ has an 'address'. A way to see this is to think of the *backward iteration* of the IFS

$$\bar{\mathbf{x}}_n = \mathbf{w}_{b_1} \circ \mathbf{w}_{b_2} \circ \cdots \circ \mathbf{w}_{b_n}(\mathbf{x}_0) \tag{2.6}$$

(note the reverse ordering of the subscripts compared with (2.3)). Since the maps are contracting on a closed bounded subset of $\mathbb{R}^m$, this process generates, for each choice of symbol sequence $\{b_k : k = 1, 2, \dots\}$, a convergent sequence $\{\bar{\mathbf{x}}_k : k = 1, 2, \dots\}$ whose limit is a point in $\mathcal{A}$ which is independent of the choice of $\mathbf{x}_0$. For any given point $\mathbf{x} \in \mathcal{A}$, any symbol sequence giving a convergent sequence under backward iteration which has $\mathbf{x}$ as its limit can be regarded as an address of $\mathbf{x}$. Each point in $\mathcal{A}$ has at least one address [2]. If the images $\mathbf{w}_b(\mathcal{A})$ are all disjoint then the address is unique and we say that $\mathcal{A}$ is *totally disconnected.* The implication of this for digital channels is that if we can at any time identify where we are on the attractor of the channel, then implicitly this gives information about the history of inputs to the channel. In the case that the channel has a totally disconnected attractor then there is a unique sequence of input symbols which produces a given channel state. Of course, it would be necessary to measure the channel state with infinite precision to get a complete history, but, as we shall see, less complete measurements nonetheless provide useful information.

2.1 An example

A simple example will be useful to illustrate the various stages of the development. The constraints imposed by the need to represent the results graphically limit this to a second-order IFS model of a binary channel. Specifically, we use equation (2.2) with

$$\mathbf{A} = \begin{pmatrix} 0.8 & -0.5 \\ 1 & 0 \end{pmatrix} \tag{2.7}$$

and $b = \pm 1$. We take $v^T = (1, 0)$. Random iteration of this model gives the attractor shown in Fig. 1 where two scales of grey have been used to label the

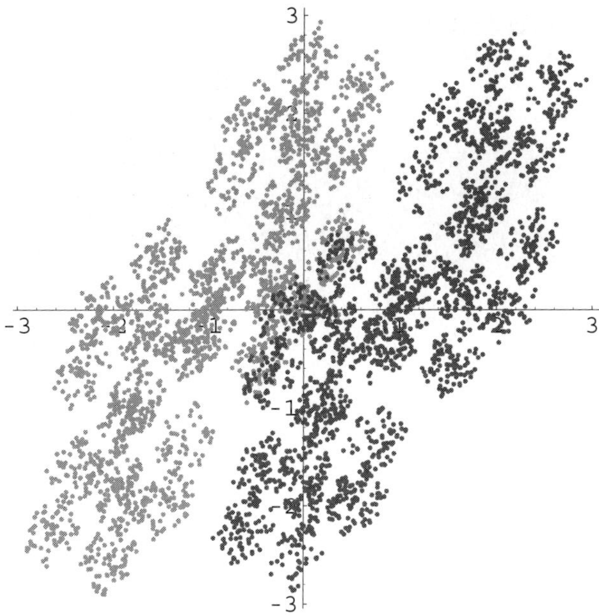

FIG. 1. Attractors of a second order linear recursive channel—see (2.2)—with $\mathbf{A}$, defined in (2.7) and $b = \pm 1$. The darker points are in the set $\mathbf{w}_{+1}(\mathcal{A})$ and the lighter points are in $\mathbf{w}_{-1}(\mathcal{A})$.

sets $\mathbf{w}_{+1}(\mathcal{A})$ and $\mathbf{w}_{-1}(\mathcal{A})$, that is, the parts of the attractor which correspond respectively to the last input being $+1$ and -1. We note that the attractor does not appear to be totally disconnected—there appears to be a region of overlap of the two differently shaded regions—and so we expect that points in the attractor will not be uniquely addressable. Despite this, the channel can be equalized by virtue of being an all-pole system. To see this we note simply that according to (2.1)

$$\mathbf{b}_{n+1} = \mathbf{x}_{n+1} - \mathbf{A}\mathbf{x}_n \tag{2.8}$$

that is, a suitable linear combination of two successive state vectors recovers the input to the channel.

We have assumed so far that the output of the channel is the first component of $\mathbf{x}$, or, in terms of the observation function introduced earlier, that $\mathbf{v}^T = (1, 0)$. A more general choice of $\mathbf{v}^T = (\cos\theta, \sin\theta)$ with $\theta \in (0, \pi)$ but not equal to $\pi/2$, will produce an output time series which is harder to invert since the linear inverse of an FIR filter is an IIR filter and hence requires an infinite history of the FIR output. Fig. 2 shows a typical output time series obtained from the pole-zero filter defined by (2.7) with $\mathbf{v}^T = (\frac{1}{2}, \frac{\sqrt{3}}{2})$, driven by an equiprobable independent sequence of inputs with $b = \pm 1$.

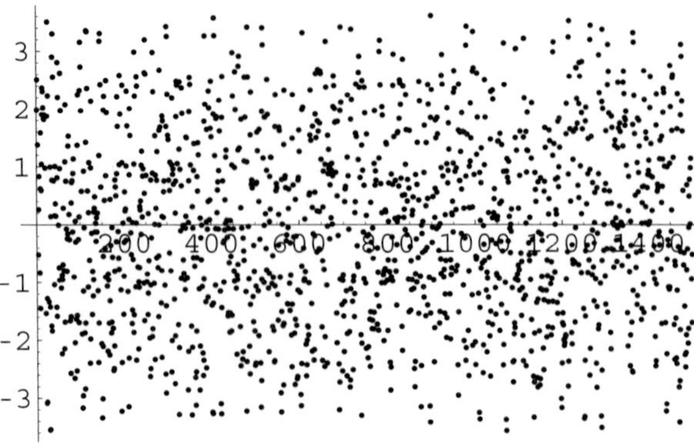

FIG. 2. Typical time series of output from the pole-zero filter defined by (2.7) with $\mathbf{v}^T = (\frac{1}{2}, \frac{\sqrt{3}}{2})$ driven by an equiprobable independent sequence of inputs with $b = \pm 1$.

3 Delay embedding of linear channels

The second part of our development is the introduction of the method of delays (for a nice description see the book by Ott, Sauer and Yorke [8]). The essentials of this are that there is a smooth dynamical system defined on a state space and a smooth measurement which is a real-valued function of the state. The basic results of this theory establish a link between the dynamical system and a construction based on time series data obtained by making successive measurements on the system. As an example, consider a dynamical system consisting of a linear map $\mathbf{A}$ which acts on a vector space $\mathbb{R}^m$. Starting with an arbitrary initial state $\mathbf{x}_0 \in \mathbb{R}^m$, repeated application of the map gives a sequence of new states, $\{\mathbf{x}_n = \mathbf{A}^n\mathbf{x}_0\}$, of the system. If at each time step we record only the projection of the state onto some fixed vector $\mathbf{v}^T$ we obtain a scalar time series $\{v_n = \mathbf{v}^T \cdot \mathbf{x}_n\}$. The construction of interest is based on the tapped delay line: that is, we consider vectors of the form $(v_n, v_{n+1}, \ldots, v_{n+d-1})^T$. This vector can be thought of as being a function of the point $\mathbf{x}_n \in \mathbb{R}^m$ since

$$(v_n, v_{n+1}, \ldots, v_{n+d-1})^T = (\mathbf{v}^T \cdot \mathbf{x}_n, \mathbf{v}^T \cdot \mathbf{A}\mathbf{x}_n, \ldots, \mathbf{v}^T \cdot \mathbf{A}^{d-1}\mathbf{x}_n)^T.$$

To be more formal, we use this tapped delay line approach to define a map $\Phi : \mathbb{R}^m \to \mathbb{R}^d$ taking points from the state space of the dynamical system to the space of states of the d-tap delay line:

$$\Phi(\mathbf{x}) = (\mathbf{v}^T \cdot \mathbf{x}, \mathbf{v}^T \cdot \mathbf{A}\mathbf{x}, \ldots, \mathbf{v}^T \cdot \mathbf{A}^{d-1}\mathbf{x})^T. \tag{3.1}$$

This map—which is clearly linear—arises in linear control theory in connection with the *observability* of a system. A well-known result from control theory asserts that Φ is full rank if all the eigenvalues of $\mathbf{A}$ are distinct and none of its eigenvectors is orthogonal to $\mathbf{v}$. Thus, if $d \geq m$, the image $\Phi(\mathbb{R}^m)$ is an

m-dimensional linear subspace of $\mathbb{R}^d$. This result is a statement of the 'usual' or 'generic' situation in the sense that special conditions must hold for it not to be the case. It provides a strong link between the original dynamical system and the tapped delay line data by showing that the information preserved in the tapped delay line representation is that which is preserved when making a change of coordinates. For example, consider the relationship $\mathbf{x}_{n+1} = \mathbf{A}\mathbf{x}_n$ and write $\mathbf{y}_n = \Phi(\mathbf{x}_n)$. Then it follows that $\mathbf{y}_{n+1} = \Phi(\mathbf{A}\Phi^{-1}(\mathbf{y}_n))$; the states of the tapped delay line evolve according to a linear map $\Phi\mathbf{A}\Phi^{-1}$ which, since it is similar to $\mathbf{A}$, has the same spectrum as $\mathbf{A}$. (Note that the inverse $\Phi^{-1}\mathbf{y}$ is meaningful whenever $\mathbf{y} \in \Phi(\mathbb{R}^m)$.)

The theorems of Aeyels [1], Takens [13] and Sauer *et al.* [11] extend this analysis to nonlinear dynamical systems and nonlinear measurement functions. In this case the map corresponding to Φ is nonlinear and an *embedding* of the state space for generic choices of the measurement function. This means that the derivative of Φ is well-defined and full rank at every point in the state space of the dynamical system and, in addition, that the map is invertible (in the linear case these properties are equivalent). Delay embedding, even in this nonlinear case, preserves the information preserved by a (nonlinear) smooth change of coordinates.

The development of these ideas to make them applicable to digital signal processing requires that we enlarge their scope to include iterated function systems. We have reported results in this direction elsewhere [12, 4] and, in [4], described how to extend IFS models to include oversampling of channels and how to exploit the resulting structure through a further development of the method of delays.

Here we shall focus on linear channels sampled at the baud rate and ask, what happens if we apply the tapped delay line idea to the output of a digital linear channel? Equation (2.4) expresses the output of the channel in terms of the sequence of input symbols and the initial state of the channel. Using this, we can define, *for each input symbol sequence*, a delay map $\Phi_\Omega : \mathbb{R}^m \to \mathbb{R}^d$ by

$$\Phi_\Omega(\mathbf{x}) = (\mathbf{v}^T \cdot \mathbf{x}, \mathbf{v}^T \cdot \mathbf{w}_{b_1}(\mathbf{x}), \ldots, \mathbf{v}^T \cdot \mathbf{w}_{b_{d-1}} \circ \mathbf{w}_{b_{d-2}} \circ \cdots \circ \mathbf{w}_{b_1}(\mathbf{x}))^T \tag{3.2}$$

where the subscript Ω labels the input sequence: $\Omega = (b_1, b_2, \ldots, b_{d-1})$. Introducing the explicit form of the $\{\mathbf{w}_b\}$ given in (2.2), reveals that the delay map is affine

$$\Phi_\Omega(\mathbf{x}) = \Phi(\mathbf{x}) + \Phi_\Omega(\mathbf{0}) \tag{3.3}$$

where Φ—which is independent of Ω—is the linear delay map defined in equation (3.1). The remaining term is a fixed offset which depends on Ω, but is independent of the channel state $\mathbf{x}$

$$\Phi_\Omega(\mathbf{0}) = (0, \mathbf{v}^T \cdot \mathbf{b}_1, \mathbf{v}^T \cdot (\mathbf{b}_2 + \mathbf{A}\mathbf{b}_1), \ldots, \mathbf{v}^T \cdot (\mathbf{b}_{d-1} + \mathbf{A}\mathbf{b}_{d-2} + \cdots + \mathbf{A}^{d-2}\mathbf{b}_1))^T \tag{3.4}$$

The fact that the observability matrix, Φ, arises naturally here makes the point that delay methods reduce to well-established theory in the special case

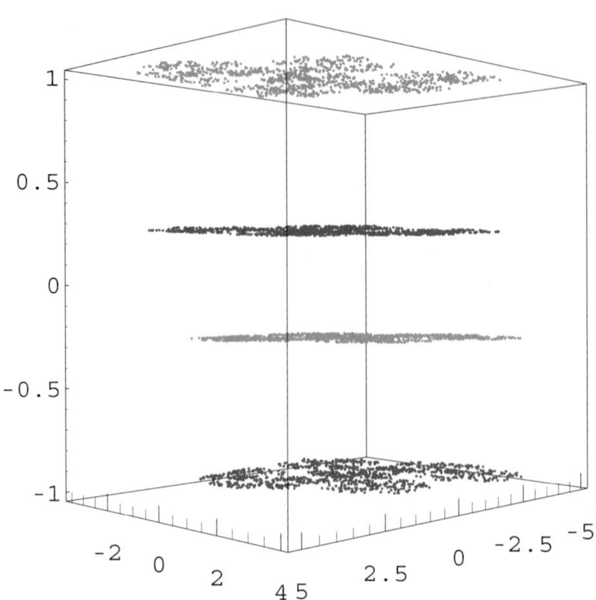

FIG. 3. The attractor shown in Fig. 1 mapped using the delay maps $\Phi_\Omega : \mathbb{R}^2 \to \mathbb{R}^3$ based on time series data shown in Fig. 2. The shading corresponds to that used in Fig. 1, for the darker points the latest symbol was $+1$ and for the lighter points the latest symbol was -1 .

of linear channels. The linear theory shows that for generic choices of $\mathbf{v}^T$, the rank of Φ is m when $d \geq m$ and, therefore, that the map Φ_Ω is an embedding for each Ω. Thus, $\Phi_\Omega(\mathcal{A})$ is $\mathcal{A}$, apart from a smooth change of coordinates. This has important implications for channel equalization since there will be a correspondence between the addresses of points in $\mathcal{A}$ and addresses of points in $\Phi_\Omega(\mathcal{A})$ which is constructed using the channel output.

The additional complication that IFSs bring to the use of delay embedding is that each sequence Ω generates a different embedding Φ_Ω. For a p symbol alphabet and using d delays there are p^{d-1} of these. The question is, therefore, are the images of the attractor under the different delay maps all disjoint? The answer is given by the following theorem.

Theorem 3.1 *If* $\mathbf{A}$ *has all distinct eigenvalues, then for generic choices of* $\mathbf{v}$ *each of the delay maps* $\Phi_\Omega : \mathbb{R}^m \to \mathbb{R}^d$, *with* $d \geq m$, *is an embedding. Moreover, if* $d > m$, *the generic case is that the images* $\Phi_\Omega(\mathbb{R}^m)$ *and* $\Phi_{\Omega'}(\mathbb{R}^m)$ *are disjoint when* $\Omega \neq \Omega'$.

The theorem has two parts. The first is just the observability condition already described. The second can be shown using a dimension-counting argument which often arises in this sort of proof. In this particular case the argument is

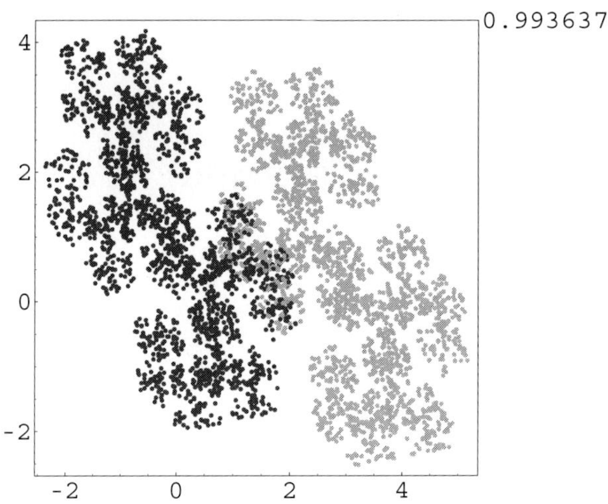

FIG. 4. A plot of the uppermost of the four sheets evident in Fig. 3. In this figure, the darker points are in the set $\Phi_{(-1,-1)}(\mathbf{w}_{+1}(\mathcal{A}))$ and the lighter points are in $\Phi_{(-1,-1)}(\mathbf{w}_{-1}(\mathcal{A}))$

based on the observation that a sufficient condition for the result to hold is that there is no $\mathbf{x} \in \mathbb{R}^m$ such that $\Phi(\mathbf{x}) = \Phi_{\Omega'}(\mathbf{0}) - \Phi_{\Omega}(\mathbf{0})$ for any pair of symbol sequences $\Omega \neq \Omega'$. Figure 3 shows what happens typically in the case of the example described in Section 2.1. The delay maps are constructed using time series data as shown in Fig. 2. Choosing $d = 3 > m = 2$ and recalling that the channel input is binary ($p = 2$), we anticipate 2^2 images of the attractor as, indeed, are seen in Fig. 3. The fact that $m = 2$ implies that each image should be a subset of a plane—that is, a displaced copy of $\Phi(\mathbb{R}^2)$. These should all be parallel because they are simply translations of one another. Again, this is evident from the figure. In Fig. 4, the image of $\mathcal{A}$ under the action of one of the Φ_Ω is shown. Since this plot is essentially of $\Phi\mathcal{A}$, it is interesting to make a comparison with the untransformed form of $\mathcal{A}$ shown in Fig. 1.

Our example also illustrates the meaning of the statement that 'generically' the images $\Phi_\Omega(\mathbb{R}^m)$ and $\Phi_{\Omega'}(\mathbb{R}^m)$ are disjoint when $\Omega \neq \Omega'$. In Fig. 5 we show how the choice of $\mathbf{v}$ changes the positions of the different image planes corresponding to the different values of Ω. This is done by calculating the points of intersection of the planes with their common normal. These are shown as a function of θ which parametrizes $\mathbf{v}$ through $\mathbf{v} = (\cos\theta, \sin\theta)$. The interpretation of 'generic' here is that at any value of θ except an exceptional set, $\mathcal{E}$, of isolated values, the four planes intersect their common normal at four different points.

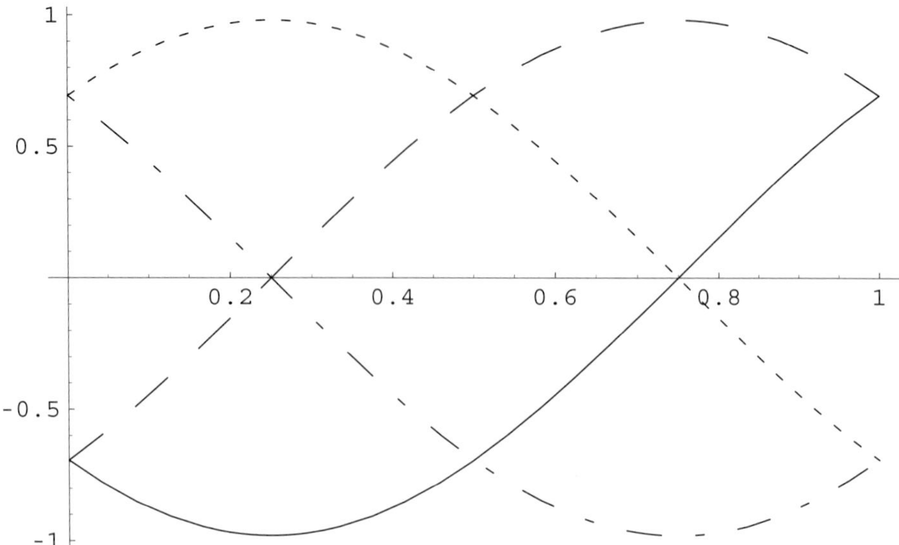

FIG. 5. The projections onto their common normal of the four parallel planes $\Phi_\Omega(\mathbb{R}^2)$ (where $\Omega = (+1,+1)$ (solid), $(-1,+1)$ (dashed), $(+1,-1)$ (dash-dotted), and $(-1,-1)$ (dotted)) plotted as a function of θ where the measurement function is $\mathbf{v} = (\cos\theta, \sin\theta)$ with $\theta \in [0, \pi]$.

The figure shows that $\mathcal{E} = \{0, \frac{\pi}{4}, \frac{\pi}{2}, \frac{3\pi}{4}, \pi\}$.

There is a subset of $\mathcal{E}$, $\{0, \frac{\pi}{2}, \pi\}$, which is benign in the sense that for θ in this set the delay vectors are essentially state space vectors of the all-pole channel model specified by (2.7). Recall that the state space dynamical system described in Section 2 is based on the structure of the tapped delay line just as is the method of delays. Therefore, a measurement function $\mathbf{v}$ that picks out a single component of the natural basis of the state space will give rise to delay vectors which are state vectors or—in the case of using $d > m$—vectors such that every m consecutive components are state vectors. It follows in these cases that all of the information about $\mathcal{A}$ is contained unambiguously in the delay plot. This is not true of the remaining points in $\mathcal{E}$ which correspond to measurements which confuse the time ordering of symbols.

4 Some remarks on the equalization of IFS channels

Given that we observe the output sequence of a digital channel and are able to construct—at least in principle—an object such as is shown in Fig. 3, how is this of any help in reconstructing the corresponding input sequence? Leaving aside any attempt to do blind equalization, let us assume that we know the input sequence which generated some part of the output. In the first instance, we can use this information to label each of the planes shown in Fig. 3. By focusing on each separately, estimates of their common normal can be obtained by, for example, forming a matrix of the differences between delay vectors in a chosen plane and computing its singular value decomposition. Alternatively, if the channel model—specified by $\mathbf{A}$, $\mathbf{b}$ and $\mathbf{v}$—is known, then there is a direct way to obtain this information using (3.4).

Assuming without loss of generality that we choose $d = m + 1$, the common normal of the planes is unique. Let us call this $\mathbf{n}$. A direct way to equalise the channel is to compute the projections of the delay vectors onto $\mathbf{n}$ and then to identify which of the projections of the image planes they correspond to. Formally, this amounts to the construction of a map from the delay space to the set of symbol sequences, $g : \mathbb{R}^d \to \{\Omega\}$. If we denote the projection of $\Phi_\Omega(\mathbb{R}^m)$ onto $\mathbf{n}$ as $\mathbf{n}_\Omega$ then $g(\mathbf{x}) = \tilde{\Omega}$ where $\tilde{\Omega}$ minimizes $|\mathbf{n} \cdot \mathbf{x} - \mathbf{n}_\Omega|$. Of course, there are more sophisticated ways of doing this which can, for example, take into account noise on the output. The point is, however, that this (nonlinear) function equalises our linear pole-zero channel using $\mathbf{x}$ which is a finite history of the channel output. If, on the other hand, we were to try to invert the channel without assuming the discreteness of the input, we would require an infinite impulse response filter.

There is another, different, way of achieving the same end. This requires that we find the inverse of the Φ_Ω. The most direct way to do this is to assume that the channel models, (2.2), are known or have been estimated using the input/output data. This allows us to compute explicitly the components of the affine form given in (3.3). The inverse of the delay maps are then simply $\Phi^{-1}(\mathbf{y} - \Phi_\Omega(0))$ (where Φ^{-1} can be computed as the pseudo-inverse of Φ). In practice, something like the g defined above will be needed to decide which of the offsets to subtract from the delay vector. The result of applying the inverses of the delay maps in our example is a set which is indistinguishable from the original attractor shown in Fig. 1. In particular the addressing of the points of the attractor is preserved by this process. In order to recover the input sequence, however, we need only apply the inverse filter given in (2.8). Moreover, it is not clear that the extra complication involved in finding the inverse of the delay maps could ever lead to a method which is superior to simply finding g.

Finally, we come to nonlinear channels. It is clear that a function like g, which maps delay vectors to symbols, can be constructed in this more general case, provided that the images $\Phi_\Omega(\mathbb{R}^m)$ are disjoint. We do not need to assume planarity of the embedded images of the channel attractor, we could, for instance, use a radial basis function expansion to fit the characteristic functions of the different sets of image points. The issue here is whether or not there is a result for nonlinear channels which is analogous to Theorem 3.1. The answer is that there is a result for nonlinear systems which is like the first part of Theorem 3.1 [12]. Indeed, it is possible to show that—generically—distinct points in the attractor do not become identified by different delay maps. However, there is a counter-example which limits our scope in the nonlinear case. It is easy to write down a hyperbolic IFS—which must be a model of some nonlinear channel—for which the images of the attractor under different Φ_Ω intersect. This property holds for any continuous measurement function and is stable in the sense that it holds also for any small perturbation of the IFS. Physically, such a 'difficult' channel is required to have a state, say $\mathbf{x}_*$, which evolves to the same new state, $\mathbf{x}'_*$, following the input of either of two different symbols. It is apparent that this is an extremely undesirable property for a communications channel to have, but unfortunately it is a possibility with nonlinear channels. For this reason we view this example as a limitation on our ability to make the simple generalization of Theorem 3.1 rather than a practical limitation on the use of delay methods for nonlinear signal processing. In this context, it is worth recalling that the approach described in [4]—which uses an oversampling technique—does not suffer from this mathematical difficulty. In this case a more detailed model of the way the channel is driven must be used. For channels which can be thought of as being driven by short pulses the issue of multiple embeddings of the attractor does not arise. Naturally, a channel which is as ambiguous as in our counter-example is likely to cause practical difficulties in this case too.

5 Conclusions

This paper has been about drawing connections between different approaches to digital signal processing. We have related the use of iterated function systems as models of digital channels to the more familiar state space models of linear channels, and we have contrasted the use of delay methods applied to IFS models with the more familiar linear methodology. Our main point has been that where the two approaches talk about the same thing there is a fundamental connection which is essentially the issue of *observability* of the channel. We have not discussed at any length the exciting prospect of a systematic and general theory of nonlinear signal processing of digital channels that the IFS work represents. This has been mentioned elsewhere [3, 4], and will be discussed in more detail in future publications.

Bibliography

[1] Aeyels D. (1981). Generic observability of differentiable systems, *SIAM J. Control Optim.*, **19**, 595–603.

[2] Barnsley M. (1988). *Fractals Everywhere*, Academic, San Diego, CA.

[3] Broomhead D.S., Huke, J.P. and Muldoon, M.R. (1998). Fractals, linear channels and delay methods. In *Mathematics in Signal Processing IV* (ed. J.G. McWhirter and I.K. Proudler), Oxford, Clarendon Press, 55–65.

[4] Broomhead. D.S., Huke, J.P. and Muldoon, M.R. (2000). Digital channels. In *Proc. IEEE 2000 Adaptive Systems for Signal Processing, Communications and Control Symp.*, pp. 123–128.

[5] Diaconis, P. and Freedman, D. (1999). Iterated random functions, *SIAM Review*, **41**, 45–76.

[6] Kailath, T., Sayed, A.H. and Hassabi, B. (2000). *Linear Estimation,*

Prentice-Hall, Englewood Cliffs, NJ.

[7] Hutchinson, J.E. (1981). Fractals and self-similarity, *Indiana University J. Math.*, **30**, 713–747.

[8] Ott, E., Sauer, T. and Yorke, J.A. (eds) (1994). *Coping with Chaos*, Wiley, New York.

[9] Packard, N., Crutchfield, J., Farmer, D. and Shaw, R. (1980). Geometry from time series, *Phys. Rev. Lett.*, **45**, 712–715.

[10] Ruelle, D. (1989). *Chaotic Evolution and Strange Attractors*, Cambridge University Press, Cambridge

[11] Sauer, T., Yorke, J.A. and Casdagli, M. (1991). Embedology, *J. Stat. Phys.*, **65**, 579–616.

[12] Stark, J., Broomhead, D.S., Davies, M.E. and Huke, J.P. (1997) Takens' embedding theorems for forced and stochastic systems, *Nonlin. Anal.*, **30**, 5303–5314, and (2001) Delay embedding for forced systems: II. Stochastic Forcing, in preparation

[13] Takens, F. (1981). Detecting strange attractors in turbulence, *Lecture Notes in Mathematics* (ed. D.A. Rand and L.-S. Young) Springer, Berlin, 366.

An Application of the Maximum Noise Fraction Method to Filtering Noisy Time Series

Markus Anderle and Michael Kirby
Department of Mathematics, Colorado State University, Fort Collins, CO 80523, USA

Abstract

We propose a tool for filtering multivariate time series that was initially developed for analysing multi-spectral satellite imagery. The basic technique, known as the maximum noise fraction (MNF) method (Green *et al.* 1988), may be used to provide a subspace decomposition of a multivariate time series in terms of basis vectors which contain maximum noise (or maximum signal). We demonstrate the utility of the method for filtering nonsmooth multivariate data that includes high variance bands such as climate data. The methodology is applied to the reduction of data on noisy manifolds. A comparison of the approach to independent component analysis (ICA) is also provided.

1 Introduction

The application of the Karhunen–Loève (KL) procedure (similarly, the singular value decomposition (SVD) or principal component analysis) to noisy data can be problematic in that the eigenvectors associated with the largest variance may contain significant amounts of noise. It is well-known that the KL eigenvectors are left unchanged by the addition of white noise while the eigenvalues are all shifted upwards by the variance of the noise. The maximum noise fraction (MNF) method proposed by Switzer [13]; see also [8], was developed as a noise removal technique for multi-spectral satellite data. Related approaches have been proposed by Allen and Smith [1] in the context of singular spectrum analysis as well as De Moor *et al.* [12] for analysing biomedical time series via QSVD.

The MNF technique produces a basis representation, the purpose of which is to separate the noise and the signal as far as possible into distinct subspaces. An interesting feature of this approach is that high-frequency components of the signal are not attenuated as a result of the filtering. Sharp features such as bursts, spikes, and non-differentiable points which may provide essential information in the time series are retained. Additionally, the method lends itself naturally to being applied locally, either in time or space.

2 Methodology

Consider the decomposition of a data matrix X into a signal matrix S and noise component N as

$$X = S + N.$$

The basic idea, due to Switzer [13], is to compute a new basis to represent the data with maximum noise fraction; the derivation here follows [10]. The optimal first basis vector, ϕ, may be written in its data-dependent form, that is, as a superposition of the data

$$\phi = \psi_1 \mathbf{x}^{(1)} + \cdots + \psi_P \mathbf{x}^{(P)} = X\psi.$$

This basis vector ϕ may also be decomposed into signal and noise components as

$$\phi = \phi_{\mathbf{n}} + \phi_{\mathbf{s}},$$

where $\phi_{\mathbf{s}} = S\psi$ and $\phi_{\mathbf{n}} = N\psi$.

The optimization problem

Now the *noise fraction* of a basis vector ϕ is defined as

$$D(\phi) = \frac{\phi_{\mathbf{n}}^T \phi_{\mathbf{n}}}{\phi^T \phi}.$$

The MNF method determines ϕ such that $D(\phi)$ is a maximum. This may now be rewritten as

$$D(\phi(\psi)) = \frac{\psi^T N^T N \psi}{\psi^T X^T X \psi}.$$

The maximization problem leads to a *symmetric definite generalized eigenproblem*

$$N^T N \psi = \mu^2 X^T X \psi. \tag{2.1}$$

Thus, given a data matrix X and the solution matrix $\Psi_m = [\psi^{(1)}| \dots |\psi^{(m)}]$ of the generalized singular vector problem, given in (2.1), the orthonormal basis for $\mathbb{R}^m$ is given by $\Phi_m = [\phi^{(1)}| \dots |\phi^{(m)}]$ where $\Phi_m = X\Psi_m$.(The basis vectors are ordered by increasing noise fraction, so a truncation of the basis corresponds to noise filtering.) Clearly we may express the data without loss as $X = \Phi_m \Phi_m^T X$. Truncating columns of Φ_m filters the noise, that is, the data may be decomposed as

$$X_D = \Phi_D B_D,$$

where the smaller matrix

$$B_D = \Phi_D^T X$$

consists of reconstruction coefficients and $D < m$.

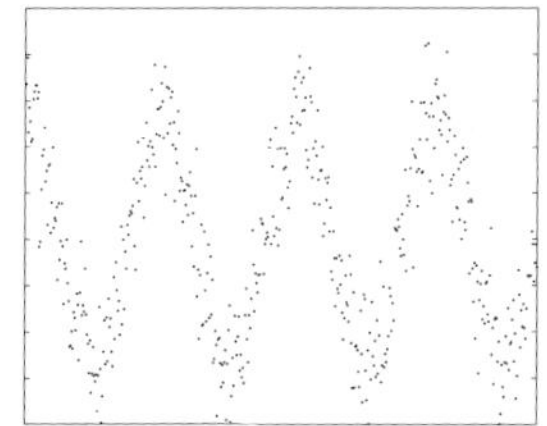
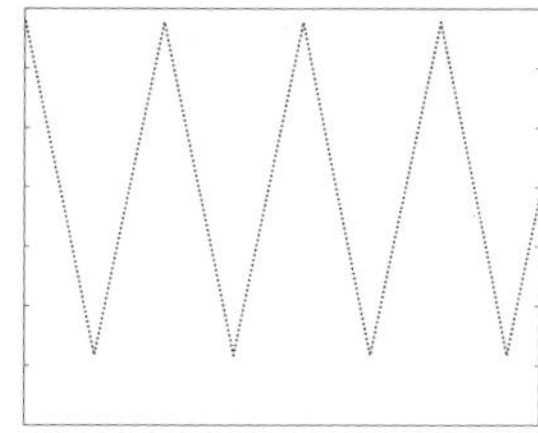

FIG. 1. (a) One of the ten original time series; (b) One term KL reconstruction of (a); (c) One term maximum noise fraction reconstruction of (a).

Estimating the covariance matrix of the noise

As suggested by Switzer [13], the covariance matrix of the noise may be estimated by shifting a time series and computing differences. One can show in this case that the covariance matrix of the differences Σ is approximately twice $N^T N$. Note that this method requires that the data be smooth, that is, $x_t \approx x_{t+1}$. It should be noted that the procedure for estimating the covariance matrix of the noise need not be absolutely perfect. The penalty of error is a modest rotation of the subspace spanning the maximum noise fraction eigenvectors.

3 Applications

We demonstrate the effectiveness of the maximum noise fraction for filtering noisy data in the context of several illustrative examples.

3.1 Filtering nonsmooth data

The data in this example was generated specifically to demonstrate the point that nonsmooth functions buried in the data may be accurately recovered. The data set consists of ten highly correlated noisy time series each consisting of 500 points. The correlation was achieved by mapping three time series (a sawtooth of amplitude one and period 2π with no noise; a pure noise signal—normally distributed with zero mean and variance 0.01; and a sinusoid squared with added noise) to $\mathbb{R}^{10}$ using a random full rank matrix. The result of applying the MNF to one of the ten time series is shown in Fig. 1. The sawtooth is in fact the first MNF basis vector and thus captures the shape of the time series without noise. The KL reconstruction is noisy as the first basis vector contains significant noise. (See, for example, [10] for a general discussion of the application of KL to data sets.)

3.2 Multivariate weather data

Here we investigate the application of the MNF method to weather data collected during the first 4 days of October 2000 with a sampling frequency of 5 mins. The time series we investigate include temperature, relative humidity, wind speed (average speed), gust speed (speed of the maximum wind) and pressure. (These data were made available by the Colorado State University Atmospheric Sci-

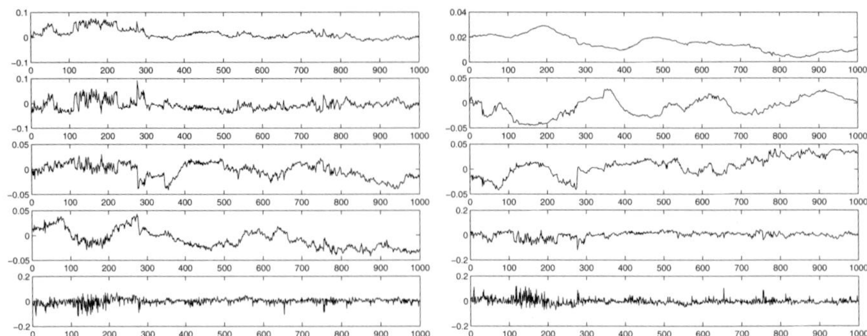

FIG. 2. Basis vectors ordered from top to bottom. (a) KL basis; (b) maximum noise fraction basis with maximum signal basis vector at the top and maximum noise basis vector at the bottom.

ence Department, Fort Collins, CO. They were collected at the Christman Field (FCC), Foothills Campus Weather Observation Station and may be viewed at www.atmos.colostate.educgi-binfcc_form.pl.)

Again, for purposes of comparison we have included both the KL reconstruction and the MNF reconstruction using the basis vectors shown in Fig. 2. Using three terms, the KL reconstruction fits each time series with comparable accuracy, see Fig. 3. Note that the wind and gust speed time series have high variance but are fit very well. The MNF fit for these time series is smoothed. Note that the temperature, relative humidity and pressure are fit more accurately using MNF than with KL. Thus, the MNF technique fits the low variance (low noise?) data with high accuracy and filters high variance (high noise?) data. This fact is made apparent by the nature of the KL and MNF eigenvectors shown in Fig. 2. The first few MNF basis vectors containing the signal are much smoother than the corresponding KL basis vectors. (Note that the data were scaled in this example such that each time series has unit variance and zero mean. This is required for the KL procedure, otherwise the dominant eigenvectors span the wind and gust speed subspaces. One advantage of the MNF method is that it is scale invariant.)

4 Reduction of noisy manifolds with MNF

If data are highly correlated in the ambient space it is often possible to construct dimensionality-reducing mappings: see [10] for a comprehensive discussion. In general, if the data matrix has full rank, then linear methods are not able to compress the dimension without loss of data. However, consider the special instance when the data matrix actually contains points that reside, at least approximately, on a manifold.

Now consider a data matrix X of size $P \times m$ where $P > m$. We proposed a new architecture for representing embedded q-dimensional manifolds of data, based on Whitney's embedding theorem, that consists of a linear reduction map-

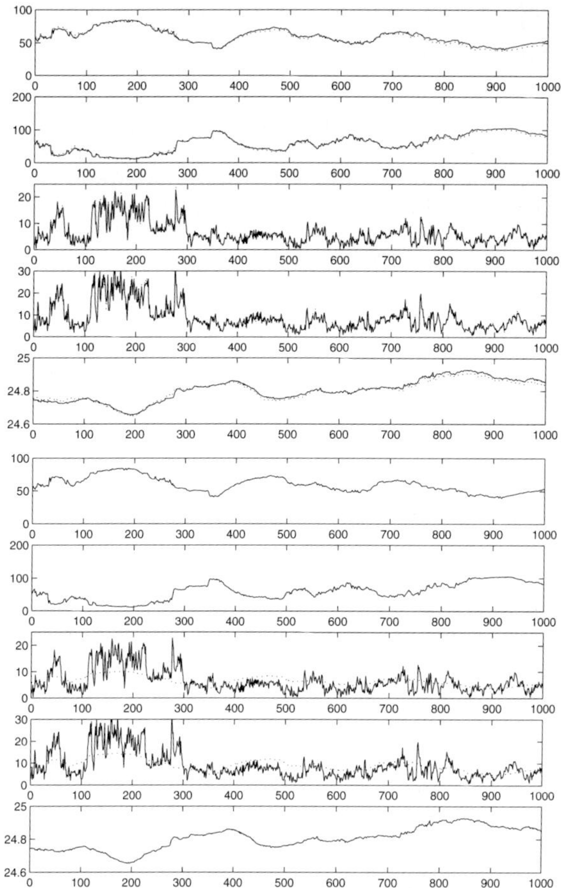

FIG. 3. Reconstructions of weather data from October 1–4, 2000 using (a) three-term KL reconstruction; (b) three-term MNF reconstruction. The five time series in each panel measure temperature, relative humidity, wind speed, gust speed and pressure. Dotted line: reconstructed data; solid line: original data.

ping and a nonlinear (radial basis function) reconstruction mapping based on [4]. Applying this architecture to noisy data is challenging and requires special modifications [3]. Here we examine the utility of MNF for use in conjunction with the Whitney Reduction Network (WRN).

4.1 The Whitney reduction network

Proposed initially in [4], the WRN provides a tool to express a data set in $\mathbb{R}^m$ globally as the graph of a function. A primary feature of the method is that it provides a good domain for the graph by optimizing on the condition number of the function. Given a data matrix X, it is reduced by projecting onto the d-dimensional domain of the graph. Whitney's theorem states that a sufficient condition for this inverse to exist is that the dimension d of this projection satisfy $d \geq 2q + 1$ where q is the dimension of the manifold on which the data lie. The theorem is an idealization that we have found to work well in practice.

The parametrization of the data is in terms of the reduced coordinates $\hat{p}$ where the representation of a point x, written $\tilde{x}$, is given by

$$\tilde{x} = U_1\hat{p} + U_2 f(\hat{p})$$

where U_1 is a basis for the domain and U_2 is a basis for the range. In this setting the data may now be viewed as the graph of a function $(\hat{p}, f(\hat{p}))$. Alternatively, the first term $U_1\hat{p}$ may be viewed as the *linear reconstruction* while the second term $U_2 f(\hat{p})$ is the *nonlinear reconstruction*: see [10] for further details and examples of this procedure.

4.2 The MNF with WRN

It has been observed [3], perhaps not surprisingly, that the presence of noise in data may impede the search for a good projection. Thus it is natural to investigate the utility of the maximum (equivalently, minimum) noise fraction method. We shall see that this approach has considerable appeal given that it naturally separates the signal and noise into orthogonal subspaces.

Recall that the MNF noise filtered data may be written

$$X_D = \Phi_D B_D.$$

We propose to compress X_D by parameterizing the last $D - d$ columns of Φ_D by the first d columns. Following [4], as a preprocessing step the basis of the row space of Φ_D is changed to improve the condition number of the reconstruction, that is,

$$\Pi = V^T \Phi_D.$$

In practice, it is the first d columns of Π that serve as the domain of the graph while the $D - d$ columns serve as the target of the domain, that is, the range. A radial basis function expansion is employed to approximate this map [4].

See Figs 4 and 5 for a summary of the decomposition and reconstruction procedures.

4.3 Predicting a change in the weather

Here we consider an application of the WRN with MNF filtering to a set of six time series measuring weather variables for the month of October, 2000 as shown in Fig. 6. (Note that the noise fraction basis was calculated over the same interval as the training data.)

We found that choosing a filter corresponding to $D = 4$ eliminated what appeared to be correlated noise. Further, we employed $d = 3$ so the radial basis function was required to empirically model a function

$$f : \Pi_3 \subset \mathbb{R}^3 \to \Pi_3^{\perp} \subset \mathbb{R}.$$

$$\begin{array}{ccccccc} X & \rightarrow & \Phi_D & \rightarrow & \Pi & \rightarrow & \Pi_d \\ & \searrow & & & & \searrow & \\ & & B_D & & & & \Pi_d^{\perp} \end{array}$$

FIG. 4. A summary of the decomposition and parameterization of a data set using the WRN with MNF.

$$\begin{array}{ccccccc} \Pi_d & \rightarrow & \tilde{\Pi} & \rightarrow & \tilde{\Phi}_D & \rightarrow & \tilde{X}_D \\ & \nearrow & & & & \nearrow & \\ f(\Pi_d) & & & & B_D & & \end{array}$$

FIG. 5. A summary of the reconstruction of a data set using the WRN with MNF based on the decomposition shown in Fig. 4. The tildes denote that the quantities are now the approximations (to the true values) as produced by the RBF fitting procedure.

We selected a period of ten days, October 9–19, for constructing the model using the method of orthogonal least squares [6].

At approximately point 475 we detect that the model residual has become significant (see Fig. 7) indicating that the underlying weather pattern is changing. Note that this point in time occurred 25 h after the end of the training data. Furthermore, at about point 510, or a day and a half after the model detected a change, there was an actual shift in the weather pattern as indicated by the nonoscillatory temperature profiles. Thus, this example suggests that the model predicted a change in the weather roughly a day before it happened.

This study is meant only to be an illustrative example of the capacity of this reduction architecture when employed with the MNF methodology.

5 Relationship to other methods

5.1 Independent component analysis

The task of filtering noise from data may also be viewed as a problem in multiple source separation. In particular, correlated noise may have significant structure, to the point that it may be interpretted as another signal. Thus, it is interesting to consider (Thanks to John McWhirter for posing this question at the IMA meeting.) how algorithms for source separation such as independent component analysis (ICA) [7] perform on the weather data.

To compare the MNF algorithm with ICA we applied both techniques to the multivariate weather data described in Section 4.2. The results are shown in Fig. 8. Perhaps surprisingly, there is considerable similarity between the resulting MNF eigenvectors and independent components. The first MNF eigenvector

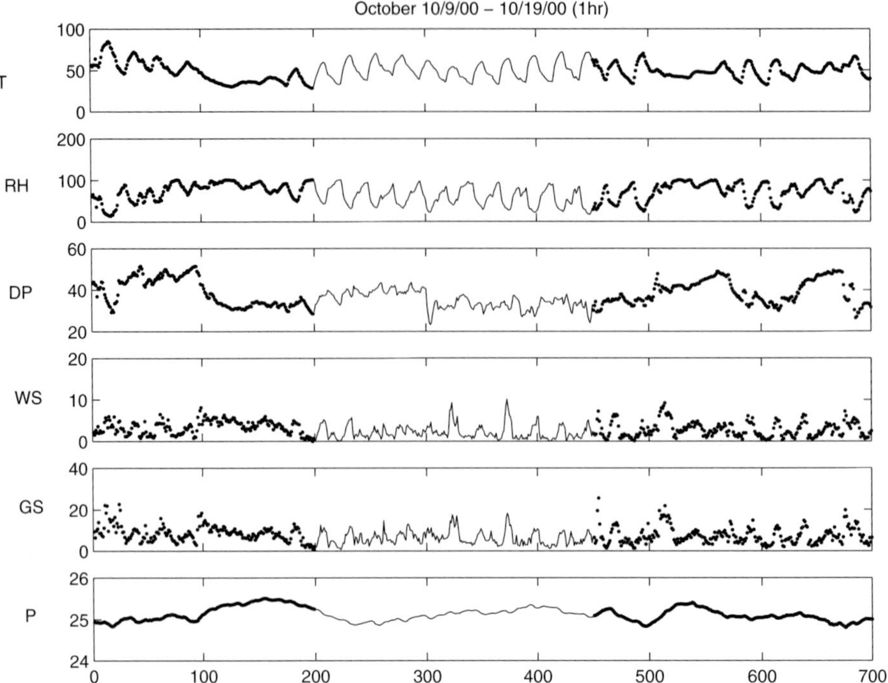

FIG. 6. Raw weather data consisting of temperature (T), relative humidity (RH), dew point (DP), wind speed (WS), gust speed (GS) and pressure (P) collected over the month of October, 2000 at hourly intervals. The dark points represent testing data while the light points from October 9–19 were used for building the radial basis function model.

corresponds closely to the pressure variable as does one of the basis vectors resulting from the FastICA routine [9]. The second MNF eigenvector is similar to independent component number two and neither correspond to a physical variable. The third MNF eigenvector provides the least correspondence to the independent components, but by elimination corresponds most to independent component number five; both manifest oscillations of similar period. The fourth MNF eigenvector corresponds closely to independent component number four; again, this is not a physical variable. The fifth MNF eigenvector corresponds closely to the wind/gust speed as well as to independent component number one. The last MNF eigenvector is the same as independent component number three and is non-physical.

Hence, we conclude that there are considerable similarities between the two methods. Indeed, the independent components appear to be a small rotation of the MNF basis. Despite these similarities, there are some fundamental differences worth noting. Firstly, the MNF method orders the resulting basis vectors.

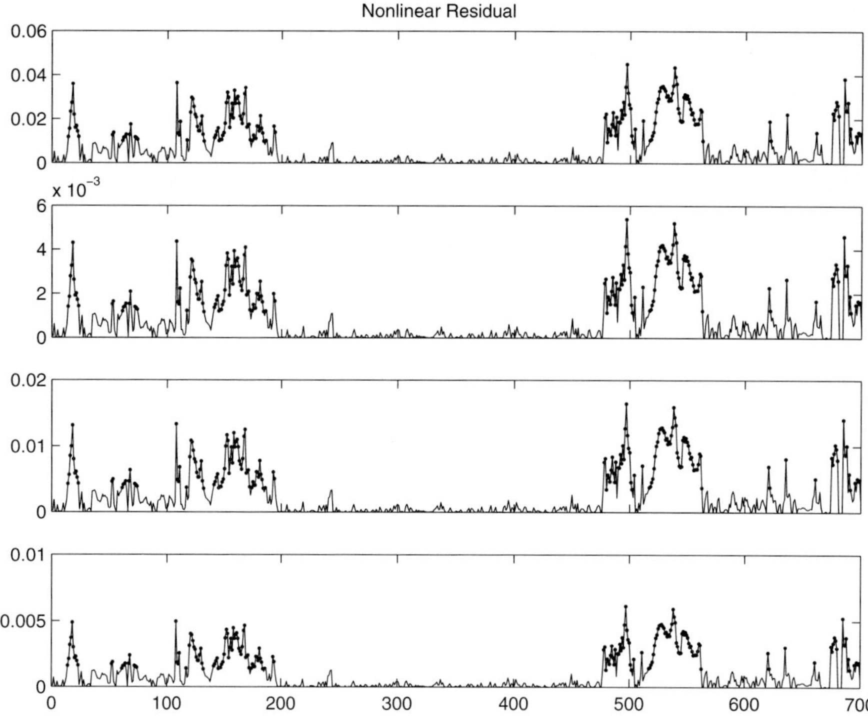

FIG. 7. Jump in nonlinear residual indicates an impending change in the weather. Points marked with a dot indicate that the magnitude of the residual exceeded the maximum residual of the training set.

Thus, it was clear in our application in Section 4 which vectors should be truncated; this is not the case with the ICA representation. Additionally, the ICA algorithm is much more involved than the solution of the generalized singular vector problem. Actually, there are many different algorithms for computing independent components. To check our work we applied the Jade algorithm [5] for ICA and the results were very similar, but not identical. Lastly, it should be noted that in general ICA cannot be used to distinguish signals from multivariate Gaussian noise, but rather to analyze independent sources and their contribution to the original signal [7].

5.2 Noise whitening

It has been suggested that this procedure may be carried out by applying a whitening transformation based on the covariance matrix of the noise [1, 11]. In this setting the resulting problem for determining the basis vectors is a standard SVD rather than a generalized SVD. While this approach is a convenient computational device, we found that the change of basis required to whiten the noise produced an unnatural coordinate system. Hence, all our results are transformed

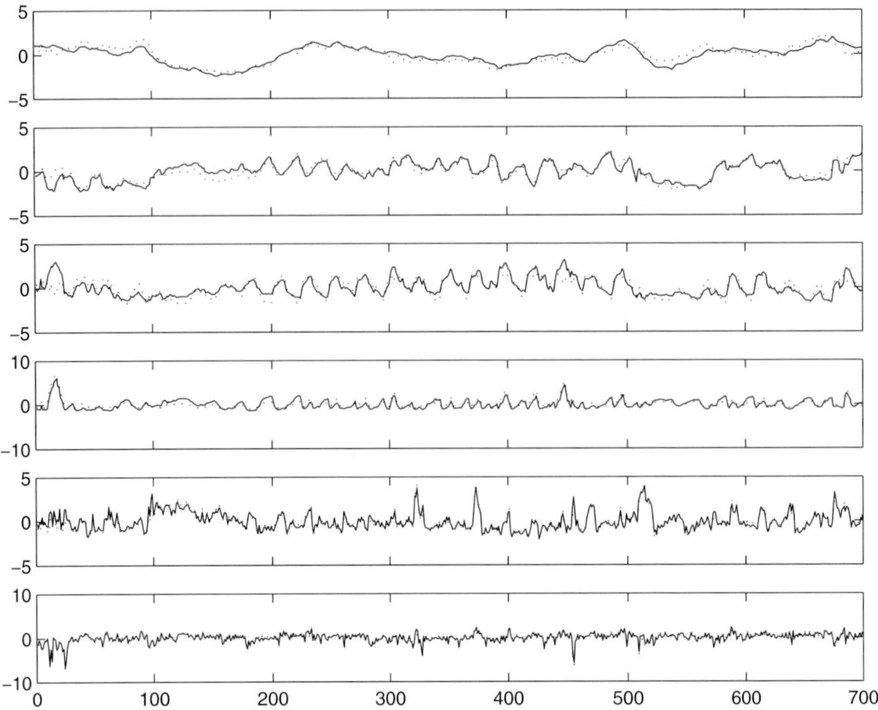

FIG. 8. A comparison of the results of applying the MNF method and ICA to the weather data. The solid lines correspond to the maximum noise fraction eigenvectors $\phi^{(i)}$, ordered from top to bottom with increasing noise. The independent components most similar to the MNF eigenvectors are plotted with dotted lines.

back to the original unwhitened coordinates.

6 Conclusions

The MNF technique provides a useful subspace decomposition for noisy as well as high-variance time series. Unlike other applications of this type of method to time series, here the covariance matrix of the noise was estimated, following Switzer [13], by computing differences of shifted vectors. The method was shown to be effective for filtering both nonsmooth data and data with high-variance bands. In addition, the MNF was integrated with the WRN [4, 3]—a tool for data reduction—as an effective means to reduce noise. Lastly, a comparison of the MNF method to ICA was presented. The results indicate, at least for the climate data considered here, that there are some surprising similarities between MNF and ICA. This will be the subject of further investigations.

Acknowldegements

The authors would like to thank the National Science Foundation for support under grant DMS-9973303. This paper is an expanded version of the 5th IMA Signal Processing Conference digest paper [2].

Bibliography

[1] Allen, M.R. and Smith, L.A. Optimal filtering in singular spectrum analysis. *Phys. Lett. A*, **234** 419–428, 1997.

[2] Anderle, M. and Kirby, M. Filtering noisy time series: Keeping the baby and most of the bathwater. In *Conference Digest, 5th IMA Int. Conf. on Mathematics in Signal Processing*, (Univeristy of Warwick, 2000).

[3] Broomhead, D.S. and Kirby, M. The Whitney reduction network: a method for computing autoassociative graphs. *Neural Comp.*, **13** 2595–2616, 2001.

[4] Broomhead, D.S. and Kirby, M. A new approach for dimensionality reduction: Theory and algorithms. *SIAM J. Appl. Math.*, **60**, 2114–2142, 2000.

[5] Cardoso, J. and Souloumiac, A. Blind beam forming for non-gaussian signals. *IEE Proc. F*, **140**, 771–774, 1993.

[6] Chen, S. Cowan, C.F.N. and Grant, P.M. Orthogonal least squares learning algorithm for radial basis function networks. *IEEE Trans. on Neural Networks*, **2**, 302–309, 1991.

[7] Comon, P. Independent component analysis, a new concept? *Signal Process.*, **36**, 287–314, 1994.

[8] Green, A.A. Berman, M. Switzer, P. and Craig, M.D. A transformation for ordering multispectral data in terms of image quality with implications for noise removal. *IEEE Trans on Geosci. Remote Sens.*, **26**, 65–74, January 1988.

[9] Hyvärinen, A. and Oja, E. A fast fixed-point algorithm for independent component analysis. *Neural Comput.*, **9**, 1483–1492, 1997.

[10] Kirby, M. *Geometric Data Analysis: An Empirical Approach to Dimensionality Reduction and the Study of Patterns.* New York, Wiley, 2001.

[11] Lee, J.B. Woodyatt, A.S. and Berman, M. Enhancement of high spectral resolution remote sensing data by a noise-adjusted principal component transform. *IEEE Trans. on Geosci. Remote Sens.*, **28**, 295–304, 1990.

[12] De Moor, B. Vandewalle, J. and Staar, J. Oriented energy and oriented signal-to-signal concepts in the analysis of vector sequences and time series. In *SVD and Signal Processing: Algorithms, Applications and Architectures*, (ed. E. Deprettere), pp. 209–232. Amsterdam, North-Holland, 1987.

[13] Switzer, P. Min/max autocorrelation factors for multivariate spatial imagery. In *Computer Science and Statistics*, (ed. L. Billard), pp. 13–16. Amsterdam, Elsevier, 1985.

Applications of Chaotic Dynamical Systems and Ergodic Theory to Spread Spectrum Sequences Design

Chi-Chung Chen, Kung Yao
Electrical Engineering Department, UCLA, Los Angeles, CA, 90095-1594 USA

Ken Umeno
CRL, Ministry of Post and Telecommunications, Koganei, Tokyo 184-9795 Japan

Ezio Biglieri
Dipartimento di Elettronica, Politecnico di Torino, Turin 10129 Italy

Abstract

This paper derives general results on the partial auto-correlation function of the optimal spreading sequences for asynchronous CDMA systems with respect to the minimization of the average error probability under the Standard Gaussian Approximation condition. A practical implementation of the optimal real-valued spreading sequence by using chaotic dynamical systems, particularly dynamical systems with Lebesgue spectrum, is provided. Based on the ergodic theory of dynamical systems, we construct a family of optimal chaotic Chebyshev spreading sequences and analyse their performance. An asynchronous CDMA system using the optimal spreading sequences allows 15 per cnet more users than when random white sequences/Gold codes are employed.

1 Introduction

In the last decade, increasing research investigations on spread spectrum communications have addressed direct sequence code division multiple access (DS-CDMA) systems, where all users transmit on the same band at the same time and are distinguished only by means of a code signature of *spreading sequence.* A DS-CDMA system has well known desirable features: universal frequency reuse, fast and accurate power control, robustness to multipath channel, graceful degradation, and ease of cellular planning, etc. These advantages result from the assignment of spreading sequences to every potential user of the system. System performance of a DS-CDMA communication system using the single-user matched filter structure critically depends on the auto-correlation and cross-correlation of the spreading sequences. As is well known, the orthogonal sequences are optimal for the downlink of wireless communication systems, where all users are synchronous, since the multiple-access interferences (MAIs) from other users are absent. It is inevitable, however, that the cross-correlations among the spreading sequences are nonzero for the uplink channel because of asynchronism. In this paper, we consider the design of the spreading sequences for an asynchronous DS-CDMA system for uplink operations.

Earlier efforts in using chaotic dynamical systems/signals for CDMA applications have been studied [1, 2, 3]. By treating the spreading sequences for an asynchronous CDMA system as random processes and assuming they are independent and stationary, Mazzini *et al.* [4] found the ensemble-averaged auto-correlation function that minimizes the expected interference-to-signal ratio under the Standard Gaussian Approximation (SGA) and proposed a nearly optimal *binary* sequence generator using a piece-wise affine map. However, this class of chaotic map generators may have practical implementational difficulties due to the need for high slope in the map as well as a finite-precision computational problem if the slope of map is some power of 2. This paper derives general results on the partial auto-correlation function of the optimal spreading sequences for an asynchronous CDMA system to minimize the average error probability under the SGA condition without the assumption of 'independence' and 'stationarity' on the spreading sequences, and also provides a practical implementation of the optimal (in a sense to be specified later) real-valued spreading sequence from a chaotic dynamical system, particularly a chaotic (ergodic) dynamical system with *Lebesgue spectrum.*

Based on the ergodic theory [5] of dynamical systems we can design a family of optimal chaotic spreading sequences and evaluate their performances analytically if the invariant measure of the dynamical system is known. The performance of a synchronous CDMA system with chaotic spreading sequences has been evaluated by Umeno *et al.* [6] using ergodic theory. Umeno also elaborates on the design of sequences to be used for Monte Carlo simulation based on various properties of ergodic theory [7]. As an application of the theory presented here, we describe a simple method to implement spreading sequences with the optimum auto-correlation property by using Chebyshev polynomials, which are exact transformations and hence ergodic transformations, and admit closed-form invariant measures.

2 System model

We consider an asynchronous CDMA system with K users and spreading factor N. The received signal is given by

$$r(t) = \sum_{k=1}^{K} s^{(k)}(t - \tau^{(k)}) + n(t), \tag{2.1}$$

where $s^{(k)}(t) = \mathrm{Re}\{b^{(k)}(t)a^{(k)}(t)\exp(j(w_c t + \phi_0^{(k)}))\}$ is the transmitted signal from the kth user, $n(t)$ is the white Gaussian noise with two-sided spectral density $N_0/2$, $b^{(k)}(t)$ and $a^{(k)}(t)$ are the data and spreading signal, respectively, w_c is the carrier frequency, and $\phi_0^{(k)}$ is the phase. The data signal $b^{(k)}(t)$ is a sequence of ± 1 rectangular pulses with a duration of T_b. The spreading sequence has a period N, and is composed of rectangular pulses with duration T_c and chip amplitude $a_j^{(k)}$ such that $\sum_{j=0}^{N-1} {a_j^{(k)}}^2 = 1$. We assume that $N = T_b/T_c$, and $\tau^{(k)}$ is the time delay.

Since we are concerned with relative phase shifts modulo 2π and relative time delays modulo T_b, there is no loss of generality in assuming $\phi^{(i)} = \phi_0^{(i)} - w_c\tau^{(i)} = 0$, $\tau^{(i)} = 0$, and $T_c = 1$ and considering only $0 \leq \phi^{(k)} < 2\pi$ (assumed to be uniformly distributed in $[0, 2\pi)$), and $0 \leq \tau^{(k)} < N$ (assumed to be uniformly distributed in $[0, N)$) for $k \neq i$.

The output of the single-user matched filter at the ith receiver is

$$\begin{aligned} Z_i &= 2\int_0^{T_b} r(t)a_i(t)\cos(w_c t)dt \\ &= b_0^{(i)} + \sum_{k\neq i}^{K} I^{(k)}(b^{(k)}, \phi^{(k)}, \tau^{(k)}) + \eta^{(i)}, \end{aligned} \tag{2.2}$$

where $\eta^{(i)}$ is the equivalent white Gaussian noise at the output of the matched filter with $E[\eta^{(i)}] = 0$ and $E[(\eta^{(i)})^2] = N_0$, and $\sum_{k\neq i}^{K} I^{(k)}(b^{(k)}, \phi^{(k)}, \tau^{(k)})$ is the MAI. The channel SNR is given by SNR $= 1/N_0$.

The interfence term $I^{(k)}(b^{(k)}, \phi^{(k)}, \tau^{(k)})$ due to the k-user has been found [8] to be

$$I^{(k)}(b^{(k)}, \phi^{(k)}, \tau^{(k)}) = (b_{-1}^{(k)} R_{k,i}(\tau^{(k)}) + b_0^{(k)} \hat{R}_{k,i}(\tau^{(k)}))\cos(\phi^{(k)}), \tag{2.3}$$

where $R_{k,i}(\tau)$ and $\hat{R}_{k,i}(\tau)$ for $0 \leq l \leq \tau < l+1 \leq N$, are given by

$$\begin{aligned} R_{k,i}(\tau) &= C_{k,i}(l-N) + [C_{k,i}(l+1-N) - C_{k,i}(l-N)](\tau - l), \\ \hat{R}_{k,i}(\tau) &= C_{k,i}(l) + [C_{k,i}(l+1) - C_{k,i}(l)](\tau - l), \end{aligned} \tag{2.4}$$

where $C_{k,i}(l)$ is the partial cross-correlation between the kth and the ith user and is defined by

$$C_{k,i}(l) \equiv \begin{cases} \sum\limits_{j=0}^{N-l-1} a_j^{(k)} a_{j+l}^{(i)}, & 0 \leq l \leq N-1 \\ \sum\limits_{j=0}^{N+l-1} a_{j-l}^{(k)} a_j^{(i)}, & 1-N \leq l \leq -1 \\ 0, & |l| \geq N. \end{cases} \tag{2.5}$$

3 Derivation of optimal sequences

The overall interference variance for the ith user from all other users can be computed [8] as

$$\begin{aligned}\sigma^2(i) &\equiv E_{\phi^{(k)},b_{-1}^{(k)},b_0^{(k)},\tau^{(k)}}\{[\sum_{k\neq i}^{K} I^{(k)}(b^{(k)},\phi^{(k)},\tau^{(k)})]^2\} \\ &= \frac{1}{6N}\sum_{k\neq i}^{K}\sum_{l=1-N}^{N-1}[2C_{k,i}^2(l)+C_{k,i}(l)C_{k,i}(l+1)]. \end{aligned} \tag{3.1}$$

With the SGA assumption, the error probability of the i-th user is given by

$$P_e(i) = Q\left(\sqrt{\frac{1}{\sigma^2(i)+N_0}}\right), \tag{3.2}$$

where $Q(.)$ is the complementary standard Gaussian probability distribution. The evaluation of the error probability of an asynchronous CDMA system based on the moment space bounding technique and its relationship to the SGA assumption were discussed in [9].

Using the following identity:

$$\sum_{l=1-N}^{N-1} C_{x,y}(l)C_{x,y}(l+n) = \sum_{l=1-N}^{N-1} C_{x,x}(l)C_{y,y}(l+n) \tag{3.3}$$

given in [10] and the trivial identity $C_x(l) \equiv C_{x,x}(l) = C_{x,x}(-l)$, (3.1) can be simplified to

$$\begin{aligned}\sigma^2(i) &= \frac{1}{6N}\sum_{k\neq i}^{K}[2C_k(0)C_i(0)+4\sum_{l=1}^{N-1}C_k(l)C_i(l)+\sum_{l=1-N}^{N-1}C_k(l)C_i(l+1)] \\ &= \frac{1}{6N}\sum_{k\neq i}^{K}[2C_k(0)C_i(0)+4\sum_{l=1}^{N-1}C_k(l)C_i(l) \\ &\quad+\sum_{l=0}^{N-1}C_k(l)C_i(l+1)+C_k(l+1)C_i(l)]. \end{aligned} \tag{3.4}$$

Since the Q-function is convex, the lower bound of average error probability can be attained by assigning the same interference variance to every user. This can be done with $C_i(l) = C_k(l) = C(l)$ for all i, k, l. With the normalization $C_i(0) = 1$, we minimize the average error probability by minimizing the interference power

$$\sigma^2 = \frac{K-1}{6N}[2 + 4\sum_{l=1}^{N-1} C^2(l) + 2\sum_{l=0}^{N-1} C(l)C(l+1)]. \tag{3.5}$$

This is a positive quadratic form in $C(l)$ whose unique minimum is achieved when $\partial\sigma^2/\partial C(l) = 0$ for $l = 1, 2, \ldots, N-1$, that is when

$$4C(l) + C(l+1) + C(l-1) = 0, \qquad \forall l = 1, 2, \ldots, N-1. \tag{3.6}$$

The solution to (3.6) is given [11] by

$$C_k(l) = (-1)^l \frac{r^{l-N} - r^{N-l}}{r^{-N} - r^N}, \qquad l = 0, 1, 2, \ldots, N-1, \quad \forall k, \tag{3.7}$$

where $r = 2 - \sqrt{3}$. Substituting (3.7) into (3.5), we obtain the minimum interference power as

$$\sigma^2_{\text{opt}} = \frac{\sqrt{3}(K-1)}{6N} \frac{r^{-2N} - r^{2N}}{r^{-2N} + r^{2N} - 2}. \tag{3.8}$$

The *ensemble-averaged* partial auto-correlation function obtained in [4] when the spreading sequences are assumed to be *stationary* and *independent random processes* is identical to the deterministic constant partial auto-correlation vector derived here. Note that when $l \ll N$, $C_k(l) \approx (-r)^l$ which decays exponentially with alternative sign. Moreover, the minimum interference variance is given by $\sigma^2_{\text{opt}} = \sqrt{3}(K-1)/6N$ as N is large, which increases by 15 per cent the number of users achieved with white sequences, that is, $(K-1)/3N$.

4 Ergodic dynamical systems

4.1 Ergodic theory

The second-order time-averaged statistic of spreading sequences is needed for sequence design and performance analysis. For a spreading sequence generated by a deterministic dynamical system, the performance can be computed analytically or numerically by using the Birkhoff individual ergodic theory which is restated as follows.

Theorem 4.1 *(4.2.4 in [5]) Let $(X, \mathcal{A}, \mu)$ be a finite measure space and $S : X \mapsto X$ be a measure-preserving and ergodic transformation. Then, for any integrable f, the average of f along the sequence generated by S, that is $\{S^{(k)}(x)\}_{k=0}^{\infty}$ for any given 'initial' $x \in X$, is equal almost everywhere to the average of f over the space X; that is,*

$$\lim_{n\to\infty} \frac{1}{n} \sum_{k=0}^{n-1} f(S^{(k)}(x)) = \frac{1}{\mu(X)} \int_X f(x)\mu(dx) \qquad a.e. \tag{4.1}$$

Use of this theorem allows us to evaluate the auto-correlation function of a sequence generated by any measure-preserving ergodic transformation. As an example of measure-preserving and ergodic transformation, consider the tent map $S(x)$ defined by

$$S(x) \equiv 1 - 2|x|, \qquad |x| \leq 1. \tag{4.2}$$

The uniform measure is invariant for this transformation; the auto-correlation function of the sequence generated by the tent map $S(x)$ can be evaluated by

$$\begin{aligned} \langle C(l) \rangle &\equiv \left\langle \frac{1}{n} \sum_{j=1}^{n} S(x_j) S(x_{j+l}) \right\rangle \\ &= \frac{1}{2} \int_{-1}^{1} x S^{(l)}(x) dx \\ &= \frac{1}{3} \delta(l), \end{aligned}$$

where $\langle\rangle$ denotes the ensemble average with respect to the initial condition x_0 under the corresponding invariant measure.

Another ergodic transformation, the nth degree Chebyshev polynomials defined by $T_n(x) \equiv \cos(n \arccos(x))$ over the interval $[-1, 1]$, has been considered as a sequence generator for synchronous CDMA system [6]. Examples of Chebyshev polynominals are given by

$$\begin{aligned} T_0(x) = 1, &\qquad T_1(x) = x, \\ T_2(x) = 2x^2 - 1, &\qquad T_3(x) = 4x^3 - 3x, \ldots. \end{aligned} \tag{4.3}$$

Adler and Rivlin [12] have shown Chebyshev polynomials of degree $n \geq 2$ are mixing and thus ergodic, and their invariant measure is given by $\rho(x)dx = \frac{dx}{\pi\sqrt{1-x^2}}$. Furthermore, by investigating the asymptotical stability of the Frobenius–Perron operator corresponding to Chebyshev transformations, Chebyshev polynomials are shown to be exact and thus mixing and ergodic transformations [5].

The Chebyshev polynominals have the orthogonality

$$\int_{-1}^{1} T_i(x) T_j(x) \rho(x) dx = \delta_{i,j} \frac{1 + \delta_{i,0}}{2}. \tag{4.4}$$

The auto-correlation functions for sequences generated by these Chebyshev polynomial are given by

$$\begin{aligned} \langle C(l) \rangle &\equiv \left\langle \frac{1}{n} \sum_{j=1}^{n} T_p(x_j) T_p(x_{j+l}) \right\rangle \\ &= \int_{-1}^{1} T_p(x) T_{p^{l+1}}(x) \rho(x) dx \\ &= \frac{1}{2} \delta(l). \end{aligned} \tag{4.5}$$

4.2 Dynamical systems with Lebesgue spectrum

One class of ergodic dynamical systems with special properties are dynamical systems with *Lebesgue spectrum* [13]. These systems, denoted as $\phi(x)$, not only have an ergodic invariant measure, but are also associated with a special set of orthonormal basis functions $\{f_{\lambda,j}(x)\}$ for Hilbert space L_2. This orthonormal basis can be split up into classes and written as $\{f_{\lambda,j}(x) : \lambda \in \Lambda, j \in F\}$, where λ labels the classes and j labels the functions within each class. The cardinality of Λ can be proven to uniquely determined and is called the *multiplicity of the Lebesgue spectrum.* If Λ is (countably) infinite, we shall speak of (countably) infinite Lebesgue spectrum. If Λ has only one element, the Lebesgue spectrum is called simple. The important property that these particular basis functions $f_{\lambda,j}$ have is

$$f_{\lambda,j} \circ \phi = f_{\lambda,j+1}, \qquad \forall \lambda \in \Lambda, j \in F. \tag{4.6}$$

That is, all the other basis functions in the same class can be generated from one of the basis function by using compositions with powers of the dynamical system $\phi(x)$. Furthermore, since the basis functions are orthogonal, every function is orthogonal both to every other function in the same class, and to every function in other classes.

An example of chaotic dynamical system with Lebesgue spectrum is the Bernoulli shift map $\phi(x)$ with invariant measure density $\rho(x) = 1$, as considered in [13, 14], and defined by

$$\phi(x) = \begin{cases} 2x, & 0 \le x \le 1/2, \\ 2x - 1, & 1/2 < x \le 1. \end{cases} \tag{4.7}$$

The associated basis functions for L_2 space are *Walsh* functions and defined by

$$\begin{aligned} w_1(x) &= 1, \\ w_{k+1}(x) &= \prod_{i=0}^{r-1} \operatorname{sgn}\{\sin^{k_i}(2^{i+1}\pi x)\}, \qquad k = 1, 2, \ldots , \end{aligned} \tag{4.8}$$

where the values of k_i, either 0 or 1, are the binary digits of k, that is, $k = \sum_{i=0}^{r-1} k_i 2^i$. Thus, this generator can produce random white binary sequences. A two-dimensional dynamical system with Lebesgue spectrum is also given [13, 14].

Another class of chaotic (ergodic) dynamical systems with Lebesgue spectrum are the Chebyshev polynomial maps as considered above. In particular, we consider the pth degree Chebyshev polynomial map, that is, $\phi(x) = T_p(x)$ where $p \ge 2$ is *prime.* The associated basis functions for $L_2([-1, 1])$ are also Chebyshev polynomials $\{T_i(x)\}_{i=0}^{\infty}$. Then, the $f_{\lambda,j}(x)$ can be defined by

$$f_{\lambda,j}(x) = T_{\lambda \cdot p^j}(x), \qquad \forall \lambda \in \Lambda, j \in F, \tag{4.9}$$

where $\Lambda = \{n | n \in \mathcal{Z},\ n \ge 0$ and relative prime to $p\}$, and F is the set of nonnegative integers. To see this, we consider the composition of ϕ with one of the basis functions:

$$f_{\lambda,j} \circ \phi(x) = T_{\lambda \cdot p^j} \circ T_p(x) = f_{\lambda,j+1}(x). \tag{4.10}$$

Note that the basis function $f_{0,j}(x) = T_0(x) = 1$ constitutes its own class and the basis function we used in (4.5) is the particular case when $\lambda = 1$.

5 Optimal chaotic spreading sequences design

Let us consider a polynomial function $G(x)$ in the Hilbert space $L_2([-1,1])$ with the form

$$G(x) \equiv \sum_{j=1}^{N} (-r)^j T_{p^j}(x), \qquad x \in [-1,1], \tag{5.1}$$

where $p \geq 2$. By using ergodic theory the average of G^2 along the sequence generated by the Chebyshev transformation $T_p(.)$ is given by

$$\langle C(0) \langle \equiv \left\langle \frac{1}{n} \sum_{i=1}^{n} G^2(x_i) \right\rangle = \frac{1}{2} \frac{r^2(1-r^{2N})}{1-r^2} \equiv A, \tag{5.2}$$

and the normalized auto-correlation function of such sequence can be evaluated by

$$\begin{aligned} \langle C(l) \rangle / A &\equiv \frac{1}{A} \left\langle \frac{1}{n} \sum_{i=1}^{n} G(x_i) G(x_{i+l}) \right\rangle \\ &= \frac{1}{A} \int_{-1}^{1} G(x) G(T_{p^l}(x)) \rho(x) dx \\ &= (-1)^l \frac{r^{l-N} - r^{N-l}}{r^{-N} - r^N}. \end{aligned} \tag{5.3}$$

Thus, with the condition $r = 2 - \sqrt{3}$, the output sequences $\{y_1, y_2, \cdots, y_N\}$ generated by

$$y_j = \frac{1}{\sqrt{A}} G(x_j), \qquad x_{j+1} = T_p(x_j), \tag{5.4}$$

are the optimal spreading sequences for asynchronous CDMA systems. Because of the property $T_{p^{j+1}}(x) = T_p \circ T_{p^j}(x)$, the function $G(x)$ in (5.1) actually is a non-causal *finite impulse response* (FIR) filter fed by the input sequence generated by Chebyshev polynomial map $T_p(x)$. This FIR filter can be easily implemented to produce the output sequence $\{y_j\}$ with certain time delay.

When the spreading factor N is large, an alternative practical design is given as follows. Since the auto-correlation function of a Chebyshev sequence is a

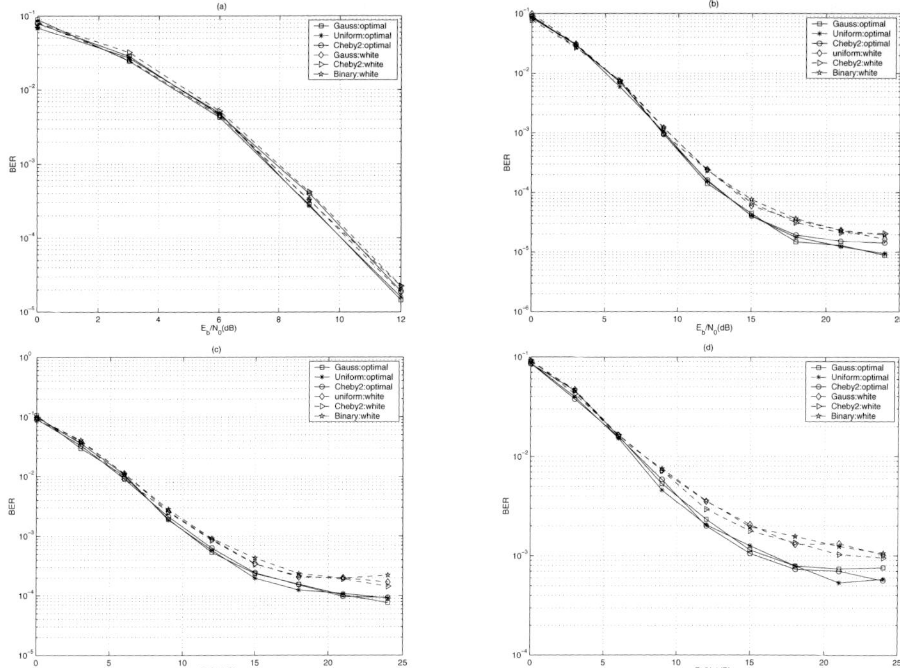

FIG. 1. Comparison of error probabilities of asynchronous CDMA system using optimal and white sequences generated by Gaussian, uniform, second-degree Chebyshev, and binary random-number generators for sequence length of 31. (a) K=3, (b) K=5, (c) K=7, and (d) K=10.

Kronecker delta function, we can design the optimal spreading sequences by passing these Chebyshev sequences through an *infinite impulse response* low-pass filter with a single pole at $(-r)$. That is,

$$y_{j+1} = -ry_j + \sqrt{2(1-r^2)}\,x_j, \qquad x_{j+1} = T_p(x_j). \tag{5.5}$$

The output sequence $\{y_1, y_2, \ldots, y_N\}$ of the filter will have an exponential autocorrelation function $(-r)^l$. Then either each non-overlapped section of the output sequences, or each sequence starting with different initial condition or choosing different Chebyshev polynomial map is assigned to a different user. The sequences designed in (5.5) are used for simulation of asynchronous CDMA systems considered in the next section.

6 Simulation results

First of all, we simulate an asynchronous CDMA communication system using optimal and white spreading sequences generated by Gaussian, uniform, Chebyshev, and binary random-number generators. The simulation results are shown

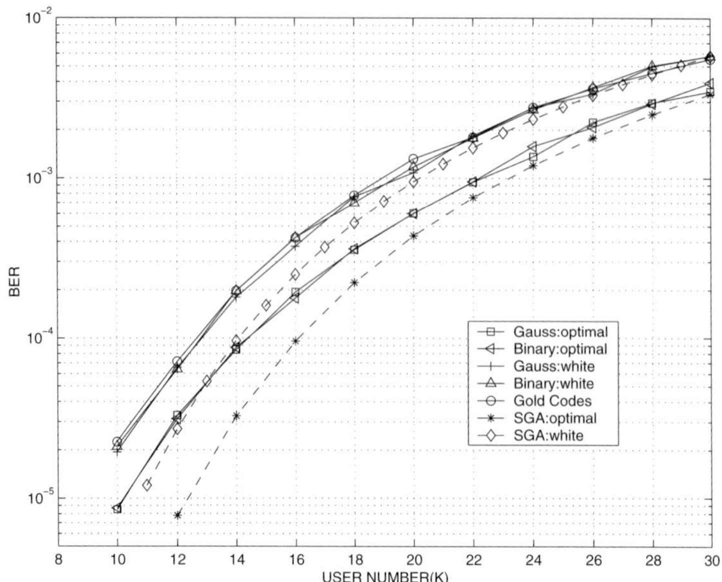

FIG. 2. Comparison of error probabilities of asynchronous CDMA system using optimal and white sequences generated by Gaussian and binary random-number generators and Gold codes ($N = 63$ and channel SNR $E_b/N_0 = 22$ dB).

in Fig. 1, where it can be seen that the error probability is independent of the distribution of the spreading sequences, and that the optimal sequences are better than random white sequences, which justifies our design. Particularly, the performances using optimal sequences generated by Chebyshev and Gaussian random-number generators are similar, consistent with our design of chaotic spreading sequences using ergodic theory. Moreover, the performance difference between optimal and random white sequences becomes more distinct when the number of users becomes larger.

In order to understand the behaviour of the sequences, we also performed simulations for different numbers of users, as shown in Fig. 2. These simulation results show that the optimal sequences are better than random white sequences by about 15 per cent in terms of allowable number of users, which is consistent with the analytical expression. We also observe that when the number of users is smaller, simulation results do not quite match with analytical results obtained under the SGA condition. This confirms the well known fact that the Gaussian approximation is not valid when the user number is small. The asynchronous CDMA system performances of optimal second- and third-degree Chebyshev sequences are shown in Fig. 3, which have the same parameters as Fig. 2, and are also better by about 15 per cent when Gold codes are employed.

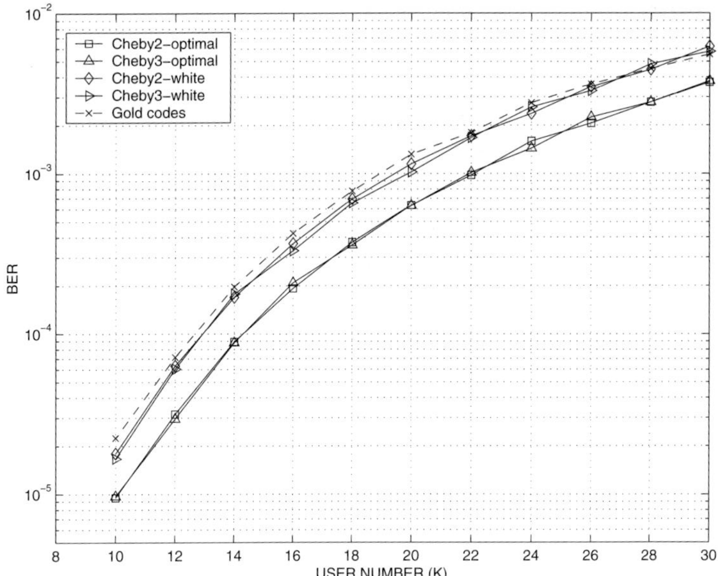

FIG. 3. Comparison of error probabilities of asynchronous CDMA system using second- and third-degree Chebyshev optimal and white sequences and Gold codes ($N = 63$ and channel SNR $E_b/N_0 = 22$ dB).

7 Conclusions

We have proposed a new design methodology for the design of optimal spread spectrum sequences for asynchronous CDMA systems with respect to minimum error probability under the SGA condition. Without any assumption on spreading sequences, the optimal partial auto-correlation function of the spreading sequences is derived. Using the ergodic theory of dynamical systems, a method to construct and to analyse such sequences based on ergodic transformations is shown. Our method generalizes some previous approaches proposed in [4, 6, 7]. Under the SGA condition, an asynchronous CDMA system using the optimal spreading sequences allows 15 per cent more users than when random white sequences/Gold codes are employed. Simulation results also show that system performances using this family of optimal chaotic Chebyshev spreading sequences are superior than and similar to Gold codes when employed in asynchronous CDMA systems.

Acknowledgements

This work is partially supported by MURI-ARO grant DAAG55-98-0269 and NASA/Dryden grant NCC2-374.

Bibliography

[1] Schweizer, J. and Hasler, M. (1996). Multiple access communications using chaotic signals, *IEEE Proc. ISCAS*, 108–111.

[2] Kodha, T. and Tsuneda, A. (1997). Statistics of chaotic binary sequences, *IEEE Trans. Information Theory*, **43**, 104–112.

[3] Abel, A., Bauer, A., Kelber, K. and Schwarz, W. (1997). Chaotic codes for CDMA applications, *Proc. ECCTD*, 306–311.

[4] Mazzini, G., Rovatti, R. and Setti, G. (1999). Interference minimization by auto-correlation shaping in asynchronous DS-CDMA systems: chaos-based spreading is nearly optimal, **35**, *Electron. Lett.*, 1054–5.

[5] Lasota, A. and Mackey, M.C. (1994). *Chaos, Fractals, and Noise: Stochastic Aspects of Dynamics*, Berlin, Springer.

[6] Umeno, K. and Kitayama, K.I. (1999). Improvement of SNR with chaotic spreading sequences for CDMA, *IEEE Information Theory Workshop.*

[7] Umeno, K. (1998). Chaotic Monte Carlo computation: a dynamical effect of random-number generations, e-print: chao-dyn/9812013 at http://xxx.lanl.gov.

[8] Pursley, M.B. (1977). Performance evaluation for phased-coded spread-spectrum multiple-access communication—part I: system analysis, *IEEE Trans. Communications*, **25**, 795–799.

[9] Yao, K. (1977). Error probability of asynchronous spread spectrum multiple access communication systems, *IEEE Trans. Communications*, **25**, 803–809.

[10] Pursley, M.B. and Sarwate, D.V. (1977). Performance evaluation for phased-coded spread-spectrum multiple-access communication—part II: code sequence analysis, *IEEE Trans. Communications*, **25**, 800–803.

[11] Chen, C.C., Biglieri, E. and Yao, K. (2000). Design of spread spectrum sequences using ergodic theory, *IEEE Proc. ISIT*, 379.

[12] Adler, R.L. and Rivlin, T.J. (1964). Ergodic and mixing properties of Chebyshev polynomials, *Proc. Amer. Math. Soc.*, **15**, 794–796.

[13] Arnold, V.I. and Avez, A. (1968). *Ergodic Problems of Classical Mechanics*, Benjamin, New York.

[14] Broomhead, D.S., Huke, J.P. and Muldoon, M.R. (1999). Codes for spread spectrum applications generated using chaotic dynamical systems, *Dynamics and Stability of Systems*, **14**, 95–105.

On the Dynamics of Some Nonhyperbolic Area-preserving Piecewise Linear Maps

Peter Ashwin and Xin-Chu Fu

School of Mathematical Sciences, Laver Building, University of Exeter, Exeter EX4 4QE, UK

Abstract

In this paper we report on an ongoing study of area-preserving piecewise linear discontinuous maps of the plane motivated by some maps that arise in signal processing. We examine some typical cases of such maps. In particular, we consider linear elliptic and parabolic maps with round-off and quantization discontinuities. In some cases such maps possess global attractors on which the map can be reduced to an almost invertible piecewise isometry. Finally, we briefly discuss some examples of global attractors in piecewise isometric systems.

1 Introduction

There are several situations in signal processing where systems are well-modelled by area-preserving discontinuous systems that are either exactly or close to being piecewise linear. In particular, the overflow oscillation problem for lossless digital filters [1], and bandpass sigma-delta modulator dynamics [10] are examples of maps that display nontrivial dynamics, where the usual techniques of smooth hyperbolic nonlinear dynamics do not work. For example, all Lyapunov exponents may be zero.

In this paper we present some general properties of area-preserving piecewise linear discontinuous maps and connections with other systems, notably piecewise isometries. We do this by examining some simple but nonetheless typical cases of such maps that arise in the study of the dynamics of signal processing systems (especially switched linear systems).

We consider cases where the linear part of the map is identical, but the translation may be different on different parts of the domain. We classify such area-preserving maps by the linear part of the map as being parabolic (two eigenvalues equal to one but only one eigenvector), elliptic (two eigenvalues are on the unit circle), or hyperbolic (no eigenvalues on the unit circle). We consider two types of discontinuity: the first due to rounding, the second due to quantization. These discontinuities are briefly compared and contrasted.

The elliptic case can give rise to a piecewise isometry on a global attracting set and, in particular, we can understand the dynamics of these maps by study of planar piecewise isometries [2]. The global attractors for parabolic maps can be decomposed into invariant straight-line segments on which the map acts as a one-parameter family of general interval translation maps. We also discuss in further details the definition and examples of global attractors in piecewise isometric maps. We do not consider hyperbolic maps any further as they can be understood to a large extent using techniques from hyperbolic dynamics: see, for example, [13]. Note that area-preserving (affine) linear maps of the plane

$$\begin{aligned} x' &= ax + by + \alpha \\ y' &= cx + dy + \beta \end{aligned} \tag{1.1}$$

where $(x, y) \in \mathbb{R}^2$, $a, b, c, d, \alpha, \beta \in \mathbb{R}, ad - bc = 1$, have very simple dynamics. It is easy to see that in nontrivial cases, either (a) all points have orbits that are bounded (elliptic case), or (b) all points (except possibly on a line through (α, β)) are unbounded (hyperbolic and parabolic cases).

1.1 Round-off discontinuity

Consider (1.1) restricted to a rectangle $B = [e, f) \times [g, h)$ by applying an appropriate translation by $(k\mu_1, l\mu_2)$ with $(k, l) \in \mathbb{Z}^2$;

$$\begin{aligned} x' &= ax + by + \alpha \ (\mathrm{mod}\ \mu_1) \\ y' &= cx + dy + \beta \ (\mathrm{mod}\ \mu_2) \end{aligned} \tag{1.2}$$

where $(x, y) \in B, \mu_1 = f - e, \mu_2 = h - g$. We say that the system (1.2) has round-off discontinuity, and the richness of its dynamical complexity is surprising given the simplicity of (1.1). This has been noted by several researchers and studied especially in the case $(a, b, c, d) = (0, 1, -1, 2\cos\theta)$, $\alpha = \beta = 0$, and $(e, f, g, h) = (-1, 1, -1, 1)$ in which case the map is referred to as the lossless digital filter overflow oscillation problem. In general, we can clearly scale the variables by $u = \frac{x-e}{\mu_1}$, $v = \frac{y-g}{\mu_2}$ so that B becomes $I = [0, 1)^2$, and the map (1.2) becomes

$$\begin{aligned} u' &= a'u + b'v + \alpha' \ (\mathrm{mod}\ 1) \\ v' &= c'u + d'v + \beta' \ (\mathrm{mod}\ 1) \end{aligned} \tag{1.3}$$

where $(u, v) \in I, a' = a, b' = \frac{\mu_2}{\mu_1} b, c' = \frac{\mu_1}{\mu_2} c, d' = d$. Note that $a'd' - b'c' = ad - bc = 1$ implying that the map (1.3) still preserves area locally. To this end we now assume (without loss of generality) that $B = I$. We distinguish between the homogeneous case $\alpha = \beta = 0$ and the nonhomogeneous case $(\alpha, \beta) \neq (0, 0)$; however, in practice both maps seem to be of similar dynamical complexity.

1.2 Quantisation discontinuity

The discontinuity of the maps discussed in Section 1.1 is induced by taking modulo 1. In this section we introduce an example with quantization discontinuity, that is, the discontinuity is induced by the function sgn$(\cdot)$ applied to the coordinates. Such a map of the form

$$\begin{aligned} x' &= ax + by + \alpha_1 \mathrm{sgn}(x) + \beta_1 \mathrm{sgn}(y) \\ y' &= cx + dy + \alpha_2 \mathrm{sgn}(x) + \beta_2 \mathrm{sgn}(y) \end{aligned} \tag{1.4}$$

where $(x,y) \in \mathbb{R}^2$ we refer to as a linear map with quantization discontinuity. Such maps can give rise to dynamics that is very similar to (1.3) after reducing to a global attractor.

2 Dynamics of discontinuous area-preserving maps

Given that $ad - bc = 1$ one can classify maps (1.2) as elliptic ($|a+d| < 2$), parabolic ($|a+d| = 2$), or hyperbolic ($|a+d| > 2$). We introduce a convenient parametrization as follows:

$$\begin{pmatrix} a & b \\ c & d \end{pmatrix} = \begin{pmatrix} \sqrt{1+A^2-B^2}+A & C^{-1}B \\ -CB & \sqrt{1+A^2-B^2}-A \end{pmatrix}, \tag{2.1}$$

where A, B, C are real quantities and we take the positive square root. It is easy to see that the trace is $2\sqrt{1+A^2-B^2}$ and so the map will be hyperbolic if $|B| < |A|$, parabolic if $|B| = |A|$, and elliptic if $|B| > |A|$. For hyperbolic and parabolic maps we can write

$$\begin{pmatrix} a & b \\ c & d \end{pmatrix} = \begin{pmatrix} \lambda^{-1}+A & \alpha^{-1}A \\ -\alpha(A+\lambda^{-1}-\lambda) & \lambda - A \end{pmatrix}, \qquad |\lambda| \geq 1, \tag{2.2}$$

where λ is an eigenvalue. For the elliptic case we can write

$$\begin{pmatrix} a & b \\ c & d \end{pmatrix} = \begin{pmatrix} \cos\theta + A\sin\theta & \alpha^{-1}(1+A^2)\sin\theta \\ -\alpha\sin\theta & \cos\theta - A\sin\theta \end{pmatrix},$$

with $\cos\theta + i\sin\theta$ an eigenvalue. We say that X^+ is a *global attractor* or *maximal invariant set* for the iterated map $F : \mathbb{R}^2 \to \mathbb{R}^2$ if X^+ is the smallest compact set such that $d(F^n(x), X^+) \to 0$ for all $x \in \mathbb{R}^2$ (in fact, there may be a zero-measure subset of X^+ which is not invariant). In Section 3 we discuss a more precise definition of global attractors for piecewise isometric systems.

It is a very interesting and apparently highly nontrivial problem to find a global attractor for general maps of the form (1.3) or (1.4), as we hint in the following sections. Note that compactness of $[0,1]^2$ implies that there is always a global attractor for (1.3); there need not be a global attractor for (1.4).

2.1 Parabolic maps

Parabolic maps seem to be a simplest class among the three classes, and their dynamical properties can be considered as an interpolating case between the hyperbolic maps and the elliptic maps. For both parametrizing methods above, the parametrization is

$$\begin{pmatrix} a & b \\ c & d \end{pmatrix} = \begin{pmatrix} 1+A & \alpha^{-1}A \\ -\alpha A & 1-A \end{pmatrix}.$$

In [5] it is shown that for all cases where α, A are rational then $\ell(X^+) > 0$ and explicit bounds on this measure are given. For several examples the expressions for the measure of the maximal invariant sets are obtained.

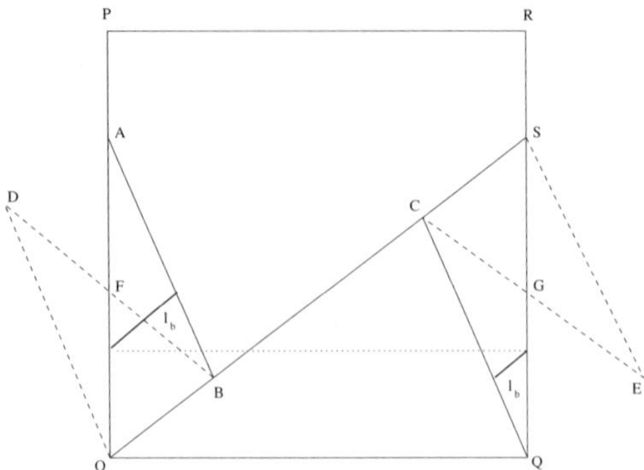

FIG. 1. Construction of the maximal invariant set X^+ consisting of the triangles $0AB$ and CQS for parabolic maps with $-1 \leq \alpha < 0$, $0 < A < 1$ (see text).

However, the question as to whether there are α and A for which $\ell(X^+) = 0$ is still open. For irrational parabolic maps, few results have been obtained yet. For the semirational (α rational) case, we show in [5] that the map may be decomposed into a one-parameter family of one-dimensional interval translation maps [3].

For illustration we consider a specific example in [5] where one can construct X^+ for $-1 \leq \alpha < 0$ and $0 < A < 1$. Figure 1 shows how the linear map maps the unit square $OPRQ$ into the maximal invariant set X^+ consisting of the union of the invariant straight line OS and the triangles OAB and CSQ, where

$$A = (0, -\alpha), \quad B = (A, -\alpha A), \quad C = (1 - A, -\alpha(1 - A)), \quad S = (1, -\alpha).$$

The union of the two triangles is invariant as the dotted images ODB and CSQ show, as is the line of fixed points OS. To see this is the maximal invariant set, note that everything in $OPRS$ decreases in its y-component unless it lands in the triangle OAB. Similarly, all points in OQS must increase in their y-components unless they land in the triangle CQS. Hence in this case we can compute

$$\ell(X^+) = |\alpha| A.$$

In fact, X^+ can be decomposed into invariant straight-line segments:

$$X^+ = \bigcup_{0 \leq b < |\alpha|} l_b, \ f(l_b) = l_b,$$

where

$$l_0 = \{(x, -\alpha x), 0 \leq x < 1\},$$

$$l_b = \{(x, -\alpha x + b), 0 \leq x < (\alpha + b)\alpha^{-1} A\}$$

$$\cup \{(x, -\alpha x + \alpha + b), Ab\alpha^{-1} + 1 \leq x < 1\},$$

where $0 < b < |\alpha|$. Let $f_b = f|_{l_b}$; as Fig. 1 shows, for $0 < b < |\alpha|, f_b$ can be transformed to a circle map. So by Denjoy's Theorem, f_b is equivalent to a pure rotation.

Therefore we can conclude that when $-1 \leq \alpha < 0$, $0 < A < 1$, the dynamics of the parabolic map on its global attractor can be reduced to a one-parameter family of one-dimensional isometries. Note that the cases $\alpha > 0$ and $\alpha < 0$ seem to be fundamentally different, and we do not have a way of conjugating one with the other.

2.2 Elliptic maps

We first consider the elliptic map with $A = -\cot\theta$, $\alpha = 1/(\sin\theta)$

$$\begin{aligned} x' &= y \\ y' &= -x + 2y\cos\theta \pmod 1 \end{aligned} \tag{2.3}$$

where $(x, y) \in I$. This map is a nonhomogeneous version of the lossless digital filter overflow, or simply the overflow map

$$\begin{aligned} x' &= y \\ y' &= -x + 2y\cos\theta \pmod 2 \end{aligned} \tag{2.4}$$

where $(x, y) \in [-1, 1)^2$. In both cases $X^+ = [0, 1]^2$ and hence by shearing the coordinates, the overflow map can be conjugated to a piecewise rotation on a rhombus: that is, fix $\theta \in [0, \pi/2]$ and consider the rhombus unit cell for a torus

$$M_\theta = \{z \in \mathbb{C} \;:\; |\mathrm{Re}(z)| \leq 1 \;\text{ and }\; |\mathrm{Re}(ze^{i\theta})| \leq 1\}.$$

We define $f : M_\theta \to M_\theta$ by

$$f(z) = e^{i\theta} z + Wk(z) \tag{2.5}$$

where $W = \frac{2i}{\sin\theta}$ is a constant and

$$k(z) = \begin{cases} +1 & \text{if } \mathrm{Re}(ze^{2i\theta}) > 1 \\ -1 & \text{if } \mathrm{Re}(ze^{2i\theta}) \leq -1 \\ 0 & \text{otherwise.} \end{cases}$$

It is obvious that, disregarding the discontinuities, the mapping is invertible.

In [6], numerical evidence of the existence of non-smooth invariant curves in an elliptic system was found, and the dynamics on these curves reduce to an interval exchange map.

We now discuss an example with quantization discontinuity, that is, the discontinuity is induced by the function $\mathrm{sgn}(\cdot)$. In the quiescent input and unity gain case, the model for a bandpass sigma-delta modulator [10] can be written

$$\begin{aligned} x' &= y \\ y' &= -x + 2y\cos\theta + \mathrm{sgn}(x) - 2\cos\theta\,\mathrm{sgn}(y) \end{aligned} \tag{2.6}$$

where $(x, y) \in \mathbb{R}^2$. By an appropriate transformation of the linearized parts into normal form, it is shown in [4] that the system (2.6) becomes a piecewise isometry of the plane, and the map is invertible on the maximal invariant set X^+, which is a compact subset of $\mathbb{R}^2$. Therefore there is an invariant disc packing of X^+ composed of periodic orbits if θ/π is rational and periodic and quasi-periodic orbits otherwise.

In [4] it is shown numerically that the maximal invariant set X^+ is a global attractor for all initial conditions in the plane. We can prove this for the special case of $\theta = \pi/2$; the more general case is discussed in Section 3. In this special case, system (2.6) becomes $(x', y') = (y, -x + \mathrm{sgn}(x))$, where $(x, y) \in \mathbb{R}^2$. We define

$$M_k = [-(k+1), k+1]^2, \quad k \geq 1, \quad \text{and} \quad M = M_0 = [-1, 1]^2.$$

Because $f^2(x, y) = (-x + \mathrm{sgn}(x), -y + \mathrm{sgn}(y))$, one can verify that

$$f^2(M_k) \subseteq M_{k-1}, \; k \geq 1$$

and moreover $f(M) = M$. Hence $X^+ = M$ is a global attractor and any point in the plane enters X^+ after a finite number of iterations.

2.3 Homogeneous and nonhomogeneous maps

The dynamics of nonhomogeneous versions of (1.3) for different (α', β') are in general not conjugate except in particular cases of symmetry, and in particular they may be different from that of the corresponding homogeneous map. For example, the parabolic map f_α:

$$\begin{aligned} x' &= x + \alpha \pmod 1 \\ y' &= 2x + y \pmod 1 \end{aligned} \tag{2.7}$$

where $(x, y) \in [0, 1)^2$ has been investigated by Marklof [11]. It is shown that when $\alpha = 0$ the map f_α is continuous and integrable but when $\alpha \neq 0$ (α rational) the dynamics of f_α on an orbit can be identified with a one-dimensional piecewise isometry. For irrational α, the map is uniquely ergodic.

Similar to the discussion in the beginning of Section 1.1, a homogeneous linear torus map on a square $[e, f) \times [g, h)$ can be transformed to a nonhomogeneous linear torus map on the unit square $[0, 1)^2$. For example, the overflow map

$$\begin{aligned} x' &= y \\ y' &= bx + ay \pmod 2 \end{aligned}$$

where $(x, y) \in [-1, 1)^2$, is equivalent to the nonhomogeneous map

$$\begin{aligned} x' &= y \\ y' &= bx + ay + \tfrac{1}{2}(1 - a - b) \pmod 1 \end{aligned}$$

where $(x, y) \in [0, 1)^2$. The elliptic sawtooth standard map is also equivalent to a nonhomogeneous map, because the former is conjugate to the lossless digital linear filter overflow oscillation system (see [6]).

3 Global attractors for piecewise isometric maps

Even though the Lyapunov exponents are zero for discontinuous parabolic [14] and elliptic [6] maps, the dynamics generated by these maps may have sensitive dependence on initial conditions due to the presence of discontinuities.

For such maps on the plane there may, however, still be compact global attractors, due to the presence of discontinuities that give rise to regions with two pre-images. For example, the maps investigated by Goetz [2] can have quite complicated but bounded global attractors for the dynamics. It has recently been recognized [12, 9] that there the definition of a global attractor is surprisingly subtle due to the presence of discontinuities of the map.

Suppose that $f : M \to M$ is a piecewise isometry with a finite partition, that is, $M = \cup_{i=1}^{N} M_i$, $M_i \cap M_j = \emptyset$ for $i \neq j$, and f restricted to each M_i is an isometry. As in [8, 9] we introduce the notion $=_0$ to mean equivalence of sets up to zero measure, that is, $U =_0 V$ means that $\ell(U \Delta V) = 0$, where $U \Delta V$ is the symmetric difference. $U \subseteq_0 V$ means that there is a $U_1 =_0 U$ such that $U_1 \subseteq V$. If $V =_0 \emptyset$ we say V is *almost empty*. We say V is *almost closed* if $V =_0 \overline{V}$, whereas V is *almost open* if $V =_0 \operatorname{int}(V)$.

Let $\Omega(U) = \overline{\cup_{x \in U} \omega(x)}$, where $\omega(x)$ is the ω-limit set of the orbit $\{f^n(x), n \geq 0\}$. We call $A \subseteq M$ an *attractor* for a piecewise isometry f, if A is invariant ($f(A) = A$) and almost closed, and there exist a neighbourhood $U \supset A$, such that $f(U) \subseteq U$, and $\Omega(U) \subseteq_0 A$. We call $A \subseteq M$ the *global attractor* for f, if A is an attractor with $\Omega(M) \subseteq_0 A$.

Remark. Here an attractor is invariant but may be not closed (only almost closed). In [12] an attractor for a piecewise isometry is defined as closed but may be not invariant but only quasi-invariant, i.e. $f(U) \subset U$ and $f(U) =_0 U$.

As an extra example in addition to that in Section 2.2, we now show that the overflow system viewed as a map (2.5) to the whole plane $f : \mathbb{C} \to \mathbb{C}$ has a global attractor. To see this, we consider the equivalent system (2.4) recalling that this is a sheared version of (2.5).

We write g to be the 'rounding' map such that $g(x) = x$ for $x \in [-1, 1)$ and $g(x+2) = g(x)$. We note that the overflow map is $f(x, y) = (y, g(-x + 2y\cos\theta))$ and so $f^2(x, y) = (g(-x + 2y\cos\theta), g(-y + 2(-x + 2y\cos\theta)\cos\theta)$. Hence the second iterate maps the whole plane onto the square $[-1, 1)^2$ after two iterates. It remains to check that this region is invariant under f (up to a set of zero measure) as we can define an inverse on the region, and hence the rhombus M_θ is the global attractor for the original map.

4 Final remarks

In this paper we have discussed some properties and examples of the dynamics of some area-preserving linear maps with round-off and quantization discontinuity. We are particularly interested in those cases where the dynamics is not hyperbolic but in fact a piecewise isometry (as in the elliptic case), or a piecewise isometry on invariant subsets (as in the parabolic case), where lack of hyperbolicity means we have to develop new techniques.

For example, for certain ranges of parameters, the parabolic map restricted to its global attractor can be reduced to a one-parameter family of one-dimensional isometries; by shearing the coordinates, the overflow map can be conjugated to a piecewise isometry on a rhombus; and the model for bandpass sigma-delta modulator can also be reduced to a piecewise isometry on a compact global attractor.

On the global attractor, the dynamics of elliptic maps dynamically define invariant disc packing problems. Moreover, the measure, geometry and dynamics of such systems can be seen as properties of these disc packings. There are some apparently very difficult conjectures concerning the dynamics and dimensions of these circle packings [6], and the riddling and invariant properties of the discontinuity set [9]. For example, it has recently been shown [7] that the overflow map has invariant disc packings that are 'loose' in the sense that have no tangencies for most parameter values.

In many cases the dynamics of invertible piecewise isometries can be characterized by finite-state symbolic dynamics. For example, when the natural and symbolic partitions are compatible in the sense that one is the refinement of the other, the maps can be symbolized over a countable alphabet. Using this we are working on techniques to predict and locate periodic orbits up to a given length, to understand the structure of some orbits, and to study bifurcations in the parameter space.

Acknowledegements

This work was supported by EPSRC grant number GR/M36335. We thank Jonathan Deane and Miguel Mendes for some helpful conversations concerning this work.

Bibliography

[1] Chua, L.O. and Lin, T. (1988). Chaos in digital filters. *IEEE Trans. CAS*, **35**, 648–658;
Davies, A.C.,(1995). Nonlinear oscillations and chaos from digital filter overflow. *Phil. Trans. R. Soc. Lond. A*, **353**, 85–99;
Kocarev, L., Wu, C.W. and Chua, L.O. (1996). Complex behaviour in Digital filters with overflow nonlinearity: analytical results. *IEEE Trans CAS-II*, **43**, 234–246;
Ashwin, P., Chambers, W. and Petkov, G. (1997). Lossless digital filter overflow oscillations; approximation of invariant fractals. *Int. J. Bifur. Chaos*, **7**, 2603–2610.

[2] Goetz, A. (1998). Dynamics of piecewise isometries, *Illinois J. Math.*, **44**, 465–478;
Buzzi, J. (2001), Piecewise isometries have zero topological entropy, to ap-

pear in *Ergodic Theory and Dynamical Systems*;
Adler, R., Kitchens, B and Tresser, C. (2000), Dynamics of nonergodic piecewise affine maps of the torus, to appear in *Ergodic Theory and Dynamical Systems.*

[3] Katok, A. and Hasselblatt, B. (1995). *Introduction to the Modern Theory of Dynamical Systems*, Cambridge University Press;
Boshernitzan, M. and Kornfeld, I. (1995). Interval translation mappings, *Ergodic Theory and Dynamical Systems*, **15**, 821–832.

[4] Ashwin, P., Deane, J.H.B., and Fu, X.-C. (2001). Dynamics of a bandpass sigma-delta modulator as a piecewise isometry, *Proc of ISCAS'2001*, pp. III-811–III-814 (Sydney, Australia).

[5] Ashwin, P., Fu, X.-C., Nishikawa, T., and Życzkowski, K. (2000). Invariant sets for discontinuous parabolic area-preserving torus maps, *Nonlinearity*, **13**, pp.819-835.

[6] Ashwin, P. (1996). Non-smooth invariant circles in digital overflow oscillations, *Proc. Workshop on Nonlinear Dynamics of Electronic Systems*, (Sevilla);
Ashwin, P. (1997). Elliptic behaviour in the sawtooth standard map, *Phys. Lett. A*, **232**, 409–416.

[7] Ashwin, P. and Fu, X.-C. (2001). Tangencies in invariant circles packings for certain planar piecewise isometries are rare, to appear in *Dynamical Systems.*

[8] Ashwin, P. and Terry, J.R. (2000). On riddling and weak attractors. *Physica D*, **142**, 87–100.

[9] Ashwin, P., Fu, X.-C. and Terry, J.R. (2001). Riddling and invariance for discontinuous maps preserving Lebesgue measure. *preprint.*

[10] Feely, O. and Fitzgerald, D. (1996). Non-ideal and chaotic behaviour in bandpass sigma-delta modulators, *Proc. Workshop on Nonlinear Dynamics of Electronic Systems*, (Sevilla).

[11] Marklof, J. and Rudnick, Z. (2000). Quantum unique ergodicity for parabolic maps, *Geom. Funct. Anal.*, **10**, 1554–1578.

[12] Goetz, A., and Mendes, M. (2000). Piecewise rotations: bifurcations, attractors and symmetries. *Proc. Conf. in Bifurcations, Symmetry and Patterns*, (Portugal).

[13] Vaienti, S. (1992). Ergodic properties of the discontinuous sawtooth map, *J. Stat. Phys.*, **67**, 251–269.

[14] Życzkowski, K. and Nishikawa, T. (1999). Linear parabolic maps on the torus, *Phys. Lett. A*, **259**, 377–386.

Nonlinear System Identification of a Broadband Subscriber Line Interface Circuitry Using the Volterra Approach

Heinz Koeppl and Gerhard Paoli
Infineon Technologies Austria, Design Center Villach, Austria

Abstract

Making use of the Volterra approach, black box modelling is applied to a large scale analogue circuitry for an ADSL (asymmetric digital subscriber line) central office application. Reducing the number of free parameters through special assumptions on the Volterra kernels, one ends up with the Hammerstein model. Taking the first few taps of the Volterra kernels and approximating the last taps through Hammerstein kernels, the Volterra–Hammerstein model is obtained. The parametrization of these models is an inverse problem and leads to an ill-conditioned linear equations system, thus regularization techniques are used. Discrete multi-tones (DMT) serve as excitation signals. For validation of the presented methods a reference model is constructed. This model consists of full factorable Volterra kernels, where the order of the nonlinearity corresponds to that occurring in the circuitry. For the circuitry, data acquisition is done by an analogue network simulator. Emphasis was put on the accuracy of the inter-modulation products, described by a generalized signal-to-noise-ratio. The Hammerstein and the Volterra–Hammerstein models are compared with each other and with measurements from the laboratory.

1 Motivation

Nonlinear distortions limit the performance of the overall ADSL system. Therefore, performance estimations of the system in the design period are desired to schedule necessary redesigns. A fast method to map the nonlinear dynamics of an analogue circuitry to a black box model is needed. In the discussed ADSL transceiver DMT multicarrier modulation [3] is utilized. The number of bits being allocated to each single carrier determines the overall bit-rate. A proper quantity to describe the possible bit-allocation is the MTPR value (multi-tone power ratio), a generalized SNR (signal-to-noise-ratio). The MTPR is defined for each carrier as

$$MTPR_i = 10 \log \left(\frac{S_i}{N_i + \sum_j D_{ij}} \right) \quad \text{in dB,} \tag{1.1}$$

where the index i denotes the ith carrier. S_i stands for the transmitted power of the ith carrier, which has to be related to the sum of the noise N_i and all the inter-modulations D_{ij}, produced from the other j carriers of the DMT signal

$$u(t) = \sum_{k=n_1}^{n_2} A_k \cos(k\omega_0 t + \varphi_k). \tag{1.2}$$

In (1.2), A_k denotes the magnitude and φ_k the phase of the kth carrier, $\omega_0 = 2\pi 4312.5\ \mathrm{s}^{-1}$. DMT signals, as given in (1.2), have different peak values, depending on their phase distribution $\{\varphi_k\}$. These peak values are described by the crest factor, which is defined as the ratio of the l_∞-norm to the l_2-norm of the signal $u(t)$.

2 Mathematical models

2.1 Volterra model

Every time-invariant nonlinear I/O operator **K** with fading memory [1,2] can be represented by a Volterra series. The doubly finite Volterra series approximates **K** [7] and is termed the Volterra filter. A discrete Volterra filter of order N and memory lengths M_n is given by

$$y(l) = h_0 + \sum_{n=1}^{N} \sum_{k_1=0}^{M_1} \cdots \sum_{k_n=0}^{M_n} h_n(k_1, \ldots, k_n) \prod_{i=1}^{n} u(l - k_i), \tag{2.1}$$

where the $h_n(k_1, \ldots, k_n)$ denote the integral kernels and $u(l)$ stands for the input sequence. A constant offset is given by h_0. The input–output relation can be expressed using tensor products. Defining

$$\mathbf{U}_i(l) = \begin{bmatrix} u(l) & u(l-1) & \ldots & u(l-M_i) \end{bmatrix}^T, \tag{2.2}$$

the Volterra filter is given by

$$y(l) = h_0 + \sum_{j=1}^{N} \mathbf{h}_j \left(\bigotimes_{i=1}^{j} \mathbf{U}_j(l) \right) \tag{2.3}$$

with $\mathbf{h}_j$ denoting the vector of kernel values of order j. The number of free parameters for each order of nonlinearity j goes with $\mathcal{O}([M_j]^j)$. Equation (2.1) is implementable only for small orders of nonlinearity N and short filter lengths M_n.

2.2 Hammerstein model

A model with fewer of free parameters is given by the Hammerstein model. It can be derived by imposing special symmetries to the general Volterra kernels $h_n(k_1, \ldots, k_n)$

$$\bar{h}_{n_H}(k_1, \ldots, k_n) = h_n(k_1, \ldots, k_n) \prod_{i=2}^{n} \delta_{k_1, k_i}, \tag{2.4}$$

with δ_{k_1,k_i} the Kronecker delta. The Hammerstein kernels $\bar{h}_{n_H}(k_1,\dots,k_n)$ given in (2.4) reduce to one-dimensional functions $h_{nH}(j)$. Using (2.4), (2.1) yields the Hammerstein model

$$y(l) = h_0 + \sum_{n=1}^{N}\sum_{j=0}^{M_n} h_{nH}(j)\,[u(l-j)]^n\,. \tag{2.5}$$

2.3 Volterra–Hammerstein model

The first part of the kernels in (2.1) govern the system response $y(l)$, and will be approximated by the Volterra–Hammerstein model, while the Hammerstein model approximates the remaining part. This can be expressed by

$$\begin{aligned} y(l) = h_0 \quad &+ \sum_{n=1}^{N}\sum_{k_1=0}^{\hat{M}_1}\cdots\sum_{k_n=0}^{\hat{M}_n} h_n(k_1,\dots,k_n)\prod_{i=1}^{n}u(l-k_i) \\ &+ \sum_{n=2}^{N}\sum_{j=\hat{M}_n+1}^{M_n} h_{nH}(j)\,[u(l-j)]^n\,, \end{aligned} \tag{2.6}$$

where $\hat{M}_n$ denote the Volterra model memory lengths. Equation (2.6) describes a more general model than the Hammerstein model, given by (2.5).

3 Parametrization

Equation (2.1), (2.5), and (2.6) are linear in the model parameters, that is, in the kernel values, thus the parameterization involves solving a linear least-square problem. All occurring combinations of the input signal for the system response $y(l)$ can be written in one vector, which is given for (2.1) as

$$\mathbf{u}(l) = [u(l) \quad \dots \quad u(l-M_1) \quad u(l)u(l-1) \quad \dots \quad u(l-M_N)^N], \tag{3.1}$$

while the corresponding parameter vector is given by

$$\mathbf{h} = [h_1(0) \quad \dots \quad h_1(M_1) \quad h_2(0,1) \quad \dots \quad h_N(M_N,\dots,M_N)]^T. \tag{3.2}$$

The system output $\mathbf{y}(l) = [y(l),\dots,y(l+S)]^T$ taken over S samples is related to the input $\mathbf{u}(l)$ by

$$\mathbf{y}(l) = \mathbf{U}(l)\mathbf{h}, \tag{3.3}$$

with the data matrix $\mathbf{U}(l) = [\mathbf{u}(l),\dots,\mathbf{u}(l+S)]^T$.

3.1 Excitation signal

For identification, the input signal $u(l)$ has to satisfy the persistence of excitation (PE) condition [5]. An input signal persistently excites a filter if the vectors $\{\mathbf{u}(l)\}$ span the whole parameter space $\mathbf{h}$. Deterministic input signals have

PE, if the smallest and the largest eigenvalues, $\lambda_{\min}$ and $\lambda_{\max}$, of the sample correlation matrix $\mathbf{U}^T(l)\mathbf{U}(l)$ satisfy

$$\rho_1 \leq \lambda_{\min} \leq \lambda_{\max} \leq \rho_2, \tag{3.4}$$

with $\rho_1, \rho_2 > 0$. Broadband DMT signals, given in (1.2), can be shown to fulfil the PE condition and are used.

3.2 Regularization

The parameterization of the above models is an inverse problem and leads to an ill-conditioning of system (3.3). Therefore, regularization [4] has to be performed. Comparison between Tikhonov regularization and regularization by conjugate gradient (CG) iterations showed the superiority of the CG iterations for the analogue circuitry under consideration. The LSQR algorithm [8], a CG-type solver, is applied.

4 Reference model

To test the applied algorithm and model structures, a reference model is constructed. It represents a full factorable Volterra system of order $N = 9$. A factorable system was chosen because of the tremendous number of kernel parameters for a general Volterra system of the same order of nonlinearity.

4.1 Factorable Volterra kernels

In general, different types of factorable kernels are possible [6,10]. A kernel is full factorable if it can be written as

$$h_n(k_1, \ldots, k_n) = \prod_{i=1}^{n} h_i(k_i). \tag{4.1}$$

Using (2.1) together with the kernel property of (4.1) results in

$$y(l) = h_0 + \sum_{n=1}^{N} \prod_{i=1}^{n} \left[\sum_{k=0}^{M_n} h_i(k) u(l-k) \right]. \tag{4.2}$$

Every kernel can be made symmetric [9] by

$$h_n(k_1, \ldots, k_n)_{\text{sym}} = \frac{1}{n!} \sum_{\text{perm}(\pi_i)} h_n(k_{\pi_i}, \ldots, k_{\pi_i}). \tag{4.3}$$

4.2 Model kernels

To construct the kernels $h_i(k_i)$ in (4.1) the following relation is used:

$$h_i(k_i) = A_i \sin(b_i k_i + \varphi_i) e^{-c_i k_i}, \tag{4.4}$$

with the model parameters b_i, φ_i and c_i shown in Table 1. For $i = 1, \ldots, N$ the magnitude is set to $A_i = 1$ and rescaled to obtain harmonics of the same order

Table 1 Configuration of the reference model

	h_1	h_2	h_3	h_4	h_5	h_6	h_7	h_8	h_9
b_i	2.3	1.3	2.0	1.6	1.3	1.1	1.2	2.3	1.5
φ_i	$\pi/20$	$\pi/7$	$\pi/13$	$\pi/5$	$\pi/4$	$\pi/4$	$\pi/6$	$\pi/12$	$\pi/4$
c_i	0.8	2.4	0.9	0.7	0.9	2.2	2.2	0.6	1.3

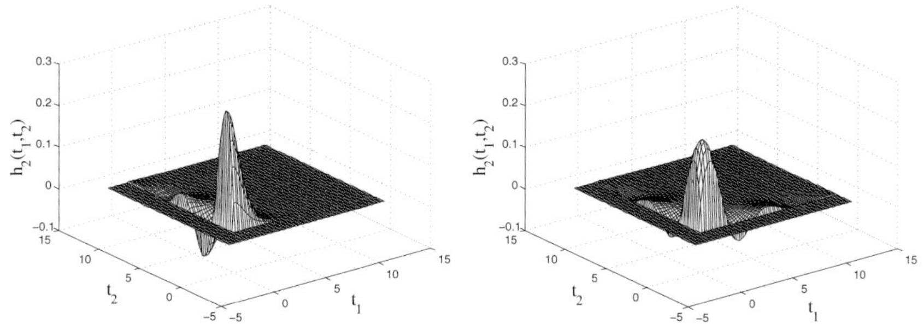

FIG. 1. Second-order Volterra kernel $h_2(t_1, t_2)$ of the reference model: generic and symmetrized.

as for the analogue circuitry. The memory length M_n in (4.2) is chosen to be $M_i = 10$ for $i = 1, \ldots, N$ and h_0 is set to zero. Using (4.1) and (4.4) one can generate higher-order kernels and symmetrize them according to (4.3). Fig. 1 illustrates the generic second-order kernel and its symmetrized version.

5 Nonlinear model of the analogue circuitry

A broadband subscriber line interface circuit (SLIC) for an ADSL central office application [11], is modeled. Its nonlinear model consists mainly of the transmit path ($K_{\mathrm{B}}(t)$), the receive path ($K_{\mathrm{s}}(t)$), and the echo compensation path ($K_{\mathrm{BF}}(t)$). Figure 2 shows the overall model of the broadband SLIC, constituted by the nonlinear operators $K_{\mathrm{s}}(t)$, $K_{\mathrm{BF}}(t)$, $K_{\mathrm{B}}(t)$, and a linear line model, given by the impulseresponses $h_{\mathrm{co2rt}}(t)$, $h_{\mathrm{rt2co}}(t)$, $h_{\mathrm{coecho}}(t)$ and $h_{\mathrm{rtecho}}(t)$. Nonlinear system identification of the broadband SLIC is done by using the Hammerstein model given in (2.5), and the Volterra–Hammerstein model, described by (2.6). Results obtained are presented for the transmit path, the most complex part of the broadband SLIC. As a frame of reference, measurements from the laboratory are used.

6 Results

In this section identification and simulation results are shown, both for the reference model and for the buffer of the broadband SLIC. A DMT signal, as given in (1.2), with n_1=33 and n_2=127 is used for the identification. The phase distribution is chosen to yield a crest factor of 3.34, which has been pointed out to be the average crest factor, obtained from several runs with random phase

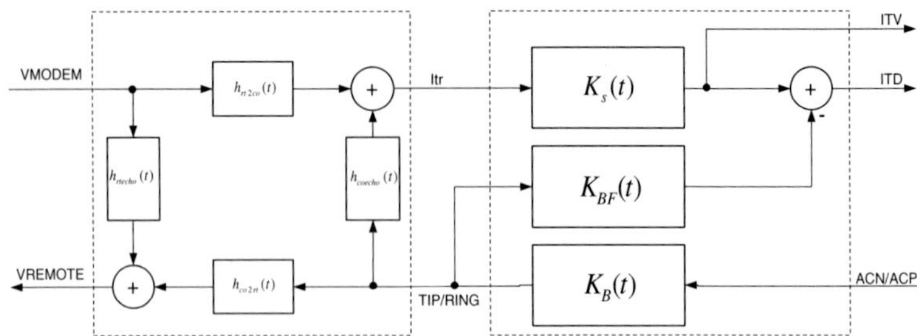

FIG. 2. Line model and nonlinear model of the broadband SLIC.

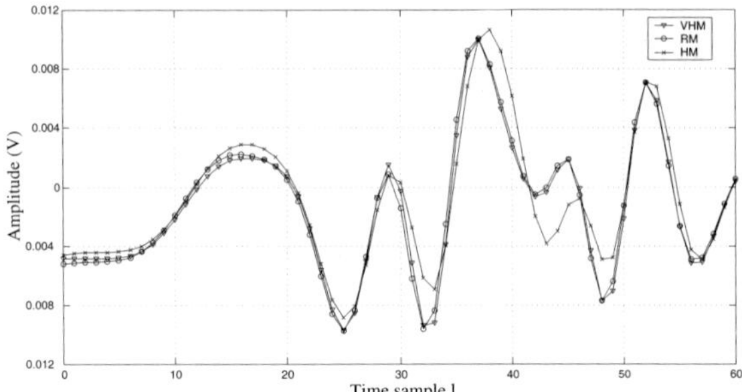

FIG. 3. Nonlinear contribution $\delta y(l)$ for the reference model: Reference, Hammerstein model and Volterra–Hammerstein model

distribution. For MTPR values, defined by (1.1), missing tone simulations for each carrier in the bandwidth of n_1=33 and n_2=127 are performed with a crest factor of 3.34. carrier are sent.

For the broadband SLIC and the reference model the linear contributions to the output are dominant. To illustrate the identification performance for the nonlinear terms, the following quantity is used:

$$\delta y(l) = y(l) - y_{ml}(l), \tag{6.1}$$

where $y(l)$ is the overall output of either the models, given by (2.6) and (2.5), or the reference systems, either the network simulator or the reference model, defined by (4.2) and (4.4). $y_{ml}(l)$ refers to the linear contribution to the model outputs of (2.5) and (2.6). The quantity $\delta y(l)$ will be denoted therefore as 'nonlinear contribution'. For brevity, the following abbreviations are used: 'VHM', Volterra–Hammerstein model; 'HM', Hammerstein model, and 'RM', reference model.

6.1 Reference model

The reference model as described in Section 4 is identified using the model structures given by (2.6), (2.5), and a DMT signal. As a quantity of measure for the identification, the mean square error (MSE) is given together with the model

Table 2 Identification results for the reference model. Memory lengths and MSE of the models. np denotes the number of model parameters and $i = 2, \ldots, 9$

	M_1	M_i	$\hat{M}_i$	np	MSE
VHM	10	10	2	111	1.44E-8
HM	10	10	0	90	7.26E-8

Table 3 Simulation results for the reference model. Average MTPR values and errors

	RM	HM	VHM
Mean MTPR (dB)	55.82	56.90	56.19
Mean MTPR error (dB)	0	1.08	0.37

Table 4 Identification results for the transmit path $K_B(t)$. Memory lengths and MSE of the models. np denotes the number of model parameters

	M_1	M_3	$\hat{M}_3$	M_5	$\hat{M}_5$	M_7	$\hat{M}_7$	M_9	$\hat{M}_9$	np	MSE
VHM	100	40	4	40	4	40	4	40	0	444	1.45E-6
HM	100	40	0	40	0	40	0	40	0	260	1.64E-6

configuration in Table 2. It is observed that the identification performance of both models are very sensitive to changes in the order of nonlinearity N and to changes in the memory length M_i. The average MTPR values and the average MTPR errors of both models compared to the reference model are shown in Table 3. For visualization of the identification performance the nonlinear contributions $\delta y(l)$ are illustrated in Fig. 3.

6.2 Analog circuitry

Single tone simulations, done with the network model of the broadband SLIC, have shown significant harmonics up to the order of 9. Due to the differential design of the broadband SLIC, only odd harmonics are present in the network simulator. The model configurations and the MSE between the output of the network simulator and the above models are given in Table 4. For the MTPR values, investigations are done in the laboratory, with the network simulator and with the discussed models. Average MTPR values and average MTPR errors are presented in Table 5, with measurements from the laboratory as the reference values. The nonlinear contribution $\delta y(l)$ for the broadband SLIC is shown in Fig. 4.

7 Conclusion

Starting from Volterra series, two models with a reasonable number of unknowns were presented and used to identify a large-scale analogue circuitry, namely a broadband SLIC, consisting of about a thousand components in BiCMOS tech-

Table 5 Simulation results for the transmit path $K_B(t)$. Average MTPR values and errors. 'LAB', Laboratory; 'SAB', SABER

	LAB	SAB	HM	VHM
Mean MTPR (dB)	64.70	67.98	70.74	66.94
Mean MTPR error (dB)	0	3.28	6.04	2.34

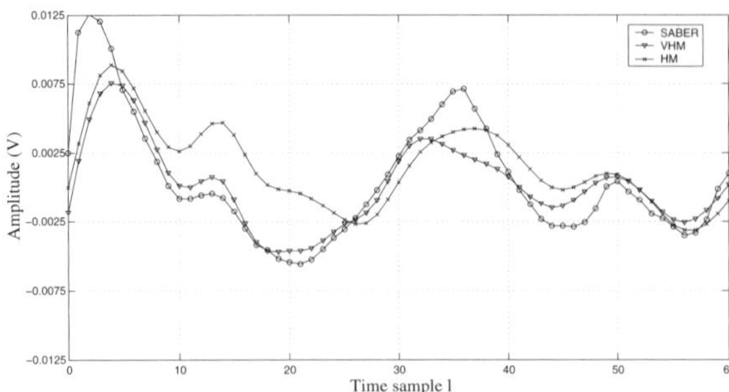

FIG. 4. Nonlinear contribution $\delta y(l)$ for the broadband SLIC: Reference, Hammerstein model and Volterra–Hammerstein model.

nology. A reference model, represented by a full factorable Volterra system, was constructed and identified for testing the reliability of the proposed models. For the identification a DMT-signal, which has PE, has been taken. The simulation results obtained from the Hammerstein model and those obtained from the Volterra–Hammerstein model were compared with each other and with measurements from the laboratory. Focus was put on the MTPR value, an integral quantity summing all errors. It was seen that both models were able to identify the behaviour of an artificial reference model with very high accuracy (0.37 dB MTPR error). Investigations for the broadband SLIC showed that the Volterra–Hammerstein model is more accurate (compared to the network simulator) than the Hammerstein model, that is, the mean MTPR errors are 1.04 dB for the Volterra–Hammerstein model and 2.76 dB for the Hammerstein model. The difference between the network simulator and the laboratory arises from the fact that the even harmonics are present on the silicon due to mismatches.

It can be concluded that the presented models are able to synthesize the nonlinearities of analogue circuitries, described by a network simulator, with high accuracy, even for large-scale circuitries.

Bibliography

[1] Boyd, S. and Chua, L.O. (1984). Analytical foundations of Volterra series. *J. Math. Control Info.*, **1**, 243–282.

[2] Chua, L.O. and Boyd, S. (1985). Fading memory and the problem of approximation nonlinear operators with Volterra series. *IEEE Trans. on Circuits and Systems*, **32**, 1150–1161.

[3] Starr, Th., Cioffi, J. and Silverman, P. (1999). *Understanding Digital Subscriber Line Technology.* Prentice-Hall, New York.

[4] Hansen, P.C. (1998). *Rank deficient and discrete ill-posed problems.* SIAM monographs on mathematical modelling and computation, Philadelphia.

[5] Novak, R.D. and Van Veen, B. (1996). Random and pseudorandom excitation sequences for identification of the truncated Volterra series. *IEEE Trans. on Signal Process.*, **1**, 213–313.

[6] Rugh, W.J. (1981). *Nonlinear System Theory.* Johns Hopkins University Press, London.

[7] Sandberg, I.W. (1992). Uniform approximation with double finite Volterra series. *IEEE Trans. on Signal Process.*, **40**, 1438–1442.

[8] Saunders, M.A. and Paige, C.C. (1982). LSQR: An algorithm for sparse linear equations and sparse least squares. *ACM Trans. on Mathematical Software*, **8**, 43–71.

[9] Schetzen, M. (1980). *The Volterra and Wiener Theories of Nonlinear Systems.* Wiley, New York.

[10] Shanmugam, K.S. and Lal, M. (1976). Analysis and synthesis of a class of nonlinear systems. *IEEE Trans. on Circuits and Systems*, **23**, 17–25.

[11] Zojer, B. *et al.* (2000). A broadband high-voltage SLIC for a splitter- and transformerless combined ADSL-lite/POTS Linecard. *IEEE J. Solid-State Circuits*, **35**, 1976–1988.

Importance Sampling Simulation and Multiple-Hyperplane Realization of the Bayesian Decision Feedback Equalizer

S. Chen and L. Hanzo
Department of Electronics and Computer Science, University of Southampton, Southampton SO17 1BJ, UK

Abstract

For the class of equalisers that employs a symbol-decision finite-memory structure with decision feedback, the optimal solution is known to be the Bayesian decision feedback equaliser (DFE). However, the complexity of the optimal Bayesian DFE increases exponentially with the length of the channel impulse response (CIR). It has been noted that, when the signal-to-noise-ratio tends to infinity, the decision boundary of the Bayesian DFE is asymptotically piecewise linear and consists of several hyperplanes. This asymptotic property can be exploited for efficient simulation and implementation of the Bayesian DFE. An importance sampling (IS) simulation technique is presented based on this asymptotic property for evaluating the lower-bound bit error rate of the Bayesian DFE under the assumption of correct decisions being fed back. A design procedure is developed, which chooses appropriate bias vectors for the simulation density to ensure asymptotic efficiency of the IS simulation. As the set of hyperplanes that form the asymptotic Bayesian decision boundary can easily be found, they can be used to partition the observation space. The resulting multiple-hyperplane detector can closely approximate the optimal Bayesian detector, at an advantage of considerably reduced decision complexity.

1 Introduction

Equalisation technique plays an ever-increasing role in combating distortion and interference in communication links [1, 2] and high-density data storage systems [3, 4]. For the class of equalisers based on a symbol-by-symbol decision with decision feedback, the Bayesian DFE [5, 6, 7] is known to provide the best performance. The complexity of this optimal Bayesian solution, however, increases exponentially with the CIR length, and this limits its practical usefulness. For example, due to its complicated structure, performance analysis of the Bayesian DFE is usually based on conventional Monte Carlo simulation, which is computationally costly even for modest signal-to-noise-ratio (SNR) conditions. To obtain a reliable BER estimate, at least 100 errors should occur during a simulation. Thus, for a BER level of 10^{-6}, at least 10^{8} data samples are needed. Investigating the Bayesian DFE under bit error rate (BER) performance better than 10^{-6} is very difficult if not impossible, using a conventional Monte Carlo simulation. In order for the Bayesian DFE to be more widely adopted in practice, it is also necessary and desired to reduce its implementation complexity without sacrificing performance too much.

Geometrically, the complexity of the Bayesian DFE is a consequence of the need to form the optimal decision boundary that is a hypersurface in the observation space [6]. It can be shown that asymptotically, as the SNR tends to infinity, the Bayesian hypersurface becomes piecewise linear and is made up of a set of hyperplanes [8]. In practice, at large rather than infinite SNR, the performance difference between Bayesian decision boundary and a piecewise linear approximation is negligible. Each of these component hyperplanes is determined by a pair of so-called dominant opposite-class channel states. This asymptotic property can be utilized for various purposes. For instance, in a previous work [9], the Bayesian equalisation solution is approximated by only using the set of the dominant signal state pairs in computation. In this paper, we exploit this asymptotic property to develop an IS simulation technique for performance evaluation of the Bayesian DFE and to implement the Bayesian DFE in a computationally very efficient multiple-hyperplane form.

Iltis [8] developed a randomized bias technique for the IS simulation of Bayesian equalisers without decision feedback. Although it can only guarantee asymptotic efficiency, as defined in [10], for certain channels, this IS simulation technique provides a valuable method in assessing the performance of the Bayesian equaliser. We extend this IS simulation technique to evaluate the lower-bound BER of the Bayesian DFE. By viewing decision feedback as a geometric translation, the Bayesian DFE is 'converted' to the Bayesian equalizer in the translated space [11], with a desired property that opposite-class channel states are always linearly separable. A design procedure is developed, which determines the set of hyperplanes that form the asymptotic Bayesian decision boundary and constructs the convex regions associated with individual states by intersecting hyperplanes that are reachable from the states concerned. This provides the appropriate bias vectors for the simulation density to ensure asymptotic efficiency.

A multiple-hyperplane partition technique for equalisation was developed by Kim and Moon [12, 13]. Their design method determines a set of hyperplanes which separate clusters of channel states. A combinatorial search and optimization process is carried out to find these hyperplanes, which is computationally very expensive. The convex regions associated with individual channel states are constructed by appropriately intersecting hyperplanes. The overall decision region is then formed from these convex regions. The decision complexity and performance of the multiple-hyperplane detector are controlled during design by a specified minimum separating distance. Although it is possible to achieve the asymptotic Bayesian solution by an appropriate choice of the minimum separating distance, this is by no means guaranteed as the combinatorial search and optimization process does not necessarily produce the set of hyperplanes which form the asymptotic Bayesian decision boundary. We propose a much simpler alternative design to explicitly realize the asymptotic Bayesian DFE.

2 The Bayesian DFE

We will assume that the channel is real-valued and the received signal sample is given by

$$y(k) = \sum_{i=0}^{n_a-1} a_i s(k-i) + e(k)\,, \tag{2.1}$$

where n_a is the CIR length, a_i are the channel taps, the Gaussian white noise $e(k)$ has zero mean and variance σ_e^2, and the transmitted symbol $s(k)$ takes values from the set $\{\pm 1\}$. A DFE uses the observation vector $\mathbf{y}(k) = [y(k) \ldots y(k-m+1)]^T$ and the past detected symbol vector $\hat{\mathbf{s}}_b(k) = [\hat{s}(k-d-1) \ldots \hat{s}(k-d-n)]^T$ to produce an estimate $\hat{s}(k-d)$ of $s(k-d)$. Without the loss of generality, the decision delay of $d = n_a - 1$, feedforward order of $m = n_a$ and feedback order of $n = n_a - 1$ are chosen, as this choice is sufficient to guarantee the linear separability [11]. The received signal vector can be expressed as

$$\mathbf{y}(k) = \mathbf{F}_1 \mathbf{s}_f(k) + \mathbf{F}_2 \mathbf{s}_b(k) + \mathbf{e}(k)\,, \tag{2.2}$$

where $\mathbf{s}_f(k) = [s(k) \ldots s(k-d)]^T$, $\mathbf{s}_b(k) = [s(k-d-1) \ldots s(k-d-n)]^T$, $\mathbf{e}(k) = [e(k) \ldots e(k-m+1)]^T$, and the $m \times (d+1)$ and $m \times n$ CIR matrices $\mathbf{F}_1$ and $\mathbf{F}_2$ are, respectively,

$$\mathbf{F}_1 = \begin{bmatrix} a_0 & a_1 & \cdots & a_{n_a-1} \\ 0 & a_0 & \ddots & \vdots \\ \vdots & \ddots & \ddots & a_1 \\ 0 & \cdots & 0 & a_0 \end{bmatrix}, \tag{2.3}$$

$$\mathbf{F}_2 = \begin{bmatrix} 0 & 0 & \cdots & 0 \\ a_{n_a-1} & 0 & \ddots & \vdots \\ a_{n_a-2} & a_{n_a-1} & \ddots & 0 \\ \vdots & \ddots & \ddots & 0 \\ a_1 & \cdots & a_{n_a-2} & a_{n_a-1} \end{bmatrix}. \tag{2.4}$$

Assuming correct past decisions, we have $\mathbf{y}(k) = \mathbf{F}_1 \mathbf{s}_f(k) + \mathbf{F}_2 \hat{\mathbf{s}}_b(k) + \mathbf{e}(k)$. Thus the decision feedback translates the original space $\mathbf{y}(k)$ into a new space:

$$\mathbf{r}(k) \triangleq \mathbf{y}(k) - \mathbf{F}_2 \hat{\mathbf{s}}_b(k)\,. \tag{2.5}$$

Let the $N_f = 2^{d+1}$ sequences of $\mathbf{s}_f(k)$ be $\mathbf{s}_{f,j}$, $1 \le j \le N_f$. The set of the noiseless channel states in the translated space is defined as

$$\mathcal{R} \triangleq \{\mathbf{r}_j = \mathbf{F}_1 \mathbf{s}_{f,j},\ \ 1 \le j \le N_f\}\,, \tag{2.6}$$

which can be partitioned into the two subsets conditioned on $s(k-d)$:

$$\mathcal{R}^{(\pm)} \triangleq \{\mathbf{r}_j \in \mathcal{R} : s(k-d) = \pm 1\}\,. \tag{2.7}$$

We point out that $\mathcal{R}^{(+)}$ and $\mathcal{R}^{(-)}$ are always linearly separable [11]. The optimal equalisation solution, however, is defined by the Bayesian decision function [6, 7]:

$$\begin{aligned} f_B(\mathbf{r}(k)) &= \sum_{\mathbf{r}_j^{(+)} \in \mathcal{R}^{(+)}} \exp\left(-\left\|\mathbf{r}(k) - \mathbf{r}_j^{(+)}\right\|^2 / 2\sigma_e^2\right) \\ &- \sum_{\mathbf{r}_j^{(-)} \in \mathcal{R}^{(-)}} \exp\left(-\left\|\mathbf{r}(k) - \mathbf{r}_j^{(-)}\right\|^2 / 2\sigma_e^2\right), \end{aligned} \tag{2.8}$$

assuming equiprobable states. The decision boundary of this Bayesian DFE

$$\mathcal{D}_B \triangleq \{\mathbf{r} : f_B(\mathbf{r}) = 0\} \tag{2.9}$$

is generally a hypersurface and cannot be realized by one hyperplane. Let us introduce the following definition. A pair of opposite-class states ($\mathbf{r}^{(+)} \in \mathcal{R}^{(+)}, \mathbf{r}^{(-)} \in \mathcal{R}^{(-)}$) is said to be *dominant* if $\forall \mathbf{r}_j \in \mathcal{R}$, $\mathbf{r}_j \neq \mathbf{r}^{(+)}$, $\mathbf{r}_j \neq \mathbf{r}^{(-)}$:

$$\|\mathbf{r}_j - \mathbf{r}_0\|^2 > \|\mathbf{r}^{(+)} - \mathbf{r}_0\|^2\,, \tag{2.10}$$

where $\mathbf{r}_0 = \left(\mathbf{r}^{(+)} + \mathbf{r}^{(-)}\right)/2$. We can now describe the asymptotic Bayesian decision boundary for SNR$\to \infty$ (or $\sigma_e^2 \to 0$).

Proposition 2.1 The asymptotic decision boundary $\mathcal{D}_B$ of the Bayesian DFE for large SNR is piecewise linear and made up of a set of L hyperplanes. Each of these hyperplanes is defined by a pair of *dominant* opposite-class states ($\mathbf{r}_l^{(+)} \in \mathcal{R}^{(+)}, \mathbf{r}_l^{(-)} \in \mathcal{R}^{(-)}$), such that the hyperplane is orthogonal to the line connecting the pair of dominant states and passes through the midpoint of the line.

Proof: See [8]. As $\sigma_e^2 \to 0$, a necessary condition for a point $\mathbf{r} \in \mathcal{D}_B$ is

$$\mathbf{r} = \frac{\mathbf{r}_l^{(+)} + \mathbf{r}_l^{(-)}}{2} + \left[\frac{\mathbf{r}_l^{(+)} - \mathbf{r}_l^{(-)}}{2}\right]^{\perp}, \tag{2.11}$$

where $\mathbf{x}^{\perp}$ denotes an arbitrary vector in the subspace orthogonal to $\mathbf{x}$; and the sufficient conditions for $\mathbf{r} \in \mathcal{D}_B$ are

$$\|\mathbf{r} - \mathbf{r}_l^{(+)}\|^2 < \|\mathbf{r} - \mathbf{r}_i\|^2,\ \forall \mathbf{r}_i \in \mathcal{R}^{(+)},\ \mathbf{r}_i \neq \mathbf{r}_l^{(+)}, \tag{2.12}$$

$$\|\mathbf{r} - \mathbf{r}_l^{(-)}\|^2 < \|\mathbf{r} - \mathbf{r}_j\|^2,\ \forall \mathbf{r}_j \in \mathcal{R}^{(-)},\ \mathbf{r}_j \neq \mathbf{r}_l^{(-)}, \tag{2.13}$$

$$\|\mathbf{r} - \mathbf{r}_l^{(+)}\|^2 = \|\mathbf{r} - \mathbf{r}_l^{(-)}\|^2\,. \tag{2.14}$$

Proposition 2.1 follows as a direct consequence. The set of all the dominant state pairs $\{\mathbf{r}_l^{(+)}, \mathbf{r}_l^{(-)}\}_{l=1}^{L}$ can easily be determined using a simple algorithm based on the conditions (2.11)–(2.14) [8, 9].

3 IS simulation method

An excellent introduction to the IS method can be found in [14]. Since the Bayesian DFE is reduced to the Bayesian equalizer in the translated space, the IS simulation technique of [8] can be extended to evaluate its lower-bound BER under the condition of correct bits being fed back, which is given by

$$\hat{P}_e = \frac{1}{N_s}\frac{1}{N_k}\sum_{i=1}^{N_s}\sum_{k=1}^{N_k} I_E(\mathbf{r}_i(k))\frac{p(\mathbf{r}_i(k)|\mathbf{r}_i)}{p^*(\mathbf{r}_i(k)|\mathbf{r}_i)}, \tag{3.1}$$

where the indicator function $I_E(\mathbf{r}(k)) = 1$ if $\mathbf{r}(k)$ causes an error, and $I_E(\mathbf{r}(k)) = 0$ otherwise; $p(\mathbf{r}_i(k)|\mathbf{r}_i)$ is the true conditional density given $\mathbf{r}_i \in \mathcal{R}^{(+)}$, and $N_s = 2^d$ is the number of states in $\mathcal{R}^{(+)}$; the sample $\mathbf{r}_i(k)$ is generated using the simulation density $p^*(\mathbf{r}_i(k)|\mathbf{r}_i)$ chosen to be

$$p^*(\mathbf{r}_i(k)|\mathbf{r}_i) = \sum_{j=1}^{L_i} p_{j,i}\frac{1}{(2\pi\sigma_e^2)^{\frac{m}{2}}}\exp\left(-\frac{\|\mathbf{r}_i(k) - \mathbf{v}_{j,i}\|^2}{2\sigma_e^2}\right). \tag{3.2}$$

In (3.2), L_i is the number of the bias vectors $\mathbf{c}_{j,i} = -\mathbf{r}_i + \mathbf{v}_{j,i}$ for $\mathbf{r}_i \in \mathcal{R}^{(+)}$, $p_{j,i} \geq 0$ for $1 \leq j \leq L_i$, and $\sum_{j=1}^{L_i} p_{j,i} = 1$. An estimate of the IS gain, which is defined as the ratio of the numbers of trials required for the same estimate variance using the Monte Carlo and IS methods, is given in [8]. To achieve asymptotic efficiency, $\{\mathbf{c}_{j,i}\}$ must meet certain conditions [10]. We present the following procedure of constructing $p^*(\mathbf{r}_i(k)|\mathbf{r}_i)$ to meet these conditions.

Each of the L dominant state pairs $\{\mathbf{r}_l^{(+)}, \mathbf{r}_l^{(-)}\}$ defines a hyperplane $H_l(\mathbf{r}) = \mathbf{w}_l^T\mathbf{r} + b_l = 0$. The weight vector $\mathbf{w}_l$ and bias b_l of the hyperplane are given by

$$\mathbf{w}_l = \frac{2\left(\mathbf{r}_l^{(+)} - \mathbf{r}_l^{(-)}\right)}{\|\mathbf{r}_l^{(+)} - \mathbf{r}_l^{(-)}\|^2}, \quad b_l = -\frac{(\mathbf{r}_l^{(+)} - \mathbf{r}_l^{(-)})^T(\mathbf{r}_l^{(+)} + \mathbf{r}_l^{(-)})}{\|\mathbf{r}_l^{(+)} - \mathbf{r}_l^{(-)}\|^2}. \tag{3.3}$$

Note that the theory of support vector machines [15, 16] has been applied to determine H_l with $(\mathbf{r}_l^{(+)}, \mathbf{r}_l^{(-)})$ as its two support vectors, and H_l is a *canonical* hyperplane having the property $H_l(\mathbf{r}_l^{(+)}) = 1$ and $H_l(\mathbf{r}_l^{(-)}) = -1$.

A state $\mathbf{r}_i \in \mathcal{R}$ is said to be *sufficiently separable* by the hyperplane H_l, if H_l can separate $\mathbf{r}_i$ correctly with $|\mathbf{w}_l^T\mathbf{r}_i + b_l| \geq 1$. Thus, if $\mathbf{w}_l^T\mathbf{r}_i^{(+)} + b_l \geq 1$ for $\mathbf{r}_i^{(+)} \in \mathcal{R}^{(+)}$, $\mathbf{r}_i^{(+)}$ is sufficiently separable by H_l and a separability index $h_{l,i}^{(+)}$ is set to 1; otherwise $h_{l,i}^{(+)} = 0$. Similarly, if $\mathbf{r}_i^{(-)} \in \mathcal{R}^{(-)}$ satisfies $\mathbf{w}_l^T\mathbf{r}_i^{(-)} + b_l \leq -1$, it is sufficiently separable by H_l and $h_{l,i}^{(-)} = 1$; otherwise $h_{l,i}^{(-)} = 0$. The *reachability* of H_l from $\mathbf{r}_i^{(+)} \in \mathcal{R}^{(+)}$ can be tested by computing

$$\mathbf{c}_{l,i} = -0.5\left(\mathbf{w}_l^T\mathbf{r}_i^{(+)} + b_l\right)\left(\mathbf{r}_l^{(+)} - \mathbf{r}_l^{(-)}\right). \tag{3.4}$$

Table 1 The separability and reachability table for the CIR of $\mathbf{a} = [-0.8\ 1.0\ -0.5]^T$. The DFE structure is defined by $m = 3$, $d = 2$ and $n = 2$

	$\mathbf{r}_1^{(-)}$	$\mathbf{r}_2^{(-)}$	$\mathbf{r}_3^{(-)}$	$\mathbf{r}_4^{(-)}$	$\mathbf{r}_1^{(+)}$	$\mathbf{r}_2^{(+)}$	$\mathbf{r}_3^{(+)}$	$\mathbf{r}_4^{(+)}$
H_1	1	1	0	1	0	0	1 (1)	0
H_2	1	0	1	1	1 (1)	1 (1)	0	1 (1)
H_3	1	1	1	1	0	1 (1)	0	0
H_4	0	1	0	0	1 (1)	0	1 (0)	1 (1)
H_5	0	0	1	0	1 (1)	1 (1)	1 (1)	1 (1)

If $\mathbf{v}_{l,i} = \mathbf{r}_i^{(+)} + \mathbf{c}_{l,i} \in \mathcal{D}_B$, H_l is said to be reachable from $\mathbf{r}_i^{(+)}$ ($\mathbf{c}_{l,i}$ is then a bias vector), and the reachability index is $\gamma_{l,i} = 1$; otherwise $\gamma_{l,i} = 0$. The process produces the following separability and reachability table:

	$\mathbf{r}_1^{(-)}$	$\cdots$	$\mathbf{r}_{N_s}^{(-)}$	$\mathbf{r}_1^{(+)}$	$\cdots$	$\mathbf{r}_{N_s}^{(+)}$
H_1	$h_{1,1}^{(-)}$	$\cdots$	$h_{1,N_s}^{(-)}$	$h_{1,1}^{(+)}$ $(\gamma_{1,1})$	$\cdots$	$h_{1,N_s}^{(+)}$ (γ_{1,N_s})
$\vdots$	$\vdots$	$\cdots$	$\vdots$	$\vdots$	$\cdots$	$\vdots$
H_L	$h_{L,1}^{(-)}$	$\cdots$	$h_{L,N_s}^{(-)}$	$h_{L,1}^{(+)}$ $(\gamma_{L,1})$	$\cdots$	$h_{L,N_s}^{(+)}$ (γ_{L,N_s})

In order to construct a convex region $\mathbf{R}_i^{(+)}$ for $\mathbf{r}_i^{(+)} \in \mathcal{R}^{(+)}$, we select those hyperplanes that can *sufficiently* separate $\mathbf{r}_i^{(+)}$ and that are *reachable* from $\mathbf{r}_i^{(+)}$ with the aid of the above table. This yields the following integer set:

$$G_i^{(+)} \triangleq \{j : h_{j,i}^{(+)} = 1 \text{ and } \gamma_{j,i} = 1\}. \tag{3.5}$$

Then $\mathbf{R}_i^{(+)}$ is the intersection of all the half-spaces $\mathcal{H}_j^{(+)} \triangleq \{\mathbf{r} : H_j(\mathbf{r}) \geq 0\}$ with $j \in G_i^{(+)}$. In fact, it is not necessary to use every hyperplanes defined in $G_i^{(+)}$ to construct $\mathbf{R}_i^{(+)}$. A subset of these hyperplanes will be sufficient, provided that every opposite-class state in $\mathcal{R}^{(-)}$ can sufficiently be separated by at least one hyperplane in the subset. If such a $G_i^{(+)}$ exists for each $\mathbf{r}_i^{(+)}$, the simulation density constructed with the bias vectors $\{\mathbf{c}_{j,i}\}$, $j \in G_i^{(+)}$, will achieve asymptotic efficiency, since all the hyperplane defined in $G_i^{(+)}$ are reachable from $\mathbf{r}_i^{(+)}$ and obviously at least one of $\{\mathbf{v}_{j,i}\}$ is the minimum rate point (as defined in [10]), and the error region $\mathcal{E}$ satisfies

$$\mathcal{E} \subset \overline{\mathbf{R}_i^{(+)}} \triangleq \bigcup_{j \in G_i^{(+)}} \mathcal{H}_j^{(-)} \tag{3.6}$$

with the half-spaces $\mathcal{H}_j^{(-)} \triangleq \{\mathbf{r} : H_j(\mathbf{r}) < 0\}$.

An example. The IS technique for the Bayesian DFE was simulated using the 3-tap CIR defined by $\mathbf{a} = [-0.8\ 1.0\ \ -0.5]^T$. The bias vectors were generated using the procedure described above. As in [8], the bias vectors were selected

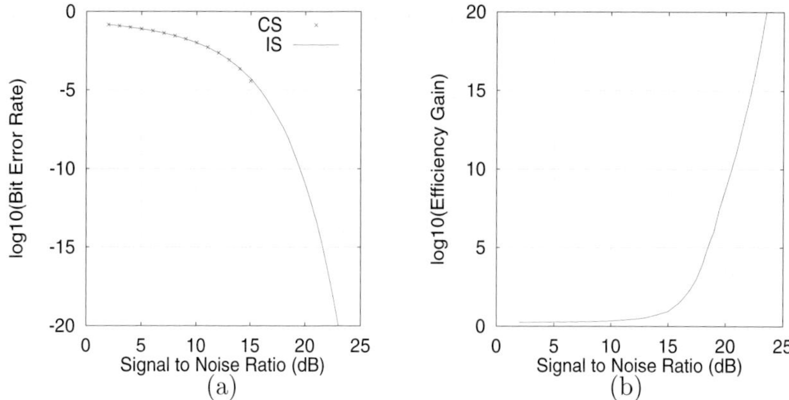

FIG. 1. The lower-bound BERs (a) and the IS gain (b) of the Bayesian DFE for the CIR of $\mathbf{a} = [-0.8\ 1.0\ -0.5]^T$ using conventional sampling (CS) and importance sampling (IS) simulation. The DFE structure is defined by $m = 3$, $d = 2$ and $n = 2$.

with uniform probability in the simulation. For all the cases, 10^5 iterations were employed at each SNR, averaging over all the possible states in $\mathcal{R}^{(+)}$. Since the channel had a length of $n_a = 3$, the DFE structure was specified by $m = 3$, $d = 2$ and $n = 2$. The asymptotic decision boundary consisted of five hyperplanes. Table 1 gives the separability and reachability table for this channel.

The states $\mathbf{r}_1^{(+)}$ and $\mathbf{r}_4^{(+)}$ require the two hyperplanes H_2 and H_4 to separate them from all the opposite-class states, and H_2 and H_4 are reachable from the both states. Thus, there are two bias vectors for $\mathbf{r}_1^{(+)}$ and $\mathbf{r}_4^{(+)}$, respectively, and $\mathcal{E} \subset \mathcal{H}_2^{(-)} \bigcup \mathcal{H}_4^{(-)}$. The state $\mathbf{r}_2^{(+)}$ is separated from $\mathcal{R}^{(-)}$ by the single reachable hyperplane H_3. Thus, there exists one bias vector for $\mathbf{r}_2^{(+)}$ and $\mathcal{E} \subset \mathcal{H}_3^{(-)}$. The state $\mathbf{r}_3^{(+)}$ is separated from $\mathcal{R}^{(-)}$ by the two reachable hyperplanes H_1 and H_5; there are two bias vectors for $\mathbf{r}_3^{(+)}$ and $\mathcal{E} \subset \mathcal{H}_1^{(-)} \bigcup \mathcal{H}_5^{(-)}$. Asymptotic efficiency of the IS simulation is therefore guaranteed for this example.

Figure 1(a) shows the lower-bound BERs obtained using the IS and conventional sampling (CS) simulation methods, respectively. It can be seen that the conventional Monte Carlo results for low SNR conditions agreed with those of the IS simulation. The estimated IS gains, depicted in Fig. 1(b), indicate that exponential IS gains were obtained with increasing SNRs. For SNR = 20 dB, the BER of the Bayesian DFE with correct bits being fed back calculated by the IS technique is 1.2×10^{-11}. The CS method could not work under the same SNR condition and, to achieve the same BER estimation accuracy, it would require approximately 4.8×10^8 times of the samples needed by the IS simulation method. As the IS method used 4×10^5 data samples, the CS method would require approximately 2×10^{14} samples to achieve a similar estimation variance.

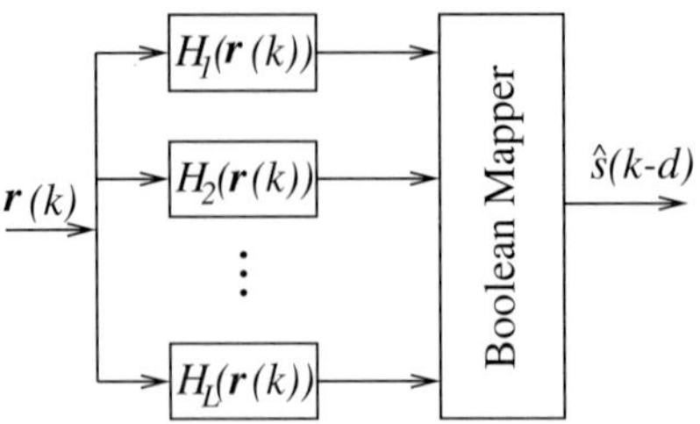

FIG. 2. Multiple-hyperplane detector for realizing the asymptotic Bayesian DFE.

4 Multiple-hyperplane detector

Since the set of the L hyperplanes that form the asymptotic Bayesian decision boundary can easily be obtained, they can be used for partitioning the observation space to form a multiple-hyperplane detector which has a structure as depicted in Fig. 2. To construct such a detector, there is no need to test whether a hyperplane H_l is reachable from each state in $\mathcal{R}^{(+)}$ and only a separability table is required. To construct a convex region $\mathbf{R}_i^{(+)}$ covering $\mathbf{r}_i^{(+)} \in \mathcal{R}^{(+)}$, select hyperplanes which can sufficiently separate $\mathbf{r}_i^{(+)}$ from the separability table and denote

$$\tilde{G}_i^{(+)} \triangleq \{l : \ h_{l,i}^{(+)} = 1\} . \tag{4.1}$$

Then $\mathbf{R}_i^{(+)}$ is obtained by the intersection of all the $\mathcal{H}_j^{(+)}$ with $j \in \tilde{G}_i^{(+)}$

$$\mathbf{R}_i^{(+)} = \bigcap_{j \in \tilde{G}_i^{(+)}} \mathcal{H}_j^{(+)} . \tag{4.2}$$

Again, a subset of the hyperplanes defined by (4.1) is enough in the construction of $\mathbf{R}_i^{(+)}$, provided that every state in $\mathcal{R}^{(-)}$ can sufficiently be separated by at least one hyperplane in the subset. The overall decision region $\mathbf{R}^{(+)}$ associated with the decision $\hat{s}(k-d) = 1$ is simply formed as the union of all the $\mathbf{R}_i^{(+)}$

$$\mathbf{R}^{(+)} = \bigcup_{i=1}^{N_s} \mathbf{R}_i^{(+)} . \tag{4.3}$$

The resulting multiple-hyperplane detector is now completely defined. Let a threshold detector output $\beta_j(\mathbf{r}(k))$ for a linear discriminant function $H_j(\mathbf{r}(k))$ have Boolean logic value 1 or 0 depending on $\mathbf{r}(k) \in \mathcal{H}_j^{(+)}$ or not. A Boolean logic value $\theta_i^{(+)}(\mathbf{r}(k))$ indicating whether $\mathbf{r}(k) \in \mathbf{R}_i^{(+)}$ or not is obtained via a logic AND operation of $\{\beta_j(\mathbf{r}(k)) : j \in \tilde{G}_i^{(+)}\}$. A Boolean logic value indicating whether $\mathbf{r}(k) \in \mathbf{R}^{(+)}$ (that is, $\hat{s}(k-d) = 1$) or not is obtained via a logic OR operation of $\{\theta_i^{(+)}(\mathbf{r}(k))\}$ for all i. This detector achieves asymptotically the

Table 2 Comparison of decision complexity for the full Bayesian and multiple-hyperplane detectors. L (usually $\ll 2^{n_a}$) is the number of hyperplanes, and n_a is the CIR length. The DFE structure is chosen to be $m = n_a$, $d = n_a - 1$ and $n = n_a - 1$

	Bayesian DFE	Multiple-hyperplane detector
Multiplications	$(n_a + 1) \times 2^{n_a}$	$n_a \times L$
Additions	$n_a \times 2^{n_a+1} - 1$	$n_a \times L$
Others	2^{n_a} exp$(\cdot)$	logic ANDs $\leq 2^{n_a-1}$
	function evaluations	a logic OR

Table 3 The separability table for the CIR of $\mathbf{a} = [0.4\ 0.7\ 0.4]^T$. The DFE structure is defined by $m = 3$, $d = 2$ and $n = 2$

	$\mathbf{r}_1^{(-)}$	$\mathbf{r}_2^{(-)}$	$\mathbf{r}_3^{(-)}$	$\mathbf{r}_4^{(-)}$	$\mathbf{r}_1^{(+)}$	$\mathbf{r}_2^{(+)}$	$\mathbf{r}_3^{(+)}$	$\mathbf{r}_4^{(+)}$
H_1	1	0	0	0	1	1	1	1
H_2	0	1	1	1	1	0	0	0
H_3	1	1	1	0	0	1	1	1
H_4	0	0	0	1	1	1	1	0
H_5	1	1	1	1	0	0	0	1

optimal Bayesian performance since it realizes exactly the asymptotic Bayesian decision boundary. Table 2 compares decision complexity for the full Bayesian DFE and the multiple-hyperplane detector. The multiple-hyperplane detector generally has much simpler decision complexity than the full Bayesian detector, since usually $L \ll N_f$.

An example. The CIR was given by $\mathbf{a} = [0.4\ 0.7\ 0.4]^T$. The structure parameters of the DFE were set to $m = 3$, $d = 2$ and $n = 2$. The asymptotic decision boundary consisted of five hyperplanes. Table 3 gives the separability table for this channel. The state $\mathbf{r}_1^{(+)}$ requires the two hyperplanes H_1 and H_2 to be separated from all the opposite-class states $\mathcal{R}^{(-)}$ and, therefore, the convex region $\mathbf{R}_1^{(+)}$ for $\mathbf{r}_1^{(+)}$ is the intersection of the two half-spaces $\mathcal{H}_1^{(+)}$ and $\mathcal{H}_2^{(+)}$. The states $\mathbf{r}_2^{(+)}$ and $\mathbf{r}_3^{(+)}$ are separated from $\mathcal{R}^{(-)}$ by the two hyperplanes H_3 and H_4. Thus $\mathbf{R}_2^{(+)} = \mathbf{R}_3^{(+)}$ is the intersection of the half-spaces $\mathcal{H}_3^{(+)}$ and $\mathcal{H}_4^{(+)}$. The state $\mathbf{r}_4^{(+)}$ is separated by the single hyperplane H_5 from all the opposite-class states, and the convex region $\mathbf{R}_4^{(+)}$ for $\mathbf{r}_4^{(+)}$ is the half-space $\mathcal{H}_5^{(+)}$ defined by H_5. The overall decision region $\mathbf{R}^{(+)}$ is the union of $\mathbf{R}_1^{(+)}$, $\mathbf{R}_2^{(+)}$ and $\mathbf{R}_4^{(+)}$.

The resulting five-hyperplane detector requires 15 multiplications and 15 additions to detect a symbol, compared with 32 multiplications, 47 additions, and eight exp$(\cdot)$ evaluations required by the full Bayesian DFE. The BERs of this multiple-hyperplane detector are compared with those of the full Bayesian DFE in Fig. 3, under different SNR conditions. The BER results were obtained with detected symbols being fed back. It can be seen from Fig. 3 that there

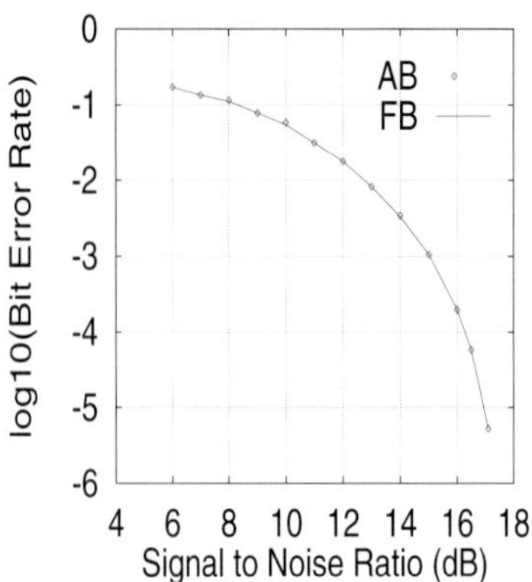

FIG. 3. Performance comparison of the multiple-hyperplane detector (AB: points) and the full Bayesian DFE (FB: solid curve) with detected symbols being fed back for the CIR of $\mathbf{a} = [0.4\ 0.7\ 0.4]^T$. The DFE structure is defined by $m = 3$, $d = 2$ and $n = 2$.

exists hardly any BER performance difference between the two equalisers for this channel.

5 Conclusions

An asymptotic property of the optimal Bayesian decision boundary has been utilized for efficient IS simulation and implementation of the Bayesian DFE. In the first application, we have extended the randomized bias technique for IS simulation of [8] to evaluate the lower-bound BER of the Bayesian DFE. A design procedure has been presented for constructing the simulation density that meets the asymptotic efficiency conditions. Although asymptotic efficiency of the IS simulation for the general channel has not rigorously been proven, we are unable to find a counter-example suggesting that the asymptotic efficiency conditions are not met. The more difficult problem of how to derive an upper-bound BER of the Bayesian DFE, taking into account error propagation, remains an open question and is still under investigation. In the second application, the set of hyperplanes that form the asymptotic Bayesian decision boundary is used to partition the observation space. The resulting multiple-hyperplane detector is guaranteed to achieve asymptotically the optimal Bayesian performance and has a much lower decision complexity than the full Bayesian DFE.

Bibliography

[1] Qureshi, S.U.H. (1985). Adaptive equalization. *Proc. IEEE*, **73**, 1349–1387.

[2] Proakis, J.G. (1995). *Digital Communications*, 3rd edition, McGraw-Hill, New York.

[3] Moon, J. (1998). The role of SP in data-storage systems. *IEEE Signal Processing Magazine*, **15**, 54–72.

[4] Proakis, J.G. (1998). Equalization techniques for high-density magnetic recording. *IEEE Signal Processing Magazine*, **15**, 73–82.

[5] Abend, K. and Fritchman, B.D. (1970). Statistic detection for communication channels with intersymbol interference. *Proc. IEEE*, **58**, 779–785.

[6] Chen, S., Mulgrew, B. and McLaughlin, S. (1993). Adaptive Bayesian equaliser with decision feedback. *IEEE Trans. Signal Processing*, **41**, 2918–2927.

[7] Chen, S., McLaughlin, S., Mulgrew, B. and Grant, P.M. (1995). Adaptive Bayesian decision feedback equaliser for dispersive mobile radio channels. *IEEE Trans. Communications*, **43**, 1937–1946.

[8] Iltis, R.A. (1995). A randomized bias technique for the importance sampling simulation of Bayesian equalizers. *IEEE Trans. Communications*, **43**, 1107–1115.

[9] Chng, E.S., Mulgrew, B., Chen, S. and Gibson, G. (1995). Optimum lag and subset selection for radial basis function equaliser. *Proc. 5th IEEE Workshop Neural Networks for Signal Processing* (Cambridge, MA, USA, Aug. 31–Sept. 2), pp. 593–602.

[10] Sadowsky, J.S. and Bucklew, J.A. (1990). On large deviations theory and asymptotically efficient Monte Carlo estimation. *IEEE Trans. Information Theory*, **36**, 579–588.

[11] Chen, S., Mulgrew, B., Chng, E.S. and Gibson, G. (1998). Space translation properties and the minimum-BER linear-combiner DFE. *IEE Proc. Communications*, **145**, 316–322.

[12] Kim, Y. and Moon, J. (1998). Delay-constrained asymptotically optimal detection using signal-space partitioning. *Proc. IEEE Int. Conf. Communications* (Atlanta, USA).

[13] Kim, Y. and Moon, J. (2000). Multi-dimensional signal space partitioning using a minimal set of hyperplanes for detecting ISI-corrupted symbols. *IEEE Trans. Communications*, to appear.

[14] Smith, P.J., Shafi, M. and Gao, H. (1997). Quick simulation: a review of importance sampling techniques in communications systems. *IEEE J. Selected Areas in Communications*, **15**, 597–613.

[15] Vapnik, V. (1995). *The Nature of Statistical Learning Theory*, Springer, New York.

[16] Chen, S., Gunn, S. and Harris, C.J. (2000). Decision feedback equalizer design using support vector machines. *IEE Proc. Vision, Image and Signal Processing*, **147**, 213–219.

Accumulated Evidence and Dimensionality Reduction

Jason F. Ralph

Department of Electrical Engineering and Electronics, The University of Liverpool, Brownlow Hill, Liverpool, L69 3GJ, UK

Abstract

In this paper we construct upper and lower bounds to the separability of two distributions using the Dempster–Shafer theory of evidential reasoning. The underlying structures of the distributions are assumed to be unknown. The bounds are constructed directly from a set of example data (a body of evidence). We examine the behaviour of these bounds when the dimensionality of the data is reduced using standard feature selection/extraction techniques.

1 Introduction

Dealing with high-dimensional data is an ever-present problem for modern image and signal processing systems. Rapid advances in sensor technology have provided high-resolution, multi-spectral imagers and high-bandwidth antenna arrays for radar and communication systems. Each has the capability to produce large amounts of data with high intrinsic dimensionality (that is, large numbers of 'features' for each data vector). Because of this, the detection of specific signals or patterns in this data can be a very complicated problem. Often the signal to be detected is not known *a priori*, either due to limited information regarding the signal and its source or because of complex interactions between the signal and its environment. In the absence of a prior distribution, an estimate of the signal distribution needs to be constructed (implicitly or explicitly) from available examples. This is the essence of statistical pattern recognition. However, in large dimensional spaces, a very large amount of data is usually required to construct an accurate estimate of the underlying distribution. As a result, and taking into consideration the computational overheads associated with manipulating high-dimensional data, many signal/image processing algorithms seek to project the data down into a lower dimensional space. Such dimensionality reduction schemes use statistical properties of the example data to determine an optimal projection, given the data. This means that the conventional statistical methods for dimensionality reduction are often reliant on sparse data, which might introduce undue bias into later processing stages.

In this paper, we examine an alternative to the standard (Bayesian) approach. Our approach is based on the Dempster–Shafer theory of evidential reasoning [2, 5, 7, 10], which is (in turn) related to the Theory of Random Sets [3]. We use the formalism of Dempster–Shafer theory rather than that of Random Sets for reasons of familiarity although they are interchangeable. In contrast to standard probability theory, evidential reasoning allocates probability only when data are available. The probability that is not allocated represents the uncertainty in the system. We use this uncertainty to derive upper and lower bounds for the underlying separability of distributions under different dimensionality reduction schemes. We assess the advantages and disadvantages of this evidential approach for several standard feature selection techniques [8, Chapter 8] applied to two example data sets: one relatively artificial example (two multivariate Gaussian distributions) and one more realistic example (two sets of simulated hyperspectral infrared data).

2 Evidential reasoning

At the heart of the theory of evidential reasoning is a function $m(A)$, which is called the *basic probability assignment* (or BPA) [2, 5, 7, 10]. This function represents the extent to which the available data, or evidence, support the hypothesis A. The main difference between this and conventional probability theory is that no probability mass is assigned to a hypothesis for which no supporting data are available. The probability that is unassigned to particular hypotheses is assigned to the whole space of hypotheses, and represents the amount of uncertainty in the system. Due to the nature of evidential reasoning, the hypothesis space for the system must be finite. Each hypothesis must correspond to one or more finite intervals of the system variables. For this we require a model for the measurement process with which the data are being collected. Typically this might include the effect of systematic measurement biases, measurement noise, and the quantization of the data to a finite set of bits (for digital representation and/or processing). For simplicity, we assume that the hypotheses correspond to finite non-overlapping intervals, whose size is determined by the number of bits used to represent the data set. That is, we assume that the dominant effect of the measurement process is the quantization of the data to a fixed number of levels. Each data point in the set is identified with a specific interval. Therefore, points drawn from the data sets will support only one of the hypotheses. With more complicated measurement models, it is possible for a piece of evidence to support more than one hypothesis, either using a set of prior probabilities to represent the measurement noise or a more general type of non-statistical measurement uncertainty [9]. However, it should be noted that using more complicated hypothesis spaces often increases dramatically the computational demands of Dempster–Shafer theory [4].

For a piece of evidence (data point) falling in an interval A, we assign a probability mass,

$$m(A) = \delta. \tag{2.1}$$

To obtain the basic probability assignment for a set of data points, we use Dempster's rule for the combination of evidence [2, 5, 7, 10]. A number of other combination rules have been suggested which aim to compensate for prior distributions [11] and the slightly conservative results obtained using Dempster's rule [12]. We choose Dempster's rule for simplicity, and because the modification of the algorithms to allow for a different combination rule is straightforward and a relatively trivial operation. For the combination of two pieces of evidence supporting the hypothesis A, as characterized by $m_1(A)$ and $m_2(A)$, Dempster's rule gives

$$m_1 \oplus m_2(A) = \frac{\sum\limits_{B \cap C = A} m_1(B) m_2(C)}{1 - \sum\limits_{B \cap C = \phi} m_1(B) m_2(C)}. \tag{2.2}$$

For our simple hypothesis space, this simplifies to expressions of the form

$$m_1 \oplus m_2(A) = \frac{m_1(A)m_2(A) + m_1(A)m_2(\Theta) + m_1(\Theta)m_2(A)}{1 - \sum\limits_{B \subseteq \bar{A}} (m_1(B)m_2(A) + m_1(A)m_2(B)} \tag{2.3}$$

where $\Theta = \bigcup A$ is the set of all intervals and their crisp complements (Θ is often called the *frame of discernment*). All of the probability mass that is not assigned to an interval or its complement is assigned to Θ, and is represented by $m(\Theta)$. Once the probability mass has been assigned based on the evidence, it is then possible to calculate two other functions: the belief function and the plausibility function. The belief function is a measure of how well the available evidence supports the hypothesis, and (for our simple hypothesis space) is equal to the corresponding BPA,

$$\text{Bel}(A) = m(A). \tag{2.4}$$

It is a superadditive measure (that is, $\text{Bel}(A \cup B) \geq \text{Bel}(A) + \text{Bel}(B)$ for any sets A and B), such that $\sum_A \text{Bel}(A) \leq 1$. The plausibility function, on the other hand, measures the extent to which the hypothesis is not contradicted by the available evidence,

$$\text{Pls}(A) = 1 - \text{Bel}(\bar{A}). \tag{2.5}$$

Plausibility is a subadditive measure (that is, $\text{Pls}(A \cup B) \leq \text{Pls}(A) + \text{Pls}(B)$ for any sets A and B), such that $\sum_A \text{Pls}(A) \geq 1$. The belief and plausibility play the role of an upper and lower probability, and should bound the Bayesian probability from above and below. The difference between the belief and plausibility, the unassigned probability mass for finite data sets, is an indication of the underlying uncertainty that remains when the available data are taken into account. (However, as data are added, care should be taken to ensure that $\delta \lesssim 1/N$, where N is the number of data points available, or the probability mass can begin to accumulate in one or other of the hypotheses. In this circumstance the belief and plausibility functions do not necessarily bound the true probability. However, in practice, it is often necessary to have $\delta > 1/N$ to obtain any non-trivial bounds).

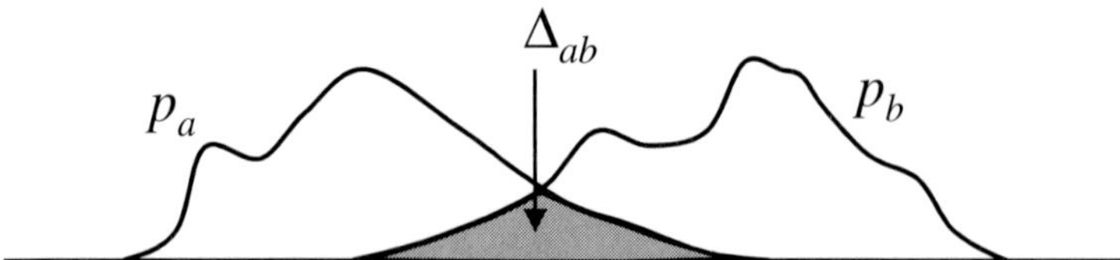

FIG. 1. Schematic diagram indicating the area of overlap Δ_{ab} in one dimension for the two arbitrary probability distributions p_a and p_b.

3 Separability

Once upper and lower bounds for the enclosed probability have been found, we can begin to construct Dempster–Shafer upper and lower bounds for the separability of two data sets. There are many different ways of measuring the separability of two probability distributions. Common methods often involve the calculation of, or more likely the estimation of, an error rate: that is, the proportion of data points that are misclassified by a pattern recognition algorithm, which can be weighted by some prior probability for the different classes or by an associated cost function [8, pp. 321–328]. Since these measures often involve the use of a specific pattern recognition technique, and are dependent on the position of the decision boundary (or boundaries), we use a simpler measure of the separability: the overlap between the two probability distributions. That is, the lower of the two probability values in each interval, summed over all intervals. (In one dimension, this becomes the area under both distributions). This has two advantages: it generates a single figure that can be used to measure the separability (zero for completely separable distributions and one for identical distributions), and it is not reliant on a specific pattern recognition technique, or particular decision boundaries. In one dimension, this overlap corresponds to the shaded region shown in Fig. 1 that falls below both of the probability distributions. Mathematically, the overlap between two (discrete) probability distributions p_a and p_b can be described by

$$\Delta_{ab} = \tfrac{1}{2} \sum_{A} [p_a(A) + p_b(A) - |p_a(A) - p_b(A)|] \tag{3.1}$$

where the sum is taken over all intervals A. The lower bound for this overlap is then defined as,

$$\Delta_{ab}^{(l)} = \tfrac{1}{2} \sum_{A} [\mathrm{Bel}_a(A) + \mathrm{Bel}_b(A) - |\mathrm{Bel}_a(A) - \mathrm{Bel}_b(A)|] \tag{3.2}$$

where $\mathrm{Bel}_a(A)$ is the belief associated with the interval A by the data set a. This represents the minimum amount of probability mass that has been assigned to

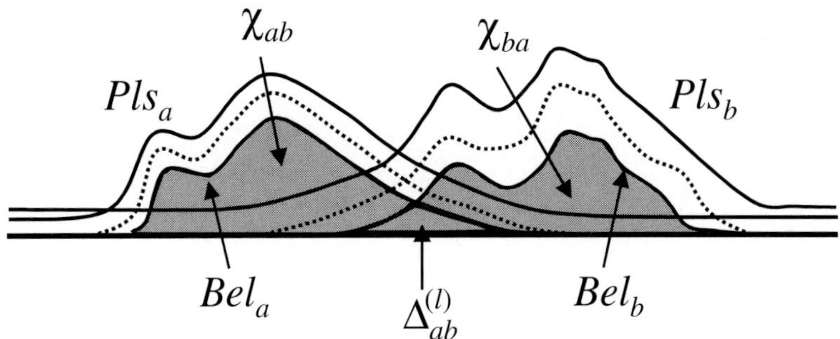

FIG. 2. Schematic diagram indicating the belief functions and plausibility functions in one dimension, the area of overlap $\Delta_{ab}^{(l)}$, and the non-overlapping areas χ_{ab} and χ_{ba}.

the overlap by the example data sets (bodies of evidence), and is indicated in Fig. 2 by the small shaded region below both of the belief curves. By contrast, the upper bound to the overlap is determined by the extent to which the belief distribution for one distribution exceeds the plausibility distribution for the other, which represents the minimum amount of probability mass that is definitely not assigned to the overlap. Hence,

$$\Delta_{ab}^{(u)} = 1 - \tfrac{1}{4}\left[(\chi_{ab} - m_b(\Theta)) + |\chi_{ab} - m_b(\Theta)| + (\chi_{ba} - m_a(\Theta)) + |\chi_{ba} - m_a(\Theta)|\right] \tag{3.3}$$

where $\chi_{ab} = \frac{1}{2}\sum_A \left[\mathrm{Bel}_a(A) - \mathrm{Bel}_b(A) + |\mathrm{Bel}_a(A) - \mathrm{Bel}_b(A)|\right]$. The areas χ_{ab} and χ_{ba} are also indicated in Fig. 2, representing the amount of probability mass that has been allocated but not assigned to the overlap. The only probability mass that cannot be assigned to the overlap is contained in these non-overlapping areas and the only probability mass that can still be added to the overlap is the probability mass that has not yet been assigned to any interval, $m_a(\Theta)$ and $m_b(\Theta)$. $\Delta_{ab}^{(u)}$ is therefore equal to the maximum possible overlap (i.e. one) minus the probability mass that is definitely not assigned to the overlap by each data set, divided by two to avoid double counting the overlap.

4 Examples

In this paper, we calculate the upper and lower bounds for two examples: the overlap between two sets of data drawn from different multivariate Gaussian distributions, and the overlap between two data sets drawn from simulated hyperspectral infrared 'images'. For the multivariate Gaussian, we fixed the covariance matrix and mean vector for each distribution and generate a large set of data vectors. For the simulated hyperspectral data, we generate a series of images in adjacent spectral bands, using a simulated blackbody source overlaid with

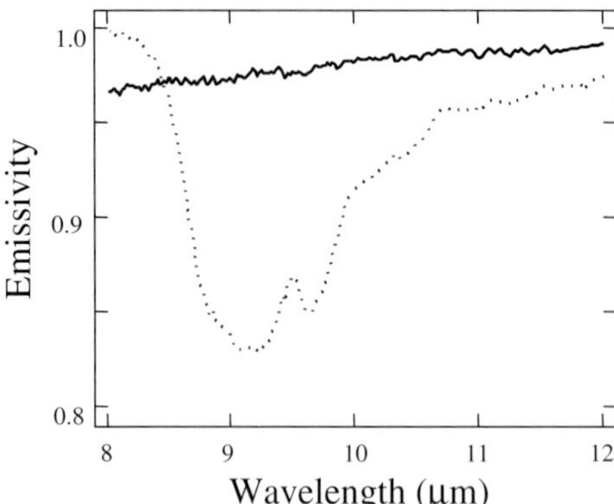

FIG. 3. Graph of thermal emissivity versus wavelength for hematite (solid line) and talc (dotted line) [1].

a spectral signature of the chosen material [1] (see Fig. 3) and a simulation of an infrared detector/imager. The spectral profile generated for individual pixels forms a set of data vectors. Using different spectral signatures for two different materials, we can generate two data sets.

The raw data vectors from each set are ten-dimensional. This can be reduced by the application of different feature selection or feature extraction algorithms. For the purposes of this paper, we use two standard processing techniques: one feature extraction technique (common principal components analysis [8, Chapter 8]) and one feature selection technique (Fisher's criterion [6]). Many more feature reduction techniques exist but, for the sake of brevity, we choose to concentrate on two. For the particular Gaussian data used in this paper, the two processing techniques produce very similar results. In the case of the simulated hyperspectral results the results of the two processing techniques are significantly different, allowing the better algorithm to be selected (that is, the one that preserves the separability of the original data), and we will also show results for a third option, which is to reduce the spectral resolution of the imager itself. We do this by averaging over larger spectral bands. After the redundant features are removed, the data vectors are quantized to a finite number of bits, each dimension being taken separately. This represents the measurement process (as discussed above). Whilst applying quantization after the dimensionality has been reduced may not be very realistic, it is necessary to ensure that the evidential hypothesis space is simple. Quantising the data vectors before applying the feature extraction procedure would not necessarily give non-overlapping intervals in the reduced space.

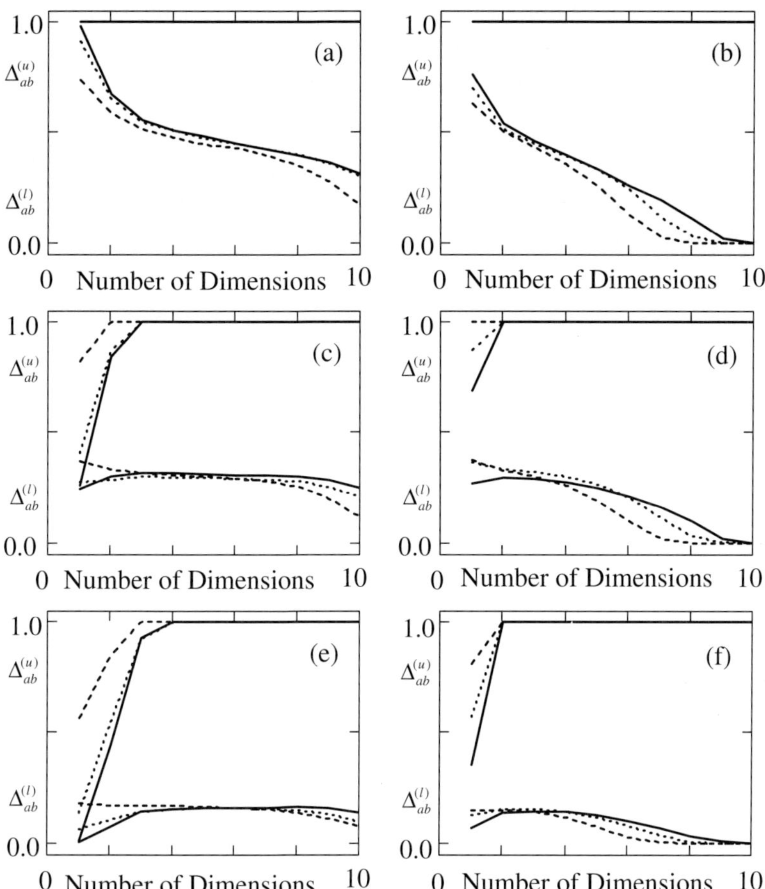

FIG. 4. Upper and lower limits to separability for Gaussian data as a function of dimension using the Fisher Ratio, $\delta = 1 \times 10^{-3}$, $N = 8000$ (solid line), $N = 4000$ (dotted line), $N = 2000$ (dashed line): (a) identical Gaussians, 2-bits, (b) identical Gaussians, 3-bits, (c) Gaussians separated by one standard deviation, 2-bits, (d) Gaussians separated by one standard deviation, 3-bits, (e) Gaussians separated by two standard deviations, 2-bits, (f) Gaussians separated by two standard deviations, 3-bits.

5 Results

Figure 4 shows the bounds generated for the first example, a multi-dimensional Gaussian distribution. For the sake of simplicity we take each dimension to represent an independent Gaussian distribution (so that the covariance matrix is diagonal). This simplifies matters because the two dimensionality reduction techniques (Fisher's criterion and common principal components analysis) are identical in the limit of infinite data sets. In our case, the difference between the two reduction techniques is small enough to be negligible. We start by considering two identical Gaussian distributions where the overlap $\Delta_{ab} = 1$ (Figs 4(a) and (b)). We see that both graphs give moderate lower bounds for most of the range, and only Fig. 4(a) approaches the true value from below, and even then only when the number of dimensions is reduced to one. It is also noticeable that doubling the number of data points from 4000 to 8000 gives a very limited improvement in the lower bounds. In Figs 4(c) and (d), we show the upper and lower bounds for two Gaussian distributions that are separated by one standard deviation along one axis. We see that the lower bounds converge fairly rapidly, and seem similar to the previous figures for dimensions above five. The main difference is that the upper bound falls below one, although it does not converge anyway near as fast as the lower bound. Similar behaviour is seen when the separation between the two distributions is increased further: Figs 4(e) and (f). The other thing to note is that the bounds shown in Figs 4(c) and (e) indicate that the probability mass is accumulating in some of the intervals to the detriment of others because we have set $\delta > 1/N$ (see above). This can be seen from the fact that the bounds for $N = 8000$ do not fall within the bounds for $N = 2000$, which is only possible when $\delta > 1/N$ because the upper and lower bounds do not strictly bound the true overlap. However, as we have stated above and due in part to the conservative nature of Dempster's rule for the combination of evidence, it is often necessary to take $\delta > 1/N$ to enable non-trivial bounds to be set at all. There is an element of compromise here but alternative approaches do exist and further work would be required to select the best compromise for a given problem.

Figures 5(a)–(f) show the bounds produced for the two simulated hyperspectral data sets. Here we look at the behaviour for the two dimensionality reduction techniques (Fisher's criterion, 5(a) and (b), and common principal components analysis, 5(e) and (f)) and the simple hardware solution (reducing the spectral resolution of the data set, 5(c) and (d)). Here we can see markedly different behaviour in the three sets of graphs. The differences between the pairs is due to the different nature of the dimensionality reduction scheme in each case. The differences within the pairs of figures comes from the different quantization and, as should be expected, the more intervals that are available, the slower the convergence. The use of Fisher's criterion clearly gives the greatest overlap, and therefore least separability. This indicates that much of the information that allows discrimination between the two minerals is highly correlated between spectral wavebands (that is, the dimensions are not independent as in the previous case). There are fewer differences between the other two reduction techniques, the main difference is that the upper bound falls dramatically for the common principal components analysis as the dimension is reduced, and falls more gradually for the reduced spectral resolution, indicating that most of the separability is contained in a few, relatively narrow spectral bands.

The principal problem to deal with when calculating the upper and lower

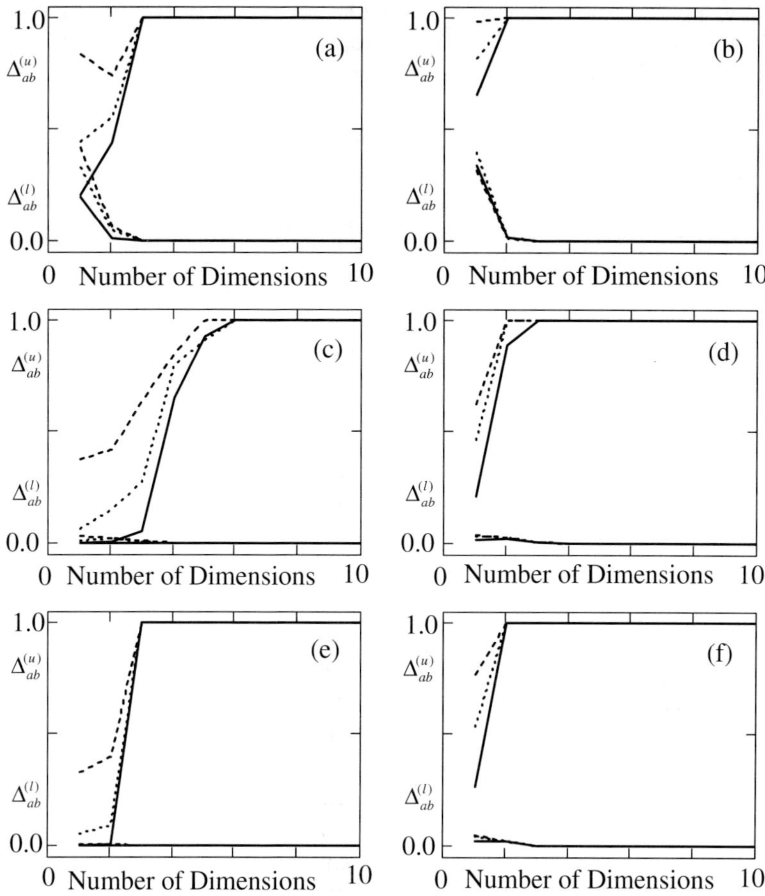

FIG. 5. Upper and lower limits to separability for the simulated hyperspectral data for talc and hematite at 300 K with Gaussian temperature variations, $\sigma = 1$ K, $\delta = 1 \times 10^{-3}$, $N = 8000$ (solid line), $N = 4000$ (dotted line), $N = 2000$ (dashed line): (a) Fisher ratio, 2-bits, (b) Fisher ratio, 3-bits, (c) reduced spectral resolution, 2-bits, (d) reduced spectral resolution, 3-bits, (e) common principal components, 2-bits, (f) common principal components, 3-bits.

bounds for the separability of two distributions in a high-dimensional space is the very large amount of data that are required to derive any meaningful (that is, non-trivial) bounds. The expressions given above for the separability are such that they are either zero or one when the amount of probability mass assigned falls below a certain level. Small data sets do not generate enough evidence to assign large probability masses to any intervals, and so lead to trivial upper and lower bounds (that is, less than one and greater than zero, respectively). For our simulated data sets, the restriction to small data sets is due to computational constraints. The data contained in this paper were generated using algorithms written in the MATLAB programming environment on a personal computer. Handling and sorting large data sets tends to be very time consuming. But, even for real data sets, there are usually restrictions on the speed at which data can be gathered and sorted, so this constraint is not entirely unrealistic. As a result, we have found that it is usually necessary to have a larger value for δ than would strictly be required, $\delta > 1/N$. Where more data are available, or the computational restrictions less of an issue, smaller values of δ could be used and would be expected to generate better upper and lower bounds. In addition, the fact that we are restricted to small data sets also restricts us to a measurement process with a small dynamic range. This is again so that we can generate meaningful upper and lower bounds. Even for standard statistical methods with Gaussian distributions, it would require a very large amount of data to generate a realistic estimate of the underlying mean square error in a large number of dimensions (in ten dimensions approximately 8.4×10^5 data points are required for a mean square error of less than 0.1 at the origin [8, p. 90]). It is therefore not surprising that a technique that is based on local properties of the data, rather than the global properties of some predefined distribution, and Dempster's rule for the combination of evidence (which has been acknowledged to give conservative estimates for the probability mass) would tend to give pessimistic bounds.

In spite of this, the evidential method does have some advantages. The main advantage being that we do not have to preselect a global distribution (or distributions) for our data. Typically, given a set of data from an unknown distribution, a series of tests would be performed to ascertain whether the distribution of the data was sufficiently Gaussian to be approximated by a multi-variate Gaussian distribution, or by another known distribution with a fixed set of parameters that could be estimated from the data. If the data does not appear to be sufficiently Gaussian, it is often analysed to see if it forms identifiable clusters that might be describable by known distributions. The evidential method described in this paper takes a different approach, in that the only assumption that is made about the statistical properties of the data corresponds to the measurement process itself. Here we have taken a simple measurement model, but more complicated and more realistic models can be used. The result is a method that makes no assumption about the global structure of the distribution. Making an assumption about the global structure, and assigning some form of probability distribution to the data, is a reasonable and often necessary step in any pattern recognition algorithm, but it is also necessary that this assumption should be acknowledged.

6 Conclusions

We have constructed expressions that can be used to set upper and lower bounds on the separability of two unknown probability distributions using the Dempster–

Shafer theory of evidential reasoning. The bounds are based on the properties of the available data (which are treated as a finite body of evidence) and the measurement process that defines the statistical properties of the data over a given length scale. As such, the bounds are only dependent on the local properties of the data sets, rather than global assumptions about the underlying distribution from which the data points were drawn. Although computational considerations restricted the number of data points that could be included in this study, we have shown that it was possible to construct upper and lower bounds for the overlap between two distributions for both of the examples considered using the expressions presented in this paper.

In addition, we have shown that the way in which the overlap is bounded from above and below is a function of the dimensionality of the data and the way in which the dimensionality is reduced. This gives information about the separability of the data in different numbers of dimensions for different reduction schemes, and provides a way of assessing the amount of uncertainty present given a finite set of examples. By comparing the bounds for different feature reduction techniques it should be possible to gauge the ability of each technique to generalize, and provide a 'best' guess of feature reduction technique and preferred number of features based on a cost function that includes overall performance and the uncertainty. Unfortunately, the fact that the bounds produced by the evidential method do not converge very fast and are computationally expensive to produce may mean that such an approach is impractical, requiring too many example data points and/or too many computational resources. There are two possible approaches that might be used to improve significantly on the bounds shown in this paper: the use of a more efficient data handling and sorting algorithm and/or the use of a less conservative rule for the combination of evidence, several of which have already been suggested in different contexts [11, 12].

The author would like to thank Mark Bernhardt (DERA Farnborough) for helpful and informative discussions in the preparation of this manuscript.

Bibliography

[1] Christensen, P.R., Bandfield, J.L., Hamilton, V.E., Howard, D.A., Lane, M. D., Piatek, J.L., Ruff, S.W. and Stefanov, W.L., (2000) A thermal emission spectral library of rock-forming mineral. *J. Geophys. Res.*, **105**, 9735–9739.

[2] Dempster, A.P. (1967), Upper and lower probabilities induced by a multi-valued mapping. *Ann. Math. Stat.*, **38**, 325–339.

[3] Goodman, I.R., Mahler, R.P.S. and Nguyen, H.T. (1997), *Mathematics of Data Fusion*, Kluwer, Dordrecht, p. 45.

[4] Gordon, J. and Shortliffe, E.H. (1985), A method for managing evidential reasoning in a hierarchial hypothesis space. *Art. Intell.*, **26**, 323–357.

[5] Klir, G.J. and Weirman, M.J. (1998), *Uncertainty-Based Information: El-*

ements of Generalised Information Theory, Physica, New York.

[6] Krishnan, S., Samudravijaya, K., Rao, P. V. S. (1996), Feature selection for pattern classification with Gaussian mixture models: a new objective criterion. *Pattern Recog. Lett.*, **17**, 803–809.

[7] Shafer, G. (1976), *A Mathematical Theory of Evidence*, Princeton University Press, Princeton, NJ.

[8] Webb, A. (1999), *Statistical Pattern Recognition*, Arnold, London.

[9] Xia, X., Wang, Z. and Gao, Y. (2000), Estimation of non-statistical uncertainty using fuzzy-set theory. *Measure. Sci. Technol.*, **11**, 430–435.

[10] Yager, R.R., Fedrizzi, M. and Kacprzyk, J. (eds) (1994), *Advances in The Dempster–Shafer Theory of Evidence*, Wiley, New York.

[11] Yen, J. (1986), A reasoning model based on an extended Dempster–Shafer theory. *AAAI-86 Proc. 5th Nat. Conf. on Art. Intell.*, **1**, 125–131.

[12] Zadeh, L. (1994), Representation, independence and combination of evidence in the Dempster–Shafer theory, in Ref. 10, Ch.3, pp. 51–69.

Information Geometric Approaches to Acoustic Signal Classification

Timothy R. Field
DERA Malvern, St. Andrews Road, Malvern WR14 3PS, UK

Abstract

A classification algorithm for acoustic signals gathered at multiple ranges is presented. The discrete wavelet transform is used to isolate the relevant impulsive parts of the signals in the time domain by reconstructing the signal at an appropriate scale. Via the spectral representation theorem we identify a complex valued vector describing the impulsive frequency spectrum, whose expected modulus squared is equal to the spectral distribution function of the impulse. Using ideas drawn from classical and quantum information geometry, we describe the discriminating features in terms of elements of the statistical and quantum Hilbert spaces. Distance-based searches enable classification, via the respective Fisher and Dirac measures applied to the impulsive spectral representations. The Dirac distance measure, based on complex amplitudes, is observed to yield improved classification performance as compared to the Fisher measure at long range. A physical basis for this result in terms of the relative phase acquired with respect to distinct frequency modes is proposed.

1 Introduction

Research examining the feasibility of automatic acoustic signal classification has been pursued for a number of years, to assess the performance of various algorithms based upon the use of a single acoustic sensor [1–4]. It was recognized that the performance of a classification system should improve as it became distributed across a network of sensors, due to the enhanced discrimination amongst frequency characteristics over range. The purpose of the paper is to assess this discriminative ability with regard to increased range, achieved by exploiting a distance measure that is sensitive to relative phase contributions acquired between distinct frequency modes.

The classification technique constitutes a distance-based search amongst certain frequency characteristics, which can be understood in the context of classical and quantum information geometry. The anticipation is that the magnitude and phase of each frequency mode contained in a signal is reasonably consistent for the same type of acoustic event recorded at a given range. Accordingly one expects well separated distances with respect to appropriate measures for acoustic signals of different types, thus providing a basis for discrimination. We compare two distance measures, both of which satisfy the metrical axioms. Firstly the statistical Fisher information distance, that considers the relative magnitude of each frequency component [4, 5], and secondly a modified technique using the quantum Hilbert space distance that is sensitive to the acquisition of relative phase. In empirical data trials, using samples of low-SNR acoustic impulses, the modified technique provided improved classification performance. (The techniques presented we expect would have wide applicability: for example, in sonar and acoustic range finding.)

Section 2 describes the application of Fourier and wavelet techniques in acoustic signal pre-processing, wherein the relevant impulsive part of the signal is isolated amongst high-amplitude noise. In Section 3 we present the classification algorithm, describing frequency feature extraction, and subsequent application of classical and quantum information theoretic measures to a distance-based search. The classification performance results are outlined in Section 4 and a brief discussion of the results and outlook given in Section 5.

2 Acoustic signal pre-processing

We consider acoustic signals in the time domain containing high-amplitude noise and a single impulsive event whose time localization, event type, and range are unknown.

2.1 Comparison of Fourier and wavelet techniques

Acoustic signals from impulsive events are highly localized in time and thus ill-suited to conventional Fourier analysis whose basis elements are sinusoids of infinite support in time. In order to extract an impulse from a large degree of environmental noise, a localized functional transform is necessary whose parameters represent location in time as well as frequency (scale) behaviour. The windowed Fourier transform has the required properties and has been used extensively in spectrogram analysis. However, the basis elements are of compact support, determined by the window independently of the frequency. For localization purposes it is more expedient to apply basis elements whose support is narrower at higher modal frequencies, such as is the case in the *wavelet decomposition* [6, 7].

2.2 Wavelet decomposition

We consider the wavelet transform of a square integrable signal $s(t)$ in the time domain. This yields a representation of the form

$$s(t) = \sum_{k\in\mathbf{Z}} \alpha_k \phi_k(t) \sum_{j\geq 0} \sum_{k\in\mathbf{Z}} \beta_{jk}\psi_{jk}(t) \tag{2.1}$$

for the scale function $\phi(t)$ and mother wavelet $\psi(t)$. Here $\phi_k(t) = \phi(t-k)$ and $\psi_{jk}(t) = 2^{j/2}\psi(2^j t - k)$, and the wavelet coefficients α and β can be calculated from $\{s, \phi\}$ and $\{s, \psi\}$ respectively [6, 7]. Since our signals are sampled in discrete time we apply the discrete wavelet transform setting $k = 2^\nu$. The resulting representation of $s(t)$ can be depicted by a plot in which the x-axis is the time translation parameter ν, the y-axis is the resolution level or scale parameter

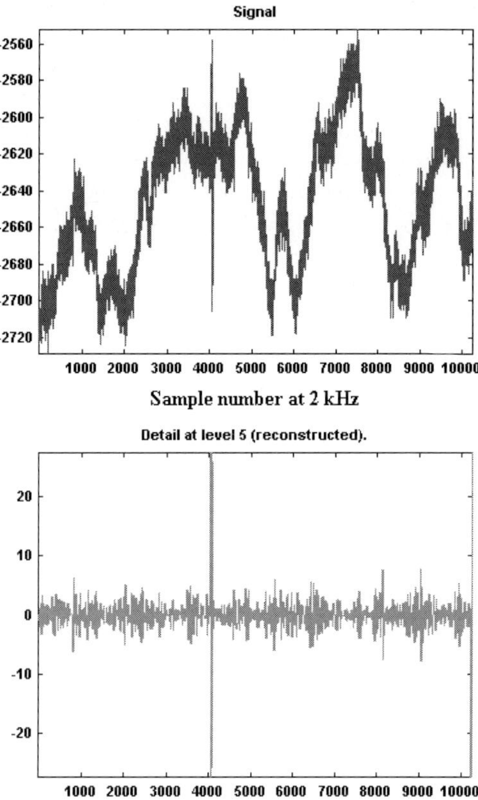

FIG. 1. Impulse localization in the time domain via wavelet reconstruction at an appropriate scale.

j and the z-axis gives the amplitude of the corresponding wavelet coefficients. This plot is (in some respects) analogous to the spectrogram of Fourier theory, where we would choose trigonometric basis functions. The wavelet transform instead represents $s(t)$ in the scale (as opposed to frequency) domain, where the wavelets themselves are basis functions. In our analysis we apply the (discrete) wavelet decomposition for the purpose of impulse localization, by reconstructing the wavelet decomposition at an appropriate scale, as shown in Fig. 1. The top panel shows that the impulse is embedded in a high degree of noise, both low frequency/high-amplitude and vice versa. (The former is likely to be due to effects of wind, while the latter constitutes general background clutter. Clutter of this type leads to bias when Fourier representations are used and it is generally recognized that wavelet representations have significant advantages.) Observe that the amplitude of the low-frequency noise sometimes exceeds that of the impulse, and the wavelet localization mechanism, selecting the time at which the

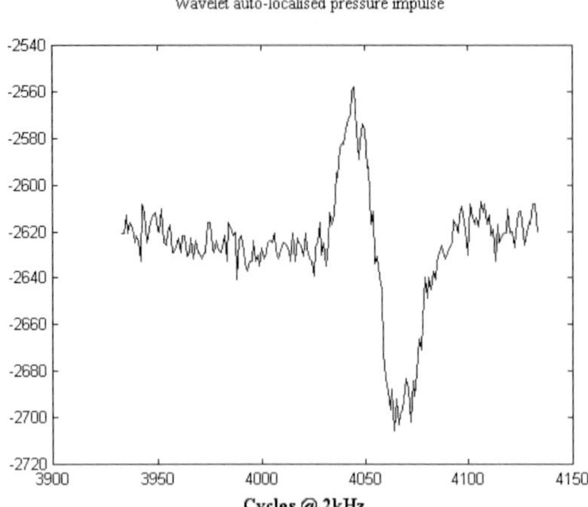

FIG. 2. Automatic localized impulse of Fig. 1(a) indicating match to Haar wavelet (square wave).

detail reconstruction is maximum in modulus, is still effective. For each event type the choice of Haar/Daubechies-1 mother wavelet was made (essentially a square wave match filter) and proved the most effective choice for localization as compared to Daubechies-n for $n > 1$. This is believed to be due to the strong match existing between the functional form of the acoustic pressure impulse as shown in Fig. 2 and the square wave profile. Moreover, since the Daubechies-n wavelets suppress the first n moments of the detail decomposition, we would expect a broader resolution to be required to achieve significant overlap between the mother wavelet and an arbitrary acoustic impulse, a property that was observed in empirical data trials [4].

3 Classification algorithm

Our classification algorithm consists of frequency feature extraction and a subsequent distance-based search. The feature extraction is based on the spectral representation theorem, which provides a means of interpreting frequency characteristics in terms of probability amplitudes. This enables the link to be made with Fisher information geometry and, by choosing a distance measure adapted to relative phase, the physics of quantum information. Of course the physics of the acoustic scenarios we are considering here is purely classical in nature, but it is nevertheless of some interest that ideas can be borrowed from the quantum domain which can be readily applied to the discrimination problem.

3.1 Frequency feature extraction

A windowed FFT taken over a characteristic number of cycles was applied to the impulse of Fig. 2, yielding the modulus frequency spectrum as shown in Fig.

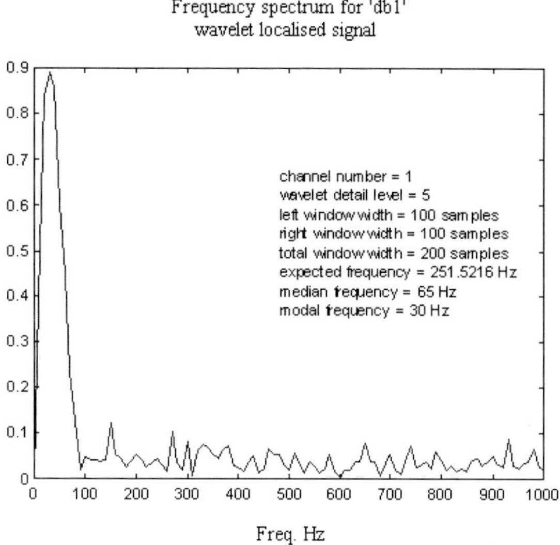

FIG. 3. Modulus frequency spectrum of impulse shown in Fig. 2, showing impulse modal frequency at approximately 30 Hz and relative low amplitude of higher frequency modes.

3. (The technique is legitimate since the modal frequency typically occurred at less than 100 Hz which is greatly exceeded by the Nyquist frequency at the sampling rate of 2k Hz.) The signal ranges were observed to skew the frequency spectrum suppressing the amplitude of higher frequencies at long range resulting in low-frequency bias. This is believed to result from the atmosphere acting as a low-pass filter owing to diffraction from objects along the line of acoustic propagation and absorption effects. Higher frequencies are also more susceptible to refraction through the atmosphere due to temperature and humidity changes [8].

The wave equation $\partial^2 p/\partial t^2 - c^2 \partial^2 p/\partial x^2 = 0$ governing the propagation of the acoustic impulse is in general nonlinear since the speed of sound is a function of wave pressure p. Thus, significant nonlinearity is present in the dependence $c^2 = c^2(p)$ in the near field due to over-pressure, approximating to a linear propagation in the far field. Thus, the peaks and troughs of the impulse propagate at different speeds and acquire relative phase at the detector. Moreover, distinct modal frequencies acquire relative phase at long range via the suppression of high-frequency modes and consequent differing average propagation speeds determined by the r.m.s. pressure (wave amplitude). The relative phase acquired with respect to a pair of distinct frequency modes propagating over large distances is the physical basis for our application of the Dirac distance measure described in Section 3.2.

In order to interpret the frequency characteristics probabilistically we observe the following result. (The author would like to thank Richard Glendinning for suggesting the spectral representation theorem as a means of interpreting measure as a correlation.)

Theorem 3.1 *Spectral representation theorem [9]. Let $X(t)$ be a complex valued random field for which $\mathbf{E}\{X(t)\} = 0$ and $\mathbf{E}\{|X(t)|^2\}$ is finite, which is continuous in mean square, and let $R(t)$ denote its covariance function. Then $R(t)$ is continuous for all t and has a representation of the form $R(t) = \int \exp(it\omega)dF(\omega)$ where $F(\omega)$ is the spectral distribution function of $X(t)$. The field $X(t)$ admits the mean square integral representation $X(t) = \int \exp(it\omega)dZ(\omega)$ where $\mathbf{E}\{Z(\omega)\} = 0$ and is related in expectation to the spectral distribution function by $\mathbf{E}\{|Z(\omega)|^2\} = F(\omega)$, $\mathbf{E}\{|Z(I)|^2\} = F(I)$ for all $I \subset \mathbb{R}$.*

Thus we make the identification $\psi(\omega)d\omega \leftrightarrow dZ(\omega)$, where ψ is the Fourier transform of the impulse, and the squared modulus frequency spectrum $p = |\psi|^2$ is identified with the spectral distribution function $F(\omega)$. (We ensure the zero-mean condition by appropriate constant translation of $X(t)$ over the impulse domain, thus removing the dc contribution in the Fourier spectrum.)

3.2 Information distance-based classification

A wide range of dissimilarity measures $d(i, j)$ between signals i, j have been used to generate classification rules of the form $\mathcal{I} \ni \arg\min_i d(i, j)$ for class $\mathcal{I}$ and test signal j. Any such distance measure is defined with respect to pairs of points on the spaces of classical or quantum states, whose geometry we describe in detail below. In the classical case a state is given by a square root probability distribution, and in the quantum case by an analogous complex-valued probability amplitude whose squared modulus is to be interpreted as a probability. The geometry on the classical and quantum state spaces is induced from the flat Euclidean geometry of the classical and quantum *Hilbert spaces* respectively. In particular, one may derive an intrinsic *metrical* geometry induced from the Euclidean metrical geometry of the ambient Hilbert space. The Fisher and Dirac measures then provide the *geodesic* separation with respect to these induced metrical geometries and accordingly the distance between a pair of states can be determined by the states themselves independently of any path joining them. More generally however, one could adopt a distance measure given by the integral of the infinitesimal metric along some specified path joining a pair of distinct states. The resulting distance measure then no longer necessarily satisfies the axioms for a metric, namely that it be symmetric and satisfy the triangle equality. An example is the Kullback–Leibler ($\mathcal{I}$-divergence) distance measure, which is akin to the Von Neumann relative entropy [10], given by

$$\mathcal{K}(\{p\}|\{q\}) = \sum_\alpha p_\alpha \log \frac{p_\alpha}{q_\alpha}. \tag{3.1}$$

This is not a metric (although its square plays the role of a Euclidean distance [11]), since it is clearly asymmetric. However, infinitesimally it is metrical. Indeed, if we set $\{q\} = \{p\} + \{\delta p\}$ then the above expression reduces to the Fisher distance [5]. In this way, infinitesimally all information distance measures coincide with the Fisher measure, since in the infinitesimal case path dependence does not arise, and the Fisher measure is the only natural choice that coincides with the (Euclidean) distance measure on the ambient Hilbert space.

In what follows, from the point of view of maximizing discrimination in the absence of prior knowledge, we exploit a pair of dissimilarity measures that are also metrics on the statistical and quantum Hilbert spaces. Both our classical and quantum Hilbert space distance measures pertain to $\psi(\omega)d\omega \leftrightarrow dZ(\omega) \in \mathbb{C}$ of Theorem 3.1 whose expected squared modulus is equal to the spectral distribution of the localized impulse.

3.2.1 *Statistical Hilbert space*

The square root normalized frequency-squared amplitudes constructed from the wavelet-localized acoustic impulses can be regarded as unit vectors $|\psi_i|/\sqrt{\int |\psi_i|^2} = \sqrt{p_i}$ (so that p_i is the spectral distribution function of the impulse) that lie on the *positive octant* of the unit sphere embedded in the statistical Hilbert space $\mathcal{H}$, as shown in Fig. 4. With respect to the *Fisher–Rao information metric* [5] the geodesic distance ϑ_{ij} between any pair of such (real-valued) feature vectors is determined by the *classical fidelity* between a pair of classical probability distributions, according to

$$\cos\vartheta_{ij} = \int \sqrt{p_i p_j} = \frac{\int |\psi_i \psi_j|}{\sqrt{\int(|\psi_i|^2)\int(|\psi_j|^2)}}, \qquad 0 \leq \vartheta_{ij} \leq \pi/2. \tag{3.2}$$

(We refer the reader to the seminal paper of Wootters [12] for a discussion of this distance and its implications for distinguishing probability distributions.) The statistical information distance-based classifier thus obtained is depicted in Fig. 4.

3.2.2 *Quantum Hilbert Space*

Expression (3.2) is the classical statistical analogue of the Dirac transition amplitude in the quantum Hilbert space $\mathcal{QH}$ in which the *complex*-valued wavefunctions ψ_i play the role of the classical $\sqrt{p_i}$ [7, p. 12 ff.]. (We could in principle regard ψ_i as wavefunctions in momentum space and thus $\psi(\omega)$ arises as the time-like component of the four-momentum of the impulse.) Explicitly we introduce the distance q via

$$\cos^2[\frac{1}{2}q(i,j)] = \frac{|\int \psi_i \bar{\psi}_j|^2}{\int(|\psi_i|^2)\int(|\psi_j|^2)}, \qquad 0 \leq q(i,j) \leq \pi. \tag{3.3}$$

Now $q(i,j)$ is the quantum angle between the two complex frequency amplitudes. This measures the geodesic distance along the quantum state space (isomorphic to $\mathbf{CP}^1$, the complex projective line or Riemann sphere) with respect to the Fubini-Study metric (see, for example, [13, 14]). (In homogeneous coordinates Z^α on $\mathbf{CP}^n$ the Fubini–Study line element is given by

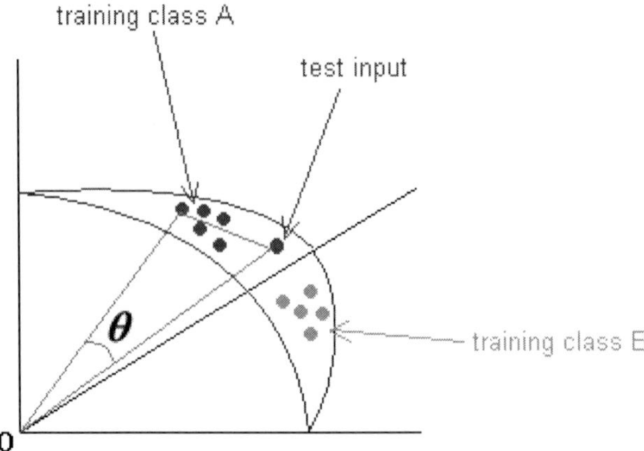

FIG. 4. Positive octant of unit sphere embedded in statistical Hilbert space $\mathcal{H}$.

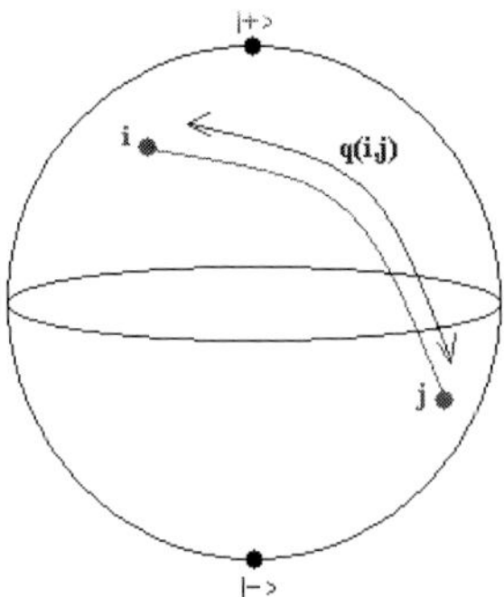

FIG. 5. Quantum projective Hilbert space $\mathcal{P}(\mathcal{QH})$.

$ds^2 = 8Z^{[\alpha}dZ^{\beta]}\bar{Z}_\alpha d\bar{Z}_\beta/(Z^\gamma \bar{Z}_\gamma)^2$.) The geometry of the Dirac quantum information distance is depicted in Fig. 5. The antipodal states $|+\rangle$ and $|-\rangle$ represent frequency amplitudes that are π out of phase with respect to each other. This geometry may also be understood in terms of the $\mathcal{Q}$-density matrix, and the theory of $\mathcal{Q}$-measurements. Consider a POVM (positive operator-valued measure) $\mathcal{P}$ [15] given as a resolution of the unit operator (in general amongst nonorthogonal states) by $\mathcal{P} \leftrightarrow \sum_\alpha \hat{E}_\alpha = \mathbf{1}$, $\hat{E}_\alpha > 0$. The transition probability to a state with density matrix $\hat{\rho}$ is then given by $p(\alpha|\hat{\rho}) = \mathrm{Tr}[\hat{E}_\alpha \hat{\rho}]$. Defining the overlap expression $F_{ij}(\mathcal{P}) = \sum_\alpha \sqrt{p(\alpha|\hat{\rho}_1)p(\alpha|\hat{\rho}_2)}$ we obtain the *quantum fidelity*, to be compared with (3.2), via

$$\cos(F_\mathcal{Q}) = \inf_\mathcal{P} F_{12}(\mathcal{P}). \tag{3.4}$$

This in turn yields the *Bures metric* [10]

$$F(\hat{\rho}_1|\hat{\rho}_2) = \mathrm{Tr}\left[\sqrt{(\sqrt{\hat{\rho}_1}\hat{\rho}_2\sqrt{\hat{\rho}_2})}\right] \tag{3.5}$$

on the space of density matrices. For a pure state, this reduces to the Fubini–Study metric. Evidently from (3.4) this metric *maximizes* the discrimination between a pair of states, in the absence of additional prior knowledge. (The author is grateful to Alonso Botero for pointing out this connection during a visit by the author to USC Columbia in April 2001.)

3.2.3 *Relative phase acquisition*

The Fisher and Dirac distance measures enable the construction of classical and quantum distance-based classifiers respectively. Contrasting the Fisher $\mathcal{H}$ and Dirac $\mathcal{QH}$ measures we observe the following result concerning their discriminative ability.

Lemma 3.2 *Under time translations* $s_i(t) \mapsto s_i(t - k_i)$ *the Fisher distance* $\vartheta(i, j)$ *is invariant. In contrast, the Dirac distance* $q(i, j)$ *acquires a relative* $U(1)$ *phase via the relation* $\psi_i\bar{\psi}_j \mapsto \exp[i(k_i - k_j)\omega]\psi_i\bar{\psi}_j$, *thus interfering the* $\{\psi_i\}$ *under (3.3) and increasing their separation with respect to* q. *More generally,* $|\psi_i\bar{\psi}_j|$ *is time dependent because the relative phase of* ψ_i *and* ψ_j *changes with time.*

Thus q, in contrast to ϑ, detects the acquisition of relative phase, a property that is significant in our application as explained in Section 3.1 above.

4 Comparative performance

In accordance with the remarks of Section 3.2.3 above we expect improved classification performance of the Dirac as compared to the Fisher measure in our specific application, and moreover that this improvement be more noticeable at longer range. These properties were observed in empirical data trials (see [4, 16] for a detailed account of classification performance) an example of which is provided in Fig. 6. (The classification algorithm was implemented in *MatLab* as detailed in [16]. Of nine neighbours, the closest to a given test signal scores 9, decreasing by 1 for subsequent neighbours. The entire data training set consisted of 278 signals, of which 10 lay in the event class tested [16].)

5 Discussion

Our study shows that the complex valued frequency amplitude and subsequent use of the Dirac distance measure provides better classification performance than the Fisher measure. This improved performance arises from the acquisition of relative phase that results from the attenuation of frequency and propagation speed with respect to range. In this way, we have exploited the skewness (kurtosis) of the frequency spectrum that occurs over distance for the purpose of range discrimination.

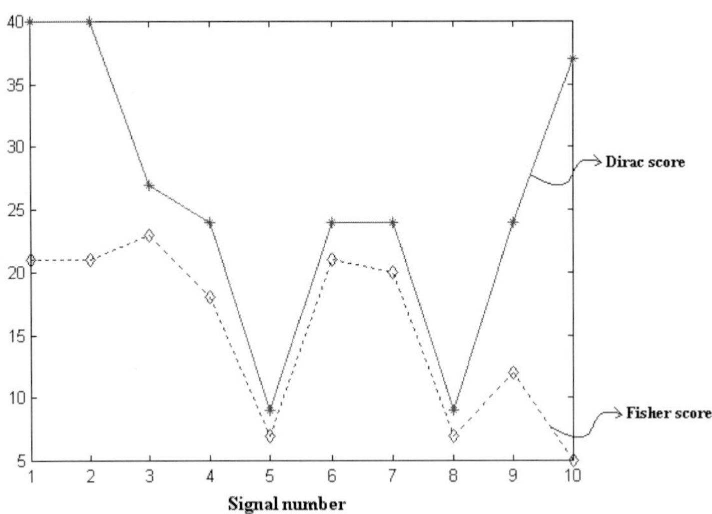

FIG. 6. Long-range classification performance improvement of Dirac over Fisher measure.

An analysis of the multiple acoustic sensor case is provided in [16] for which the Dirac measure again yields improved performance. It would be informative to investigate the physical origin of multi-path signals where (approximate) copies of a signal in the time domain overlap. This could be achieved by manually superposing small-time translations of a given signal and analyzing the effect on the wavelet transform. Further use of information from the discrete wavelet decompositions, beyond impulse localization, is recommended in the enhancement of classifier performance. The determination of refined range information may also benefit from semi-parametric modelling [17].

Bibliography

[1] Meyer, N. and Tuthill, T. (1995). Bayesian classification of ultrasound signals using wavelet coefficients. *Proc. IEEE Conf. on National Aerospace and Electronics (NAECON95)*, Vol 1, pp. 240–243.

[2] *Proc. IEEE*, **84** (1996).

[3] Field, T.R., Glendinning, R.H. and Goode, A.J. (2000). Signal classification using wavelets, *IEE Time-Scale and Time-Frequency Analysis and Applications*, ISSN 0963-3308, Ref. 2000/019. Institution of Electrical Engineers.

[4] Field, T.R., Glendinning, R.H. and Goode, A.J. (2000). Acoustic Signal Classification. *DERA Technical Report*, DERA/S&P/SPI/TR000240.

[5] Amari, S. (1985). *Differential-Geometrical Methods in Statistics*, Berlin, Springer.

[6] Daubechies, I. (1992). *Ten Lectures on Wavelets*. SIAM, Philadelphia.

[7] Kaiser, G. (1994). *A Friendly Guide to Wavelets*. Basle, Birkhäuser.

[8] Kinsler, L.E., Frey, A.R., Coppens, A.B. and Sanders, J.V. (1982). *Fundamentals of Acoustics* (3rd edn). ISBN 0-471-02933-5. Wiley.

[9] Adler, R.J. (1981). *The Geometry of Random Fields.* New York, Wiley.

[10] Vedral, V., Plenio, M.B. and Knight, P.L. (2000). Entanglement quantification. In *The Physics of Quantum Information*, (eds, D. Bouwmeester, A. Ekert and A. Zeilinger). Berlin, Springer.

[11] Gersch, W. (1978). Nearest neighbour rule classification of stationary and non-stationary time series. In *Applied Time Series Analysis II. (ed. D. F. Findley*, 221–270. Academic, New York.

[12] Wootters, W.K. (1981). *Phys. Rev. D*, **23**, 357.

[13] Kobayashi, S and Nomizu, K (1996). *Foundations of Differential Geometry*, Vol. 2, Chapter IX, Section 6. Wiley.

[14] Field, T.R. and Hughston, L.P. (1999). The geometry of coherent states, *J. Math. Phys.*, **40**, 2568–2583.

[15] Peres, A. (1995). *Quantum Theory: Concepts and Methods.* Dordrecht, Kluwer.

[16] Field, T.R. (2001). Multiple sensor acoustic signal classification. *DERA Technical Report*, DERA/S&P/SPI/TR010563.

[17] [17] Goode, A.J. and Calloway, V.C. (2001). Techniques for detection and identificaion of transient signals, *DERA Technical Report*, DERA/S&P/SPI/652/CRP/TG10/SPATS/5_4.

Using Stochastic Vector Quantizers to Characterize Signal and Noise Subspaces

S. P. Luttrell

Signal and Information Processing Department, Defence Evaluation and Research Agency, St. Andrews Rd, Malvern WR14 3PS, UK

Abstract

In this paper a stochastic generalization of the standard Linde-Buzo-Gray (LBG) approach to vector quantizer (VQ) design is presented, in which the encoder is implemented as the sampling of a vector of code indices from a probability distribution derived from the input vector, and the decoder is implemented as a superposition of reconstruction vectors. This stochastic VQ (SVQ) is optimized using a minimum mean Euclidean reconstruction distortion criterion, as in the LBG case. Numerical simulations are used to demonstrate how this leads to self-organization of the SVQ, where different stochastically sampled code indices become associated with different input subspaces.

1 Introduction

In vector quantization a code book is used to encode each input vector as a corresponding code index, which is then decoded (again, using the codebook) to produce an approximate reconstruction of the original input vector [1, 2]. The purpose of this paper is to generalize the standard approach to vector quantizer (VQ) design [3], so that each input vector is encoded as a *vector* of code indices that are stochastically sampled from a probability distribution that depends on the input vector, rather than as a *single* code index that is the deterministic outcome of finding which entry in a code book is closest to the input vector. This will be called a stochastic VQ (SVQ), and it includes the standard VQ as a special case.

One advantage of using the stochastic approach is that it automates the process of splitting high-dimensional input vectors into low-dimensional blocks before encoding them, because minimizing the mean Euclidean reconstruction error can encourage different stochastically sampled code indices to become associated with different input subspaces [4, 5]. Another advantage is that it is very easy to connect SVQs together, by using the vector of code index probabilities computed by one SVQ as the input vector to another SVQ [6].

SVQ theory will be extended to the case of encoding noisy (or distorted) data, with the intention of subsequently reconstructing an approximation to the noiseless data. This theory is then applied to the problem of encoding data vectors which are a superposition of a 'large' jammer and a 'small' signal, where the

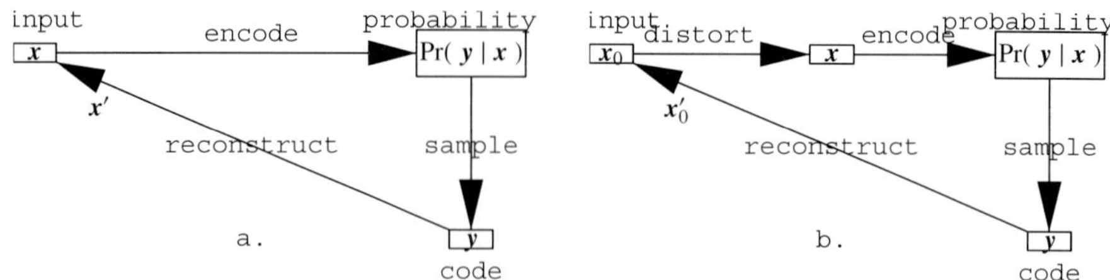

FIG. 1. (a) A SVQ in which an input vector $\boldsymbol{x}$ is encoded as a code index vector $\boldsymbol{y}$ that is drawn from a conditional probability $\Pr(\boldsymbol{y}|\boldsymbol{x})$, which is then decoded as a reconstruction vector $\boldsymbol{x}'$ drawn from the Bayes' inverse conditional probability $\Pr(\boldsymbol{x}'|\boldsymbol{y})$. (b) A SVQ in which an input vector $\boldsymbol{x}_0$ is first distorted into $\boldsymbol{x}$, which is then encoded as a code index vector $\boldsymbol{y}$ that is drawn from a conditional probability $\Pr(\boldsymbol{y}|\boldsymbol{x})$, which is then decoded as a reconstruction vector $\boldsymbol{x}_0{}'$ drawn from the Bayes' inverse conditional probability $\Pr(\boldsymbol{x}_0{}'|\boldsymbol{y})$.

signal is regarded as a distortion superimposed on the jammer, rather than the other way around. The reconstruction is then an approximation to the jammer, which can thus be subtracted from the original data to reveal the underlying signal of interest.

In Section 2 the underlying theory of SVQs is developed together with its extension to the encoding of noisy data, and in Section 3 some simulations illustrating the application of SVQs to the nulling of jammers are presented.

2 Theory

In Section 2.1 the basic theory of SVQs is summarized, in Section 2.2 SVQ theory is extended to the case of encoding noisy or distorted data with the intention of eventually recovering the undistorted data, in Section 2.3 this extended theory is applied to the problem of encoding data that contain unwanted 'nuisance degrees of freedom', in Section 2.4 some constraints on the optimization of the encoder are introduced to encourage the encoder to disregard the nuisance degrees of freedom (that is, discover invariances), and finally in Section 2.5 this invariant encoder theory is applied to the problem of encoding and subsequently nulling 'large' jammers that obscure 'small' signals.

2.1 Stochastic vector quantisers

The basic building block of the encoder/decoder model used in this paper is the folded Markov chain (FMC) [7], which is equivalent to the SVQ discussed in Section 1, and analysed in detail in [8]. The operations that occur in a SVQ are summarized in Fig. 1(a). Thus an input vector $\boldsymbol{x}$ is encoded as a code index vector $\boldsymbol{y}$, which is then subsequently decoded as a reconstruction $\boldsymbol{x}'$ of the input vector. Both the encoding and decoding operations are allowed to be probabilistic, in the sense that $\boldsymbol{y}$ is a sample drawn from $\Pr(\boldsymbol{y}|\boldsymbol{x})$, and $\boldsymbol{x}'$ is a sample drawn from $\Pr(\boldsymbol{x}'|\boldsymbol{y})$, where $\Pr(\boldsymbol{y}|\boldsymbol{x})$ and $\Pr(\boldsymbol{x}|\boldsymbol{y})$ are Bayes' inverses of each other, as given by

$$\Pr(\boldsymbol{x}|\boldsymbol{y}) = \frac{\Pr(\boldsymbol{y}|\boldsymbol{x})\Pr(\boldsymbol{x})}{\int d\boldsymbol{x}'\Pr(\boldsymbol{y}|\boldsymbol{x}')\Pr(\boldsymbol{x}')},$$

and $\Pr(\boldsymbol{x})$ is the prior probability from which $\boldsymbol{x}$ was sampled. Because the chain of dependences in passing from $\boldsymbol{x}$ to $\boldsymbol{y}$ and then to $\boldsymbol{x}'$ is first-order Markov (that is, it is described by the directed graph $\boldsymbol{x} \longrightarrow \boldsymbol{y} \longrightarrow \boldsymbol{x}'$), and because the two ends of this Markov chain (that is, $\boldsymbol{x}$ and $\boldsymbol{x}'$) live in the same vector space, it is called a *folded* Markov chain [7].

In order to ensure that the SVQ encodes the input vector optimally, a measure of the reconstruction error must be minimized. There are many possible ways to define this measure, but one that is consistent with many previous results, and which also leads to many new results, is the mean Euclidean reconstruction error measure D, which is defined as

$$D \equiv \int d\boldsymbol{x} \, \Pr(\boldsymbol{x}) \sum_{\boldsymbol{y}} \Pr(\boldsymbol{y}|\boldsymbol{x}) \int d\boldsymbol{x}' \, \Pr(\boldsymbol{x}'|\boldsymbol{y}) \, \|\boldsymbol{x} - \boldsymbol{x}'\|^2 \tag{2.1}$$

where $\boldsymbol{y} = (y_1, y_2, \cdots, y_n)$, $1 \le y_i \le M$, $\|\boldsymbol{x}-\boldsymbol{x}'\|^2$ is the Euclidean reconstruction error, and $\int d\boldsymbol{x} \, \Pr(\boldsymbol{x}) \sum_{\boldsymbol{y}} \Pr(\boldsymbol{y}|\boldsymbol{x}) \int d\boldsymbol{x}' \, \Pr(\boldsymbol{x}'|\boldsymbol{y})$ averages over all possible states of the SVQ.

Using the Markov chain property $\boldsymbol{x} \longrightarrow \boldsymbol{y} \longrightarrow \boldsymbol{x}'$, Bayes' theorem may be written in the form $\Pr(\boldsymbol{x})\Pr(\boldsymbol{y}|\boldsymbol{x})\Pr(\boldsymbol{x}'|\boldsymbol{y}) = \Pr(\boldsymbol{y})\Pr(\boldsymbol{x}|\boldsymbol{y})\Pr(\boldsymbol{x}'|\boldsymbol{y})$, which may then be used to express D in a form in which $\boldsymbol{x}$ and $\boldsymbol{x}'$ appear symmetrically. This allows the $\boldsymbol{x}'$ integrals to be expressed in terms of the corresponding $\boldsymbol{x}$ integrals, and after using Bayes' theorem in the form $\Pr(\boldsymbol{x})\Pr(\boldsymbol{y}|\boldsymbol{x}) = \Pr(\boldsymbol{y})\Pr(\boldsymbol{x}|\boldsymbol{y})$ this yields the following expression for D:

$$D = 2 \int d\boldsymbol{x} \, \Pr(\boldsymbol{x}) \sum_{\boldsymbol{y}} \Pr(\boldsymbol{y}|\boldsymbol{x}) \, \|\boldsymbol{x} - \boldsymbol{x}'(\boldsymbol{y})\|^2 \tag{2.2}$$

where the reconstruction vector $\boldsymbol{x}'(\boldsymbol{y})$ is defined as $\boldsymbol{x}'(\boldsymbol{y}) \equiv \int d\boldsymbol{x} \, \Pr(\boldsymbol{x}|\boldsymbol{y}) \, \boldsymbol{x}$. Because of the quadratic form of the objective function, it turns out that $\boldsymbol{x}'(\boldsymbol{y})$ may be treated as a vector of free parameters, whose optimum value (that is, the solution of $\frac{\partial D}{\partial \boldsymbol{x}'(\boldsymbol{y})} = 0$) is $\int d\boldsymbol{x} \, \Pr(\boldsymbol{x}|\boldsymbol{y}) \, \boldsymbol{x}$, as required.

2.2 Noisy data

By analogy with the results reported in [9], the SVQ approach can be generalized to the problem of encoding noisy or distorted data, with the intention of eventually recovering the undistorted data. The operations that occur in his type of SVQ are summarized in Fig. 1(b). The input vector is $\boldsymbol{x}_0$, which is converted into the distorted input vector $\boldsymbol{x}$ by a distortion process $\Pr(\boldsymbol{x}|\boldsymbol{x}_0)$, which is then encoded as a code index vector $\boldsymbol{y}$, which is then subsequently decoded as a reconstruction $\boldsymbol{x}_0{}'$ of the original input vector. This is described by the directed graph $\boldsymbol{x}_0 \longrightarrow \boldsymbol{x} \longrightarrow \boldsymbol{y} \longrightarrow \boldsymbol{x}_0{}'$.

The mean Euclidean reconstruction error measure D becomes (compare (2.1))

$$D = \int d\boldsymbol{x}_0 \, \Pr(\boldsymbol{x}_0) \int d\boldsymbol{x} \, \Pr(\boldsymbol{x}|\boldsymbol{x}_0) \sum_{\boldsymbol{y}} \Pr(\boldsymbol{y}|\boldsymbol{x}) \int d\boldsymbol{x}' \, \Pr(\boldsymbol{x}_0'|\boldsymbol{y}) \, ||\boldsymbol{x}_0 - \boldsymbol{x}_0'||^2 \tag{2.3}$$

The Bayes' inverse probability $\Pr(\boldsymbol{x}_0'|\boldsymbol{y})$ may be integrated out of this expression for D to yield (compare (2.2))

$$D = 2 \int d\boldsymbol{x}_0 \, \Pr(\boldsymbol{x}_0) \int d\boldsymbol{x} \, \Pr(\boldsymbol{x}|\boldsymbol{x}_0) \sum_{\boldsymbol{y}} \Pr(\boldsymbol{y}|\boldsymbol{x}) \, ||\boldsymbol{x}_0 - \boldsymbol{x}_0'(\boldsymbol{y})||^2 \tag{2.4}$$

where the reconstruction vector $\boldsymbol{x}_0'(\boldsymbol{y})$ is defined as $\boldsymbol{x}_0'(\boldsymbol{y}) \equiv \int d\boldsymbol{x}_0 \, \Pr(\boldsymbol{x}_0|\boldsymbol{y}) \, \boldsymbol{x}_0$, which may be treated as a vector of free parameters. Bayes' theorem $\Pr(\boldsymbol{x}_0)\Pr(\boldsymbol{x}|\boldsymbol{x}_0) = \Pr(\boldsymbol{x})\Pr(\boldsymbol{x}_0|\boldsymbol{x})$ may be used to integrate out $\boldsymbol{x}_0$ to yield

$$D = 2 \int d\boldsymbol{x} \, \Pr(\boldsymbol{x}) \sum_{\boldsymbol{y}} \Pr(\boldsymbol{y}|\boldsymbol{x}) \, ||\boldsymbol{x}_0(\boldsymbol{x}) - \boldsymbol{x}_0'(\boldsymbol{y})||^2 + \text{constant} \tag{2.5}$$

where $\boldsymbol{x}_0(\boldsymbol{x})$ is defined as $\boldsymbol{x}_0(\boldsymbol{x}) \equiv \int d\boldsymbol{x}_0 \, \Pr(\boldsymbol{x}_0|\boldsymbol{x}) \, \boldsymbol{x}_0$.

It is much more difficult to optimize this version of the objective function than the version in (2.2), because the $\boldsymbol{x}_0(\boldsymbol{x})$ term is in general a nonlinear function of $\boldsymbol{x}$. Worse still, the expression for $\boldsymbol{x}_0(\boldsymbol{x})$ involves $\Pr(\boldsymbol{x}_0|\boldsymbol{x})$, which depends on the unknown $\Pr(\boldsymbol{x}_0)$, so $\boldsymbol{x}_0(\boldsymbol{x})$ cannot be computed analytically anyway. However, progress can be made by splitting $\boldsymbol{x}$ into 'signal' and 'noise' subspaces, as will be shown in Section 2.3.

2.3 Nuisance degrees of freedom

For convenience, split up the input space into (possibly non-orthogonal) subspaces as $(\boldsymbol{x}_0, \boldsymbol{x}_\perp)$, where the undistorted input lives entirely in the $\boldsymbol{x}_0$ subspace, and the distortion lives principally in the $\boldsymbol{x}_\perp$ subspace, and any distortion that lies in the $\boldsymbol{x}_0$ subspace is regarded as part of the undistorted input. For simplicity, the part of the distortion that lies in the $\boldsymbol{x}_0$ subspace is ignored throughout the rest of this paper. The directed graph becomes $(\boldsymbol{x}_0, \boldsymbol{0}) \longrightarrow (\boldsymbol{x}_0, \boldsymbol{x}_\perp) \longrightarrow \boldsymbol{y} \longrightarrow (\boldsymbol{x}_0', \boldsymbol{0})$ as shown in Fig. 2(a). The expression for D becomes (compare (2.4))

$$D = 2 \int d\boldsymbol{x}_0 \, \Pr(\boldsymbol{x}_0) \int d\boldsymbol{x}_\perp \, \Pr(\boldsymbol{x}_\perp|\boldsymbol{x}_0) \sum_{\boldsymbol{y}} \Pr(\boldsymbol{y}|\boldsymbol{x}_0, \boldsymbol{x}_\perp) \, ||\boldsymbol{x}_0 - \boldsymbol{x}_0'(\boldsymbol{y})||^2. \tag{2.6}$$

Now consider the related optimization problem in which an attempt to reconstruct $(\boldsymbol{x}_0, \boldsymbol{x}_\perp)$ is made, as shown in Fig. 2(b). The corresponding objective function may be obtained by modifying (2.6) (omitting the cross-term arising

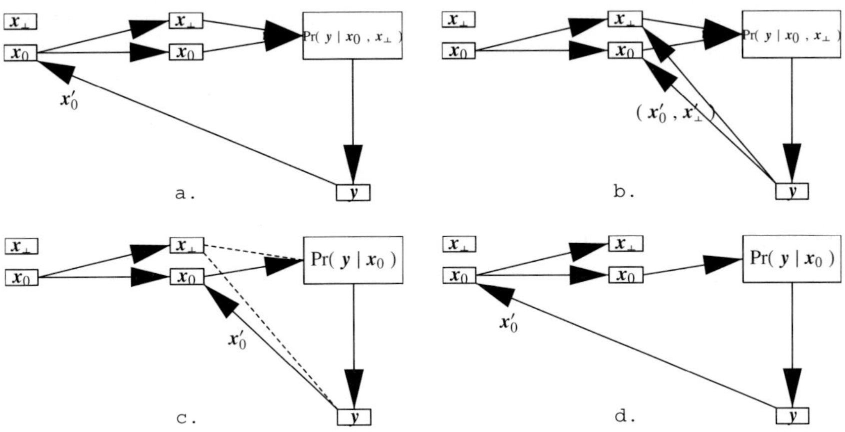

FIG. 2. The input $(\boldsymbol{x}_0, \boldsymbol{x}_\perp)$ is represented as a pair of channels, with $\boldsymbol{x}_\perp$ lying above $\boldsymbol{x}_0$ in each of the diagrams. (a) A SVQ in which an input vector $(\boldsymbol{x}_0, \mathbf{0})$ is first distorted into $(\boldsymbol{x}_0, \boldsymbol{x}_\perp)$, which is then encoded as a code index vector $\boldsymbol{y}$ that is drawn from a conditional probability $\Pr(\boldsymbol{y}|\boldsymbol{x}_0, \boldsymbol{x}_\perp)$, which is then decoded as a reconstruction vector $\boldsymbol{x}_0{}'$ drawn from the Bayes' inverse conditional probability $\Pr(\boldsymbol{x}_0{}'|\boldsymbol{y})$. (b) Modified version of Fig. 2(a) in which the reconstruction link is switched from the original undistorted signal $(\boldsymbol{x}_0, \mathbf{0})$ to the distorted signal $(\boldsymbol{x}_0, \boldsymbol{x}_\perp)$. (c) Modified version of Fig. 2(b) in which the encoder (and reconstruction) links from (and to) the distortion subspace are deleted (as indicated by the dashed lines). (d) Alternative version of Fig. 2(c) in which the reconstruction link is moved to an equivalent position, ignoring differences that arise due to the part of the distortion that lies in the $\boldsymbol{x}_0$ subspace.

from $\boldsymbol{x}_0.\boldsymbol{x}_\perp \neq 0$, which will later be shown to be zero) to yield

$$\begin{aligned} D &= 2 \int d\boldsymbol{x}_0 \Pr(\boldsymbol{x}_0) \int d\boldsymbol{x}_\perp \Pr(\boldsymbol{x}_\perp|\boldsymbol{x}_0) \sum_{\boldsymbol{y}} \Pr(\boldsymbol{y}|\boldsymbol{x}_0, \boldsymbol{x}_\perp) \\ &\quad \times (||\boldsymbol{x}_0 - \boldsymbol{x}_0{}'(\boldsymbol{y})||^2 + ||\boldsymbol{x}_\perp - \boldsymbol{x}_\perp{}'(\boldsymbol{y})||^2). \end{aligned} \tag{2.7}$$

Assume for now that some of the links in Fig. 2(b) are broken as shown in Fig. 2(c). Because the distortion subspace is not involved in the computations in Fig. 2(c), it may be redrawn as shown in Fig. 2(d). This is the same as Fig. 2(a), except that the encoder now disregards (or is invariant with respect to) the nuisance degrees of freedom. Note that the simplification of Fig. 2(c) to Fig. 2(d) ignores differences that arise due to the part of the distortion that lies in the $\boldsymbol{x}_0$ subspace.

The following argument explains the conditions under which the links in Fig. 2(c) may be broken. Firstly, it is necessary to *assume* that the encoder is independent of $\boldsymbol{x}_\perp$, so that $\Pr(\boldsymbol{y}|\boldsymbol{x}_0, \boldsymbol{x}_\perp) = \Pr(\boldsymbol{y}|\boldsymbol{x}_0)$ (this assumption will be discussed in detail later). The $||\boldsymbol{x}_\perp - \boldsymbol{x}_\perp'(\boldsymbol{y})||^2$ term in (2.7) needs to simplify to a constant, which requires that $\int d\boldsymbol{x}_0 \ \Pr(\boldsymbol{x}_0) \int d\boldsymbol{x}_\perp \ \Pr(\boldsymbol{x}_\perp|\boldsymbol{x}_0) \sum_{\boldsymbol{y}} \Pr(\boldsymbol{y}|\boldsymbol{x}_0, \boldsymbol{x}_\perp) \ ||\boldsymbol{x}_\perp - \boldsymbol{x}_\perp'(\boldsymbol{y})||^2 =$ constant. However, to obtain this constant result, it is sufficient to assume $\int d\boldsymbol{x}_\perp \ \Pr(\boldsymbol{x}_\perp|\boldsymbol{x}_0) \ ||\boldsymbol{x}_\perp - \boldsymbol{x}_\perp'(\boldsymbol{y})||^2 =$ constant independent of $\boldsymbol{x}_0$ and $\boldsymbol{y}$, and to guarantee that this constant is indeed independent of $\boldsymbol{x}_0$ and $\boldsymbol{y}$, it is sufficient to assume $\Pr(\boldsymbol{x}_\perp|\boldsymbol{x}_0) = \Pr(\boldsymbol{x}_\perp)$ and $\boldsymbol{x}_\perp'(\boldsymbol{y}) =$ constant. Note that the result $\boldsymbol{x}_\perp'(\boldsymbol{y}) =$ constant follows from the assumptions $\Pr(\boldsymbol{y}|\boldsymbol{x}_0, \boldsymbol{x}_\perp) = \Pr(\boldsymbol{y}|\boldsymbol{x}_0)$ and $\Pr(\boldsymbol{x}_\perp|\boldsymbol{x}_0) = \Pr(\boldsymbol{x}_\perp)$ which imply $\Pr(\boldsymbol{x}_\perp|\boldsymbol{y}) = \Pr(\boldsymbol{x}_\perp)$, so it may be omitted as a separate assumption.

These assumptions may be summarized as

$$\begin{aligned} \Pr(\boldsymbol{y}|\boldsymbol{x}_0, \boldsymbol{x}_\perp) &= \Pr(\boldsymbol{y}|\boldsymbol{x}_0) \\ \Pr(\boldsymbol{x}_\perp|\boldsymbol{x}_0) &= \Pr(\boldsymbol{x}_\perp) \end{aligned} \tag{2.8}$$

which allow the objective function D (see (2.7)) to be replaced by the equivalent objective function (dropping a constant term)

$$D = 2 \int d\boldsymbol{x}_0 \Pr(\boldsymbol{x}_0) \sum_{\boldsymbol{y}} \Pr(\boldsymbol{y}|\boldsymbol{x}_0) \ ||\boldsymbol{x}_0 - \boldsymbol{x}_0'(\boldsymbol{y})||^2. \tag{2.9}$$

This is the standard SVQ objective function (compare (2.2)) for encoding and reconstructing the undistorted input, for which the directed graph is $\boldsymbol{x}_0 \longrightarrow \boldsymbol{y} \longrightarrow \boldsymbol{x}_0'$.

Note that, under the stated assumptions, the simplification leading from (2.7) to (2.9) occurs even if $\boldsymbol{x}_0.\boldsymbol{x}_\perp \neq 0$, because the potential cross-term $\int d\boldsymbol{x}_\perp \ \Pr(\boldsymbol{x}_\perp|\boldsymbol{x}_0) \ (\boldsymbol{x}_0 - \boldsymbol{x}_0'(\boldsymbol{y})).(\boldsymbol{x}_\perp - \boldsymbol{x}_\perp'(\boldsymbol{y}))$ in (2.7) is zero.

In summary, the encoder $\Pr(\boldsymbol{y}|\boldsymbol{x}_0, \boldsymbol{x}_\perp)$ has access only to the distorted signal $(\boldsymbol{x}_0, \boldsymbol{x}_\perp)$ (see Fig. 2(b) and (2.7)), but the assumptions in (2.8) force the encoder to disregard the distortion (see Fig. 2(d) and (2.9)). In practice, it is not possible to satisfy the assumptions in (2.8) in general, because it is not known in advance how to separate the distorted signal into signal and distortion $(\boldsymbol{x}_0, \boldsymbol{x}_\perp)$, given examples of only the distorted signal. However, it turns out that the first assumption in (2.8) is true if D (as defined in (2.7)) is minimized under certain constraints, in which case Fig. 2(d) and (2.9) follow automatically from Fig. 2(b) and (2.7), respectively. These constraints are discussed in Section 2.4.

This type of encoder, in which the signal is encoded whilst the distortion is ignored, can be used as the basis of a so-called 'residual vector quantizer' [10], in which (quoting from [10]) 'the quantizer has a sequence of encoding stages, where each stage encodes the residual (error) vector of the prior stage'. The key identification needed to make this connection is that the reconstruction error discussed in this paper is the same as the residual (error) vector discussed in [10]. Note that a residual vector quantizer is a special case of the type of multistage encoder discussed in [6].

2.4 Optimisation constraints

Henceforth, only the scalar code case will be considered, so the vector $\boldsymbol{y}$ is now replaced by the scalar y, $1 \leq y \leq M$. Also, it will be assumed that the input manifold has its centroid at the origin. In order to implement a practical optimization procedure for minimizing D (as defined in (2.7)) it is necessary to introduce a variety of assumptions and constraints.

Because $\Pr(y|\boldsymbol{x})$ is a probability it satisfies $\Pr(y|\boldsymbol{x}) \geq 0$ and $\sum_{y=1}^{M} \Pr(y|\boldsymbol{x}) = 1$, which are guaranteed to be true if $\Pr(y|\boldsymbol{x})$ is written as

$$\Pr(y|\boldsymbol{x}) = \frac{Q(y|\boldsymbol{x})}{\sum_{y'=1}^{M} Q(y'|\boldsymbol{x})} \tag{2.10}$$

where $Q(y|\boldsymbol{x}) \geq 0$ for all y, and $Q(y|\boldsymbol{x}) > 0$ for at least one y. This removes the need to explicitly impose the constraint $\sum_{y=1}^{M} \Pr(y|\boldsymbol{x}) = 1$ during optimization. The $Q(y|\boldsymbol{x})$ can be viewed as the unnormalized likelihood of sampling code index y from the code book.

However, $Q(y|\boldsymbol{x})$ itself needs to be described by a finite number of parameters in order that the values that minimize D may be derived from a finite amount of training data. It can be shown that the optimal form of $\Pr(y|\boldsymbol{x})$ must have a piecewise linear dependence on $\boldsymbol{x}$ [6], and that for training data that lie on smooth curved manifolds this dependence is well approximated by a $Q(y|\boldsymbol{x})$ that has the following simple piecewise linear dependence on $\boldsymbol{x}$ [5]:

$$Q(y|\boldsymbol{x}) = \begin{cases} \boldsymbol{u}(y) \cdot \boldsymbol{x} + v(y) & \boldsymbol{u}(y) \cdot \boldsymbol{x} + v(y) \geq 0 \\ 0 & \boldsymbol{u}(y) \cdot \boldsymbol{x} + v(y) \leq 0 \end{cases} \tag{2.11}$$

However, $Q(y|\boldsymbol{x})$ needs to approximate this behaviour only in the vicinity of the data manifold, so it can be allowed to depart from this behaviour elsewhere. A convenient functional form that achieves this is the sigmoid function

$$Q(y|\boldsymbol{x}) = \frac{1}{1 + \exp(-\boldsymbol{w}(y) \cdot \boldsymbol{x} - b(y))}. \tag{2.12}$$

This reduces the problem of minimizing D to one of finding the optimal values of the $\boldsymbol{w}(y)$, $b(y)$, and $\boldsymbol{x}'(y)$. This may be done by using the gradient descent procedure described in [8].

If the input is an *undistorted* signal (that is, $\boldsymbol{x} = (\boldsymbol{x}_0, \boldsymbol{0})$) which lies on a smooth curved manifold, then the sigmoids can co-operate in encoding this input as illustrated in Fig. 3, where the sigmoid threshold planes $\boldsymbol{w}(y) \cdot \boldsymbol{x} + b(y) = 0$ are shown slicing pieces off a curved one-dimensional manifold embedded in a two-dimensional space. When D is minimized, the sigmoids adjust themselves so that they partition the one-dimensional curved manifold into a set of non-overlapping pieces [5], as illustrated in Fig. 3(a). If these pieces are required to overlap, then it is sufficient to introduce an additional constraint which pulls the sigmoid threshold planes inwards, as illustrated by the dotted lines in Fig. 3(b). This

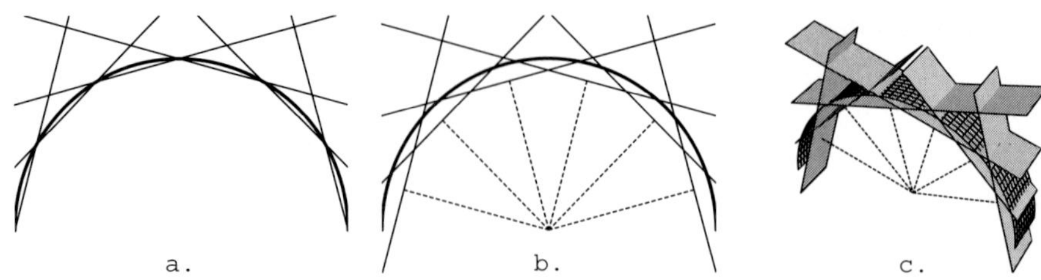

FIG. 3. (a) Illustration of how a number of sigmoids slice non-overlapping pieces off a curved signal manifold. (b) Illustration of how a threshold constraint can be used to force a number of sigmoids to slice overlapping pieces off a curved signal manifold. (c) Illustration of how a number of sigmoids slice overlapping pieces off a curved signal manifold thickened by nuisance degrees of freedom.

type of constraint is easily introduced by choosing $b(y) = -\theta \, ||\boldsymbol{w}(y)||$, so that

$$Q(y|\boldsymbol{x}) = \frac{1}{1 + \exp(-(\hat{\boldsymbol{w}}(y) \cdot \boldsymbol{x} - \theta) \, ||\boldsymbol{w}(y)||)} \tag{2.13}$$

where $||\boldsymbol{w}(y)|| = \sqrt{\boldsymbol{w}(y) \cdot \boldsymbol{w}(y)}$ and $\hat{\boldsymbol{w}}(y) = \frac{\boldsymbol{w}(y)}{||\boldsymbol{w}(y)||}$. The sigmoid threshold planes are then $\hat{\boldsymbol{w}}(y) \cdot \boldsymbol{x} = \theta$, which lie at a distance θ from the origin.

If the input is a *distorted* signal (that is, $\boldsymbol{x} = (\boldsymbol{x}_0, \boldsymbol{x}_\perp)$) then it lies on a 'thickened' version of the smooth curved manifold of Fig. 3(b), which is shown in Fig. 3(c). The thickness represents the nuisance degrees of freedom, which can lie both in the $\boldsymbol{x}_0$ subspace and the $\boldsymbol{x}_\perp$ subspace, but only the latter are represented in Fig. 3(c). If the threshold constraint $b(y) = -\theta \, ||\boldsymbol{w}(y)||$ is used (see (2.13)), as illustrated by the dotted lines in Fig. 3(c), then the sigmoids slice pieces off the curved manifold in a way that disregards the nuisance degrees of freedom.

The threshold constraint in (2.13) thus implements the invariant behaviour required by $\Pr(\boldsymbol{y}|\boldsymbol{x}_0, \boldsymbol{x}_\perp) = \Pr(\boldsymbol{y}|\boldsymbol{x}_0)$ in (2.8), and is analogous to the threshold trick that was introduced in [11] to encourage the emergence of neural responses that were invariant with respect to various degrees of freedom in their input.

In practice, for numerical efficiency and to encourage the optimization procedure to locate the global minimum of D, it is useful to introduce two additional constraints. Firstly, in problems where the input manifold is sliced into overlapping pieces as shown in Fig. 3(c), optimal solutions approximately satisfy $\boldsymbol{x}'(y) \propto \boldsymbol{w}(y)$, so good solutions are obtained if each reconstruction vector $\boldsymbol{x}'(y)$ is *forced* to lie parallel to the corresponding weight vector $\boldsymbol{w}(y)$. Secondly, the norm of the weight vectors $||\boldsymbol{w}(y)||$ can be constrained as $||\boldsymbol{w}(y)|| = w_0$, in order to avoid situations where they grow to rather large values which make $Q(y|\boldsymbol{x})$ (and hence $\Pr(y|\boldsymbol{x})$) depend very strongly on $\boldsymbol{x}$ in some regions. Both of these constraints speed up convergence to the global minimum of D, and can finally be lifted in the vicinity of an optimal solution to obtain complete convergence.

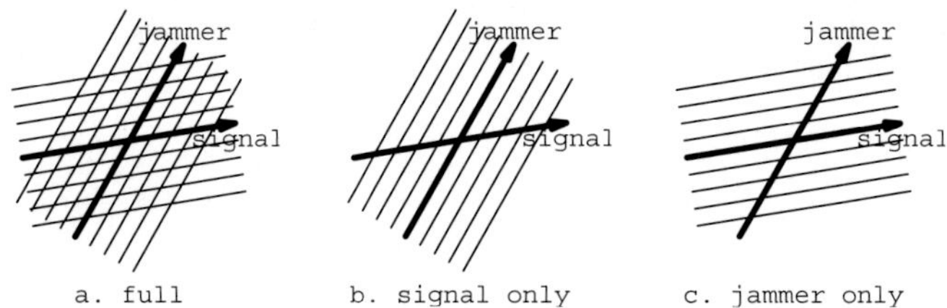

FIG. 4. Examples of the response (represented as a contour plot) of $\Pr(y|\boldsymbol{x})$ to signal and jammer subspaces. (a) The 'full' case shows a pair of $\Pr(y|\boldsymbol{x})$, one of which responds only to the jammer and the other only to the signal. (b) The 'signal' case shows a single $\Pr(y|\boldsymbol{x})$ responding only to the signal. (c) The 'jammer' case shows a single $\Pr(y|\boldsymbol{x})$ responding only to the jammer.

2.5 Jammer nulling

A number of examples of typical behaviours of $\Pr(y|\boldsymbol{x})$ are shown in Fig. 4. The signal and jammer degrees of freedom generate a pair of non-orthogonal subspaces, whose axes are indicated in bold (only one dimension of each of these subspaces is illustrated in Fig. 4). The response contours of a variety of possible $\Pr(y|\boldsymbol{x})$ are shown. The 'full' case in Fig. 4(a) shows the contours of each of a pair of $\Pr(y|\boldsymbol{x})$, one of which responds only to the signal subspace, and the other only to the jammer subspace. The 'signal' case in Fig. 4(b) shows the contours of a $\Pr(y|\boldsymbol{x})$ that responds only to only the signal subspace, and is thus invariant over the jammer subspace. The 'jammer' case in Fig. 4(c) is the converse of 'signal' case, so that $\Pr(y|\boldsymbol{x})$ responds to only the jammer subspace, and is thus invariant over the signal subspace. This argument may readily be generalized to any number of $\Pr(y|\boldsymbol{x})$.

If it is assumed that the jammer is the 'large' degree of freedom and the signal is the 'small' degree of freedom, then the signal and jammer subspaces may be separated by adjusting the threshold parameter θ in (2.13) so that in Fig. 4 the 'jammer' case (that is, Fig. 4(c)) is obtained, in which case $\Pr(y|\boldsymbol{x})$ for $y = 1, 2, \ldots, M$ will all become invariant over the signal subspace. The jammer subspace is then spanned by the set of gradient vectors $\boldsymbol{\nabla}\Pr(y|\boldsymbol{x})$ for $y = 1, 2, \ldots, M$, which can thus be used to construct a projection operator $\boldsymbol{J}$ onto the jammer subspace, and a projection operator $\mathbf{1} - \boldsymbol{J}$ onto the signal subspace. This definition of the projection operator may also be used in cases where the jammer and signal subspaces are curved, so that the directions of their axes are functions of $\boldsymbol{x}$, and all of the straight lines in Fig. 4 are replaced by curves defining a curvilinear coordinate system and its coordinate surfaces. The projection operator then becomes a function of the input vector $\boldsymbol{x}$, so that $\boldsymbol{J} = \boldsymbol{J}(\boldsymbol{x})$. Note that curved subspaces are the norm rather than the exception.

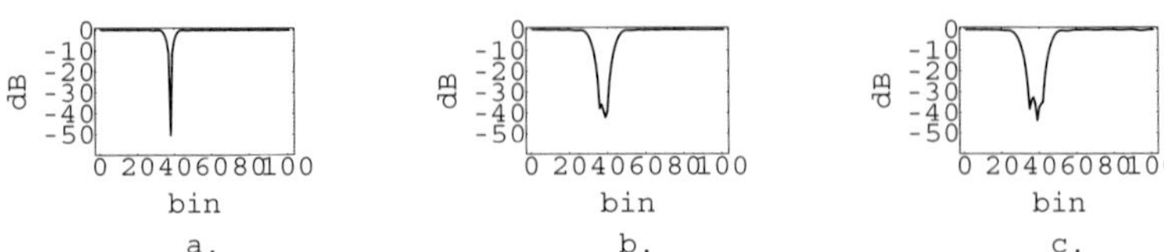

FIG. 5. Plot of degree of nulling against jammer location, for jammer locations that are spread over the intervals (2 units is approximately a resolution cell) (a) $[38, 38]$ using $M = 2$, (b) $[36, 40]$ using $M = 4$, and (c) $[34, 42]$ using $M = 6$.

3 Simulations

The optimization of the encoder may be done by minimizing D in (2.2) using gradient descent [8], using the threshold θ in the sigmoid function in (2.13) to constrain the optimization so that it encodes only the jammer subspace, as illustrated in Fig. 4(c), and discussed in Sections 2.4 and 2.5.

In these simulations the input vector $\boldsymbol{x}$ is 100-dimensional so that $\boldsymbol{x} = (x_1, x_2, \cdots, x_{100})$, and each vector in the training set is independently generated as a superposition of a pair of 'sinc' response functions

$$x_i = a_s \frac{\sin(\frac{i-i_s}{\sigma})}{\frac{i-i_s}{\sigma}} + a_j \frac{\sin(\frac{i-i_j}{\sigma})}{\frac{i-i_j}{\sigma}} \tag{3.1}$$

where a_s is the signal amplitude that is uniformly distributed in the interval $[-\sqrt{10^{-3}}, \sqrt{10^{-3}}]$ (this corresponds to a signal level of -30 dB), a_j is the jammer amplitude that is uniformly distributed in the interval $[-1, 1]$ (this corresponds to a jammer level of 0 dB), i_s is the signal location that is chosen to be 50, i_j is the jammer location that is uniformly distributed in the interval $[38 - \Delta, 38 + \Delta]$ ($\Delta = 0, 2, 4$ is used in the simulations), and σ is the width of the response function that is chosen to be 2. The peak and the first zero of the 'sinc" response function are separated by $\pi\sigma$, which defines the resolution cell size. The mean jammer location $\langle i_j \rangle$ and the signal location satisfy $i_s - \langle i_j \rangle = 12$, which corresponds to a separation of $\frac{12}{\pi\sigma} \approx 2$ resolution cells. Random noise uniformly distributed in the interval $[-\sqrt{10^{-5}}, \sqrt{10^{-5}}]$ (this corresponds to a noise level of -50 dB) is also added to each component of the training vector.

In Fig. 5 an encoder is trained on three different jammer scenarios $\Delta = 0, 2, 4$. After training, the encoder is tested for how well it can be used to null a pure jammer (that is, with no signal or noise added), where the degree of nulling is defined as the ratio of the squared lengths of the nulled input vector and the original input vector. This is a good test of the ability of the encoder to learn the shape of the jammer manifold which is generated by sweeping this profile over the interval $[38 - \Delta, 38 + \Delta]$. When $\Delta = 0$ there is a sharp minimum at the jammer location $i_j = 38$, as expected. When $\Delta = 2$ the minimum becomes spread over jammer locations i_j in the interval $[36, 40]$ (that is, approximately

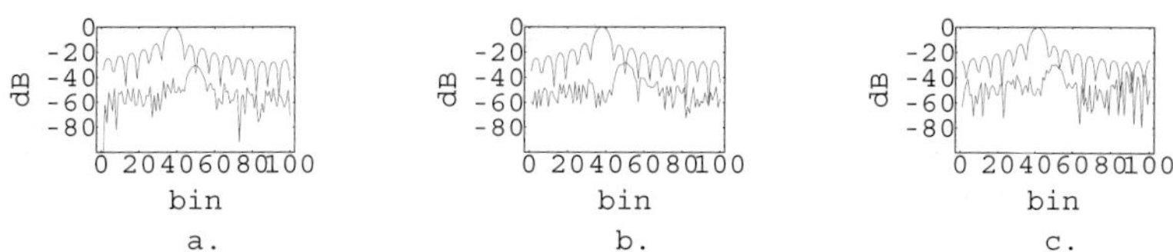

FIG. 6. Plot of a typical input vector before (upper curve) and after (lower curve) jammer nulling, for jammer locations that are spread over the intervals (2 units is approximately a resolution cell) (a) [38, 38] using $M = 2$, (b) [36, 40] using $M = 4$, and (c) [34, 42] using $M = 6$. These correspond to the degree of nulling plots in Fig. 5.

two resolution cells), and when $\Delta = 4$ the minimum becomes spread even more broadly over jammer locations i_j in the interval [34, 42] (that is, approximately four resolution cells). All of these results are as expected.

In Fig. 6 typical examples of an input vector together with how it appears after jammer nulling are shown for each of the jammer scenarios considered in Fig. 5. In every case the signal is clearly revealed at its correct location after nulling the jammer.

In all of these training scenarios, one could envisage further constraining some of the properties of the encoder, in order to introduce prior knowledge of the form of the jammer and/or signal subspaces, and to thereby reduce the computational complexity of the jammer nulling. For instance, the signal subspace could be predefined, as in conventional algorithms which hold constant the response in a predefined 'look direction'. Similarly, the jammer subspace could be built out of predefined subspaces which are optimized so as to maximally null the jammer(s), as in conventional algorithms in which a number of jammer 'templates' are used to remove the jammer(s). In general, by choosing appropriate additional constraints, the SVQ approach to jammer nulling can be made backwardly compatible with conventional approaches.

4 Conclusions

The theory of SVQs [8] has been extended to allow the quantizer to develop invariances, so that only 'large' degrees of freedom in the input vector are represented in the code. This has been applied to the problem of encoding data vectors which are a superposition of a 'large' jammer and a 'small' signal, so that only the jammer is represented in the code. This allows the jammer to be subtracted from the total input vector (that is, the jammer is nulled), leaving a residual that contains only the underlying signal. Several numerical simulations have shown how that idea works in practice, even when the jammer location is uncertain so that the jammer subspace is curved.

The main advantage of this approach to jammer nulling is that little prior knowledge of the jammer is assumed, because these properties are automatically discovered by the SVQ as it is trained on examples of input vectors. Provided that the signal is much weaker than the jammer, the SVQ acquires an internal representation of only the jammer manifold, so it is invariant with respect to the signal manifold. In a sense, the SVQ regards the 'large' jammer as the normal type of input that it expects to encode, whereas it regards the 'small' signal as an anomaly that is not to be encoded.

Acknowledgement

This work was carried out as part of Technology Group 10 of the MoD Corporate Research Programme.

Bibliography

[1] Gray, R.M. (1984). Vector quantization. *IEEE Acoustics, Speech, Signal Process. Mag.*, April 4–29.

[2] Gersho, A. and Gray, R.M. (1992). *Vector Quantization and Signal Processing*, Dordrecht Kluwer.

[3] Linde, Y., Buzo, A. and Gray, R.M. (1980). An algorithm for vector quantizer design. *IEEE Trans. on Communications*, **28**, 84–95.

[4] Luttrell, S.P. (1997). Self-organization of multiple winner-take-all neural networks. *Connection Science*, **9**, 11–30.

[5] Luttrell, S.P. (1999). Self-organized modular neural networks for encoding data. In: *Combining Artificial Neural Nets: Ensemble and Modular Multi-Net Systems*, (ed. A.J.C. Sharkey), Berlin, Springer, 235–263.

[6] Luttrell, S.P. (1999). An adaptive network for encoding data using piecewise linear functions. In: *Proc. Ninth Int. Conf. on Artificial Neural Networks (ICANN99)*, Edinburgh, 7–10 September 1999, pp. 198–203.

[7] Luttrell, S.P. (1994). A Bayesian analysis of self-organizing maps. *Neural Comput.*, **6**, 767–794.

[8] Luttrell, S.P. (1997). A theory of self-organizing neural networks. In: *Mathematics of Neural Networks: Models, Algorithms and Applications*, (eds, S.W. Ellacott, J.C. Mason and I.J. Anderson), Dordrecht, Kluwer, 240–244.

[9] Ephraim, Y. and Gray, R.M. (1988). A unified approach for encoding clean and noisy sources by means of waveform and autoregressive model vector quantization. *IEEE Trans. on Information Theory*, **34**, 826–834.

[10] Barnes, C.F., Rizvi, S.A. and Nasrabadi, N.M. (1996). Advances in residual vector quantization: a review. *IEEE Trans. on Image Processing*, **5**, 226–262.

[11] Webber, C.J.S. (1994). Self-organization of transformation-invariant detectors for constituents of perceptual patterns. *Network: Computat. Neural Systems*, **5**, 471–496

Multiresolution Gaussian Mixtures for Image Analysis

Roland Wilson
University of Warwick, Coventry CV4 7AL, UK

Abstract

This paper introduces a generalization of scale-space and pyramids, which combines statistical modelling with a spatial representation. The representation uses the familiar concept of multiple resolutions, but applied to a Gaussian mixture density representation of the image. It is shown that MGMM can approximate any probability density and can accommodate the effects of smooth motions. After a brief presentation of the theory, examples are used to show how MGMM can be applied to problems such as segmentation.

1 Introduction

Multiresolution image representations have been successfully applied to many problems in image analysis [14, 2] because they trade off spatial against frequency domain resolution and show a degree of symmetry with respect to translations and dilations. The most commonly used forms, such as pyramids and orthonormal wavelets, sacrifice symmetry for compression [3]. While this is useful in applications such as data compression, it is arguably not the best approach in more general problems, especially where the effects of image motion are significant.

Another interesting development in recent years has been the use of Gaussian mixture models to cope with statistical problems for which no simple parametric model exists [11, 15, 9]. While it is well known that algorithms such as Expectation-Maximization (EM) can lead to effective approximations in terms of a finite number of components, the general problem of mixture modelling is difficult when the number of components is unknown and methods based on likelihood alone, such as EM, are known to have weaknesses. This has led a number of authors to explore the use of Bayesian methods [11, 9]. In their application to image analysis, such models tend to be applied to intensity or colour co-ordinates, while spatial interactions are dealt with using one of the many varieties of random field model, such as Markov random fields (MRFs) [4, 5, 6]. While effective in limited classes of problem, such models often require intensive computation.

This paper describes a new representation which combines mixture modelling with multiresolution representation and computation: multiresolution Gaussian mixture models (MGMMs). It shares important features with the Classification and Regression Trees (CART) system of [1] and may also be seen as a generalization of scale space [14], in that it uses Gaussian functions and includes spatial co-ordinates, but unlike a conventional basis set, they represent the data statistically. Moreover, they are defined in a space whose dimension reflects the inference problem, not the data. Thus, in dealing with colour images, a five-dimensional space is required (two spatial and three colour dimensions); for inferring three-dimensional structure from motion, typically nine dimensions are required (three spatial, three colour and three motion axes). In its use of Bayesian methods, MGMM resembles the approach of Richardson and Green on mixture analysis [9]. Its goal, however, is different: it is to model an arbitrary density as a mixture of Gaussian components, not to estimate a mixture density with an unknown number of components. While the former makes no assumptions about the shape of the density, the latter assumes it to be a sum of an unknown number of known components. The next section of the paper contains a brief outline of the theory of MGMM. This is followed by some simple experiments, which illustrate how it may be used in image analysis segmentation. The paper is concluded with some remarks on the new approach.

2 The properties of MGMM

Three main elements dictate the form of representation called MGMM: ability to approximate *any* probability density in a space of arbitrary dimension, closure under affine motions, and a multiresolution structure, which can be used for computational efficiency.

To define the MGMM approach in a general way, we start by observing that any smooth probability density function can be approximated to an arbitrary precision by a set of Gaussian functions: it is well known that the Gaussian functions are a complete set on $L^2(R^n)$. However, we require a stronger result—namely that the approximation need only involve *positive* coefficients in the expansion. To this end, we state the folowing theorem.

Theorem 2.1 *Let $f(.) : R^n \to R$ be any nonnegative integrable function on R^n with*

$$\int_{R^n} d\boldsymbol{x}\; f(\boldsymbol{x}) = 1. \tag{2.1}$$

Then for any $\delta > 0$ there exists an approximation of $f(.)$ by a strictly positive, finite sum of Gaussian functions of the form

$$\hat{f}(\boldsymbol{x}) = \sum_i f_i g_{\Sigma_i}(\boldsymbol{x} - \boldsymbol{\mu}_i) \tag{2.2}$$

of means $\boldsymbol{\mu}_i$ and covariances Σ_i, such that $f_i > 0, \forall i$ and

$$\delta > \int_{R^n} d\boldsymbol{x}\; |f(\boldsymbol{x}) - \hat{f}(\boldsymbol{x})|. \tag{2.3}$$

Proof: We shall sketch the proof here. It hinges on two observations:

(1) The convolution of $f(.)$ with a Gaussian function of sufficiently small scale, that is, $|\boldsymbol{\Sigma}| < \epsilon$, satisfies

$$\int_{R^n} |f(\boldsymbol{x}) - f(\boldsymbol{x}) * g_\Sigma(\boldsymbol{x})| < \frac{\delta}{2} \tag{2.4}$$

which follows from the observation that the Gaussian functions tend uniformly to an impulse function as the scale decreases.

(2) The convolution, which is smooth, can be similarly approximated by a sum,

$$f(\boldsymbol{x}) * g_\Sigma(\boldsymbol{x}) \approx \sum_i f_i g_\Sigma(\boldsymbol{x} - \boldsymbol{\mu}_i) \tag{2.5}$$

where the means $\boldsymbol{\mu}$ are appropriately chosen, such that there is an integral absolute approximation error less than $\frac{\delta}{2}$.

Hence, putting the two together, we arrive at an arbitrarily good approximation of $f(.)$ using n-dimensional Gaussians.

The other property, a crucial one for motion analysis, is the closure of the set of n-dimensional Gaussian functions G^n under affine maps $\boldsymbol{A} : R^n \mapsto R^n$

$$\boldsymbol{A}\boldsymbol{x} = \boldsymbol{L}\boldsymbol{x} + \boldsymbol{a} \tag{2.6}$$

where $\boldsymbol{L}$ is an invertible matrix and $\boldsymbol{a}$ a translation. Again, it is obvious that the action of $\boldsymbol{A}$ on G^n is closed, since

$$g_\Sigma(\boldsymbol{A}^{-1}(\boldsymbol{x} - \boldsymbol{\mu})) = g_{\Sigma_A}(\boldsymbol{x} - \boldsymbol{\mu}_A) \tag{2.7}$$

where

$$\boldsymbol{\mu}_A = \boldsymbol{\mu} - \boldsymbol{a} \tag{2.8}$$

and

$$\boldsymbol{\Sigma}_A = \boldsymbol{L}^T \boldsymbol{\Sigma} \boldsymbol{L}. \tag{2.9}$$

Now we are in a position to prove a rather interesting result.

Theorem 2.2 *Let $f(.) \geq 0$ be an integrable function $f : R^n \to R$, as above and let $\mathbf{t}(.) : R^n \to R^n$ be a smooth, invertible map from R^n to itself, having bounded first and second derivatives. Then for any $\delta > 0$, f has an approximation as a Gaussian mixture of the form of (2.2), with integral absolute error no greater than $\delta/2$ and there exists an approximation of the transformed function $\boldsymbol{T}f$*

$$\boldsymbol{T}f(\boldsymbol{x}) = f(\mathbf{t}^{-1}(\boldsymbol{x})) \tag{2.10}$$

of the same form, where each Gaussian component g_{Σ_i} is transformed according to a local affine approximation of the flow field $\mathbf{t}$, with error no greater than δ.

Proof: Again, we confine ourselves to a sketch. The proof is based on the smoothness of the mapping and the local nature of the MGMM approximation. First, observe that *any* smooth flow can be approximated using

$$\boldsymbol{t}(\boldsymbol{x}) = \boldsymbol{t}(\boldsymbol{x}_0) + \nabla_x(\boldsymbol{t})(\boldsymbol{x} - \boldsymbol{x}_0) + O(\|\boldsymbol{x} - \boldsymbol{x}_0\|^2) \tag{2.11}$$

via Taylor's theorem, where $\nabla_x(.)$ is the gradient. Now, let $g_\Sigma(\boldsymbol{x} - \boldsymbol{\mu})$ be an arbitrary Gaussian function on R^n. It follows from (2.7) and (2.11) that by scaling Σ by a factor $\sigma(\delta)$, say, we can achieve an approximation of the transformed Gaussian $g_{\sigma(\delta)\Sigma}(\boldsymbol{t}(\boldsymbol{x} - \boldsymbol{\mu}))$ with an error which is no greater than $\delta/2$. But from Theorem 1, we know that we can approximate an arbitrary function $f(.)$ by a mixture of Gaussians, with arbitrary low error. Setting this error to be no greater than $\delta/2$, we achieve the result.

The proofs, which are straightforward, are detailed in [13]. The significance of these two results is that if we can find an efficient way to compute MGMM representations, then we will have a powerful and general way of statistically modelling data, in any number of dimensions, which deals directly with the effects of motion on the representation.

3 MGMM

The key to applying these ideas is to use a sequential approach, which leads to a multiresolution tree structure: MGMM. We start from the natural assumption that we do not know the density of interest, but wish to estimate it from some data, such as an image or set of images. For example, in the simplest case, a grey-level image is modelled as a set of three-dimensional samples from an unknown density. If these data are denoted $\boldsymbol{X}_i, 1 \leq i \leq N$, we can compute the sample mean and covariance [12] and thence infer a single multivariate Gaussian model $g_{\Sigma_0}(\boldsymbol{x} - \boldsymbol{\mu}_0)$, where $\boldsymbol{\mu}_0, \Sigma_0$ are the sample (maximum likelihood) estimates for the data. Now, with luck, this will model the data adequately. If not, then we split the data into two parts and model each with a Gaussian. This can be done using the Markov Chain Monte Carlo sampling technique described in [10], which treats the inference as one containing *hidden variables*, namely the class Z_i to which each datum belongs; it samples from the posteriors for the population size, means, and covariances, assuming *conjugate* priors, whose parameters are simply those of the population as a whole. Thus the prior for the means of the two classes are Gaussian, while the covariances are drawn from a Wishart density and the population sizes from a Dirichlet distribution. Sampling for (i) the hidden variables and (ii) the corresponding population densities gives the two Gaussians, based on the posterior estimates from the sampler $g_{\Sigma_{1j}}(\boldsymbol{x} - \boldsymbol{\mu}_{1j}), j \in [0, 1]$. Clearly, this leads to a binary tree-structured recursion:

(1) Select class j and test its normal density approximation for 'goodness-of-fit'. If the fit is adequate, terminate, *else*

(2) Split class j into two components: class $j0$ and class $j1$, by sampling the hidden variables $Z_{ji}, 1 \leq i \leq N_j$ and thence obtaining a Bayesian estimate

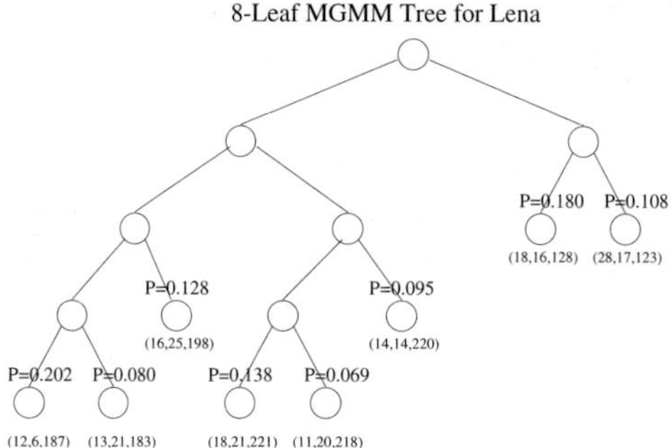

FIG. 1. MGMM tree for Lena image, showing component weights and mean parameters for each vertex in the tree.

of the class means and covariances by sampling from the posteriors, given the prior $g_{\Sigma_j}(\boldsymbol{x} - \boldsymbol{\mu}_j)$.

A description of the sampling procedure is given in the Appendix. Thus, provided we can find a measure of 'goodness-of-fit', we obtain a tree of Gaussians of decreasing variance, which has the property of approximating the density of the data as we desired. The question of goodness of fit is not trivial, but a simple rule is to 'charge' an appropriate amount for each split, since it increases the complexity of the description by a fixed number of parameters, giving the rule

$$\ln \frac{P(\boldsymbol{Y}|\theta_{j0}, \theta_{j1})}{P(\boldsymbol{Y}|\theta_j)} > \gamma(n, N_j, N_{j0}) \tag{3.1}$$

where $\gamma(n, N, N_0) > 0$ depends in general on the size of the data set N, that of the partition N_0 and the dimension of the space n. With appropriate choice of γ, the common complexity measures, including the Bayes Factor, Akaike's Information Criterion and the Bayesian Information Criterion, can be implemented. The parameters θ_i define the Gaussian components for the MGMM and the log-likelihood ratio refers to the pre- and post-split Gaussian models. This is the scheme used in the experiments.

We conclude by observing that an alternative view of the MGMM description is as a patchwork of affine models, each leaf node being the result of a linear regression on the data [1]. For example, in the case of a grey-level image, the MGMM description gives for each class a Gaussian model, which is directly related to a least-squares approximation of the form

$$z_i(\boldsymbol{x}) = \boldsymbol{A}_i(\boldsymbol{x} - \boldsymbol{x}_i) + z_{i0} + \nu_i(\boldsymbol{x}) \tag{3.2}$$

where $z_i(.)$ is the grey-level as a function of the spatial coordinate $\boldsymbol{x}$ for the ith class and $\nu_i(.)$ is the *residual.* The matrix $\boldsymbol{A}_i$ is easily found from the covariance matrix Σ_i for that class and $\boldsymbol{x}_i, z_{i0}$ from the mean.

FIG. 2. Classification of Lena image from MGMM representation with eight leaves.

FIG. 3. Reconstruction corresponding to previous image.

4 Experiments

To illustrate MGMM, Figs 1–4 show its use in the segmentation of a well known image—the 256×256 pixel 'Lena'. In this case, the data are three-dimensional: two spatial dimensions and one for intensity. Thus Figs 2 and 3 shows the representation of the 'Lena' image in eight regions, each corresponding to a leaf of the MGMM tree, which is illustrated in Fig. 1. The classification of Fig. 2 is just the MAP classification of pixels based on all three coordinates (spatial and

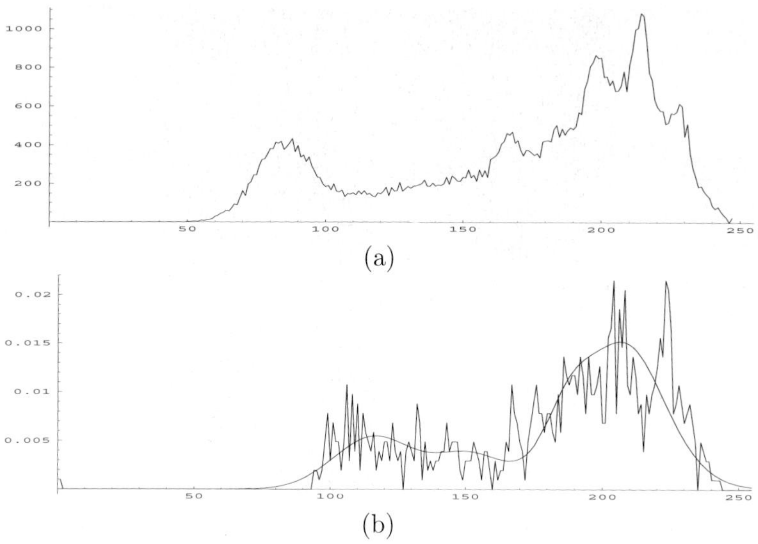

FIG. 4. Comparison of histogram with MGMM approximation: (a) histogram from full 256×256 pixel 'Lena' image, (b) histogram and MGMM approximation from level 3 of grey-level pyramid.

intensity), while the reconstruction of Fig. 3 uses the least-squares approximation based on the estimated mean and covariance associated with the MAP class at that pixel, as in (3.2). Figure 4 compares the approximation of the intensity probablility density based on the MGMM with the intensity histogram on level 3 of a grey-level pyramid, which has just 1024 pixels. This is a trivial computation from the MGMM representation, which clearly gives a better approximation of the density than can be obtained by simple histogramming.

The last experiment was designed to test the motion properties of MGMM, using the 'Miss America' sequence. In the example shown in Fig. 5, the MGM model of one frame of a sequence is moved using local affine motion estimates and used to approximate the next frame of the sequence. In this example, the model had 66 leaves in total; the signal–noise ratio was 20 dB, for both frames, demonstrating that the approximation does not have to be recomputed *de novo* for each frame of a sequence.

5 Conclusions

In this paper, a novel and versatile statistical image representation has been presented. The theory of MGMM as a form of statistical approximation was outlined and its use in describing images and sequences illustrated. The key properties which mark MGMM out as a representation are as follows: although it is a statistical model, it incorporates spatial relationships directly through spatial co-ordinates, rather than through neighbour correlations (like MRF models); it is 'auto-scaling' because it classifies data based on their likelihood, rather than on

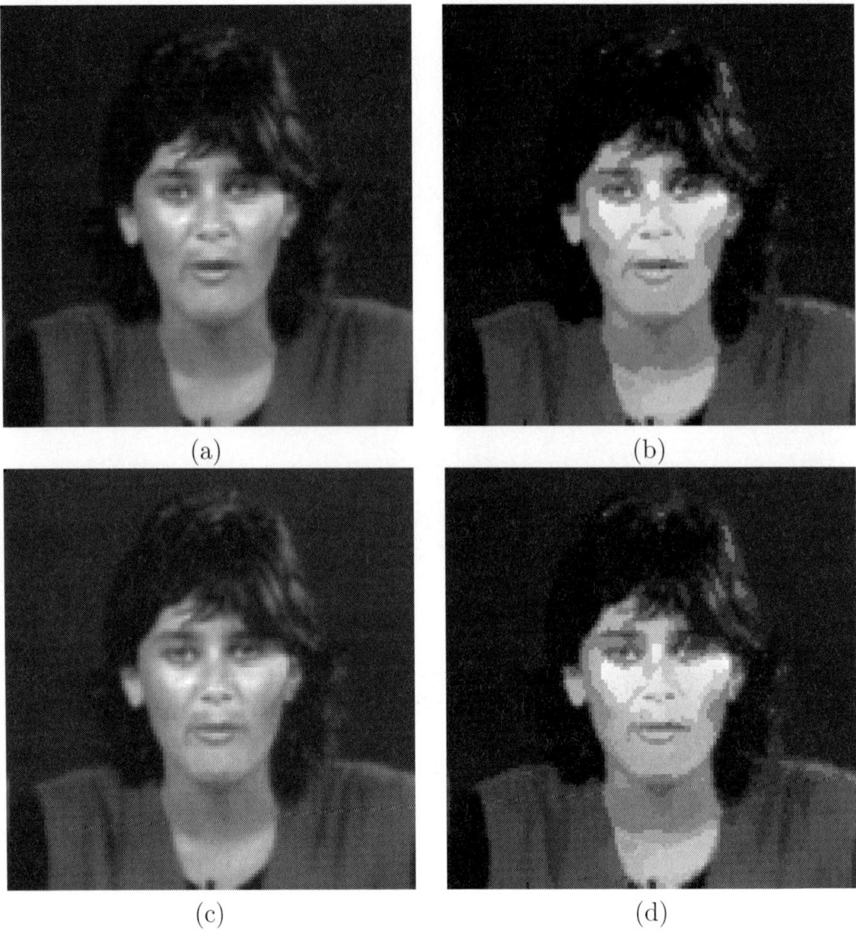

FIG. 5. (a) Frame 15 of Miss America sequence and (b) reconstruction from MGMM tree using 66 classes. (c) Original frame 16 and (d) reconstruction based on moved MGMM from frame 16.

Euclidean distance; because it is multiresolution, it allows efficient computation; because of the closure of the Gaussian functions under affine motions, it deals directly with the problem of image motion. This combination of properties makes MGMM a unique approach to image modelling, which we have used in the analysis of colour images, stereopsis and motion estimation, with promising results. While there are some important problems yet to be solved, the MGMM approach offers a new way of attacking the problems of inference which are common in image analysis and computer vision.

Acknowledgement

This work was supported by the UK EPSRC.

Bibliography

[1] Breiman, L., Friedman, J.H., Ohlsen, R.A. and Stone, C.J. *Classification and Regression Trees.* Wadsworth, 1984.

[2] Burt, P.J. and Adelson, E.H. The Laplacian pyramid as a compact image code. *IEEE Trans. Commun.*, COM(31), 1983.

[3] Daubechies, I. Orthogonal bases of compactly supported wavelets. *Comm. on Pure and Appl. math.*, XLI, 1988.

[4] Geman, S. and Geman, D. Stochastic relaxation, Gibbs distributions, and the Bayesian restoration of images. *IEEE Trans. Patt. Anal. Machine Intell.*, **6**, 1984.

[5] Hassnew, M. and Sklansky, J. The use of Markov random fields as models of texture. In: *Image Modelling*, (ed., A. Rosenfeld). Academic, 1981.

[6] Mao, J. and Jain, A. Texture classification and segmentation using multiresolution simultaneous autoregressive models. *Patt. Recog.*, **25**, 1992.

[7] O'Hagan, A. *Kendall's Advanced Theory of Statistics.* Vol. 2B. London, Arnold, 1994.

[8] Press, S.J. *Applied Multivariate Analysis.* Florida, Robert E. Krieger, 1982.

[9] Richardson, S. and Green, P.J. On Bayesian analysis of mixtures with unknown numbers of components. *J. R. Statist. Soc.*, **59**, 1997.

[10] Robert, C.P. Mixtures of Distributions: Inference and Estimation. In: *Markov Chain Monte Carlo in Practise*, (eds, S. Richardson, W. R. Gilks, D. J. Spiegelhaler), Chapman and Hall, 1996.

[11] Roberts, S.J., Husmeier, D., Rezek, I. and Perry W. Bayesian approaches to Gaussian mixture modelling. *IEEE Trans. Patt. Anal. Machine Intell.*, **20**, 1998.

[12] Seber, G.A.F. *Multivariate Observations.* New York, Wiley, 1984.

[13] Wilson, R. Multiresolution Gaussian mixture models: theory and applications. Research report RR404, Department of Computer Science, University of Warwick, UK, December 1999.

[14] Witkin, A. Scale-Space Filtering. IN *Proc. of IEEE ICASSP-84*, 1984.

[15] Zhuang, X., Huang, Y., Palaniappan, K. and Zhao, Y. Gaussian mixture density modelling, decomposition and applications. *IEEEIP*, **5**, 1293–1301, 1996.

Appendix: A Gibbs sampler for normal-inverse Wishart variates with hidden variables

In using the MVN distribution as a *universal approximator*, it is possible to make use of the properties of this widely studied distribution. The development here follows that in [7] and [8]. A Bayesian analysis starts from the choice of priors and in the present application, it is obvious to use the so called *natural conjugate prior* distribution: the Normal-Inverse Wishart (NIW) distribution, for multivariate normal data with unknown mean and covariance. This is defined for n-dimensional data as

$$f(\boldsymbol{\mu}, \boldsymbol{\Sigma}) = k^{-1} \, |\boldsymbol{A}|^{\frac{d}{2}}| \, |\boldsymbol{\Sigma}|^{-\frac{d+n+2}{2}} \exp\left[-\frac{c}{2}(\boldsymbol{\mu} - \boldsymbol{a})^T \boldsymbol{\Sigma}^{-1}((\boldsymbol{\mu} - \boldsymbol{a}) - \frac{1}{2}\mathrm{tr}(\boldsymbol{\Sigma}^{-1}\boldsymbol{A})\right] \tag{5.1}$$

or, briefly, $f(\boldsymbol{\mu}, \boldsymbol{\Sigma}) = NIW(\boldsymbol{A}, d, \boldsymbol{a}, c)$, a function of the parameters $d > n+1$, the *degrees of freedom*, c, the prior weight on the mean, $\boldsymbol{a}$, the prior mean and $\boldsymbol{A}$, the prior covariance. Increasing c and d attaches relatively more weight to the prior, for a given sample size N. The normalizing constant k is given by

$$k = 2^{\frac{n(d+1)}{2}} \pi^{\frac{n(n+1)}{4}} c^{\frac{n}{2}} \prod_{i=1}^{n} \Gamma\left(\frac{d+1-i}{2}\right). \tag{5.2}$$

The likelihood for the data $\boldsymbol{X}_i, 1 \leq i \leq N$, is then simply a product of normals, assuming that data are independently sampled, that is,

$$f(\boldsymbol{X}|\boldsymbol{\mu}, \boldsymbol{\Sigma}) = (2\pi)^{-\frac{nN}{2}} |\boldsymbol{\Sigma}|^{-\frac{N}{2}} \exp\left[-\frac{1}{2}\sum_{i=1}^{N} (\boldsymbol{X}_i - \boldsymbol{\mu})^T \boldsymbol{\Sigma}^{-1}(\boldsymbol{X}_i - \boldsymbol{\mu})\right] \tag{5.3}$$

and because the NIW is the natural conjugate prior, the posterior is also NIW, with parameters [7]

$$\boldsymbol{a}^* = \frac{c\boldsymbol{a} + N\bar{\boldsymbol{X}}}{c+N} \tag{5.4}$$

where $\bar{\boldsymbol{X}} = N^{-1}\sum_i \boldsymbol{X}_i$ is the sample mean and

$$\boldsymbol{A}^* = \boldsymbol{A} + N\boldsymbol{S} + \frac{cN}{c+N}(\boldsymbol{a} - \bar{\boldsymbol{X}})(\boldsymbol{a} - \bar{\boldsymbol{X}})^T \tag{5.5}$$

where $\boldsymbol{S} = N^{-1}\sum_i(\boldsymbol{X}_i - \bar{\boldsymbol{X}})((\boldsymbol{X}_i - \bar{\boldsymbol{X}})^T$ is the sample covariance. It is not hard to see that in the posterior, the scalar parameters are $c^* = c + N, d^* = d + N$.

This sampler follows exactly the model proposed by Robert [10]. It uses NIW as the prior for the sample data and a Dirichlet prior for the population size (experience has shown that this is not always helpful, however). Thus, given the estimate $\boldsymbol{\mu}_j, \boldsymbol{\Sigma}_j$ for the mean and covariance at a node j, the sampler performs the following steps:

(1) Sample $\boldsymbol{\mu}_{ji}, \boldsymbol{\Sigma}_{ji}, i = 1, 2, \sim \mathrm{NIW}(\boldsymbol{\mu}_j, \boldsymbol{\Sigma}_j, c_i^*, d_i^*)$.

(2) Sample for the population size using the Dirichlet distribution.

(3) Sample the hidden variables $Z_k \in [1, 2], 1 \leq k \leq N_j$ using a Gibbs sampler and the current estimates of $P, \boldsymbol{\mu}_{ji}, \boldsymbol{\Sigma}_{ji}, i = 1, 2$.

(4) Calculate the posterior NIW parameters, based on the classified data and (5.4)–(5.5).

This is repeated, starting with $c_i^* = c_i = 1, d_i^* = d_i = n+2$, the minimum values, to keep the prior vague.

It may be noted here that a deterministic algorithm, based on $2-means$, but using the log-likelihood rather than simple Euclidean distance, works adequately on some simpler problems, for example where classes are convex. In general, however, it is little cheaper to implement $2 - means$ and much more likely to settle in a local minimum.

Mathematics in Biomedical Signal Processing

Sabine Van Huffel
Katholieke Universiteit Leuven, ESAT Laboratory, Division SISTA/COSIC, Belgium

Abstract

In biomedical signal processing, the aim is to extract clinically, biochemically or pharmaceutically relevant information (e.g metabolite concentrations in the brain) in terms of parameters out of low-quality measurements in order to enable an improved medical diagnosis. Three case studies in different biomedical areas are investigated. For each topic, the applied computational techniques—making use of advanced linear algebra, optimization, signal processing, and system identification—-are explained.

1 Introduction

Biomedical data are typically affected by large measurement errors, largely due to the noninvasive nature of the measurement process or the severe constraints to keep the input signal as low as possible for safety and bio-ethical reasons. Accurate and automated quantification of this information requires an ingenious combination of the following four issues:

(1) an adequate pretreatment of the data,

(2) the design of an appropriate model and model validation,

(3) a fast and numerically robust model parameter quantification method, and

(4) an extensive evaluation and performance study, using *in vivo* and patient data, up to the embedding of the advanced tools into userfriendly user interfaces to be used by clinicians.

Signal preprocessing implies the use of advanced signal processing tools (using generalized singular value decomposition (SVD), total least squares (TLS), multichannel adaptive filtering, subspace algorithms, maximum entropy methods) for achieving the best signal separation (noise elimination, filtering out the relevant information such as one metabolite concentration). The next issue of *modelling and model validation* requires data mining techniques (such as principal component and canonical correlation analysis, higher-order statistics) for data exploration (retrieval of biomedically relevant information and relationships), as well as the use of so-called grey-box model fitting techniques which fully exploit the available prior knowledge (biochemical, patient-related, expert-based), such as a combination of constrained optimization with subspace-based methods or neural networks combined with a Bayesian-based input selection scheme. *Model parameter quantification* strikes the right compromise between computational speed and quality of the computed estimates (numerical robustness). Diverse optimization algorithms (nonlinear least squares, weighted subspace algorithms) are combined in an optimal way, tailored to the specific biomedical needs, and improved in efficiency by exploiting the matrix structure (e.g. structured TLS, displacement rank). Finally, the clinical *performance needs to be evaluated*, using patient data and statistical tools commonly used in biomedicine, such as the Receiver Operating Characteristic curve, and implemented in a userfriendly way for use in a clinical environment. The following biomedical applications and underlying computational problems, making use of linear algebra, signal processing, and system identification, are considered in this paper:

- Quantification of the kidney impulse response in renography (Section 2),
- Quantification of metabolite concentrations and images using *in vivo* Magnetic Resonance Spectroscopic data (Section 3),
- Quantification of tactile sensation through oral implants using Trigeminal Somatosensory Evoked Potentials (Section 4).

2 Renography

The goal is to determine via deconvolution the so-called renal retention function of the kidney, which in system theoretic terms corresponds to the impulse response of the system. The renal retention function visualizes the mean whole kidney transit time of one unit of a tracer injected into the patient and enables a physician to evaluate the renal function and dysfunction severity after transplantation. A radioactive tracer is injected in an artery of the patient. The measured input of the system (u) is the noise-perturbed arterial concentration of the radioactive tracer as a function of time. This concentration is measured by means of a gamma camera, and thus in discretized time $u(k)$ represents the number of counts registered in the vascular region at the entrance of the kidney under study during the kth sampling interval. The measured output $y(k)$ (the so-called renogram) represents the (noise-perturbed) number of counts registered in the whole kidney region by the gamma camera in the kth sampling interval. Deconvolution analysis of the renogram is based on modelling the kidney as a linear time-invariant causal system with zero initial state and reduces to solving an overdetermined set of linear equations, say $Ax \approx y$, where the $u(k)$ are arranged in $A \in \Re^{m \times n}$ in Toeplitz form with zero upper-triangular block and y contains the outputs. The current deconvolution techniques based on backsubstitution or Gaussian elimination are not reliable: they are sensitive to the inaccuracies present in the data. A more reliable, stable and efficient deconvolution technique based on the SVD [7] and TLS [20] has been proposed and is shown to outperform the current one [19]. Exploiting the structure by formulating the kernel problem as a Structured TLS (STLS) problem further improves the statistical accuracy of the TLS estimator. To this end, we used a simulation example as described in [19] imitating a real *in vivo* measurement and performed a Monte Carlo study. Table 1 shows that the relative errors of the STLS estimator are

Table 1 Relative error $\frac{\|x-x_0\|_2}{\|x_0\|_2}$ for the TLS and STLS estimator, averaged over 100 runs, for different noise standard deviations σ_ν

σ_ν	0.071	0.1	0.224	0.316	0.707	1	2.24	3.16
TLS	0.0015	0.0021	0.0046	0.0064	0.0138	0.0204	0.0454	0.0660
STLS	0.0013	0.0018	0.0041	0.0057	0.0123	0.0180	0.0402	0.0601

Table 2 Ratio of the number of Matlab flops needed by STLS algorithm [13] to that of the fast STLS algorithm [11] for computing the (Q)R factorization of $A_{m\times n}$ (only that part is different)

$m \times n$	300×20	600×20	300×10
Flop ratio	1327.67	4949.15	2290.15

9 to 14 per cent lower than those of the TLS estimator, confirming its statistical superior performance. The STLS algorithm, as described in [13], solves, as the kernel problem in each iteration step, a least squares problem involving a higher-dimensional structured and sparse data matrix. However, this algorithm, requiring $O\left((m+n)^3\right)$ operations, does not exploit these matrix features. We developed a computationally faster algorithm of $O(mn+n^2)$ operations [11] based on the generalized Schur algorithm [9]. The improved efficiency, as shown in Table 2, is obtained by fully exploiting the low displacement rank of the involved matrices (displacement rank 5) and the sparsity of the corresponding generators throughout all computations. In summary, the main computational issues are:

- SVD enables one to filter out noise from the raw input/output curves;
- TLS outperforms Gaussian elimination for impulse response estimation;
- Structured TLS outperforms TLS for impulse response estimation;
- Displacement rank algorithms enhance the computational efficiency of the STLS estimator with a factor 100 to 5000.

3 Magnetic resonance spectroscopy

This area of research concerns the improved quantitative analysis of time-domain data of humans and animals acquired from *in vivo* Medical Magnetic Resonance Spectroscopy (MRS) measurements [18].

Introduction. MR Imaging (MRI) enables one to noninvasively view the inside of a human body (*in vivo*). As a result, MRI has become an important tool in medicine. The high-quality images displayed lately in the literature originate from the ubiquitous water in the human body. Apparently, this water can be made to serve as a medium that reveals anatomical details. But MRI can do more. It appears possible to view also molecules other than water. In fact, when MRI is fine-tuned, it can distinguish between most molecules of different chemical composition. This is achieved by exploiting the powerful spectroscopic capabilities of MR. Thus, it becomes possible to view different metabolites in a human body separately. It is clear that images of metabolite concentrations over an extended region provide very useful additional (biochemical or pharmaceutical) information to a physician and enable a detailed medical diagnosis. In addition, it is possible to acquire detailed spectroscopic information from a selected trouble spot in a patient. Note that the concentration of metabolites in the human body is more than 1000 times lower than that of water, or, on various locations, than that of fat. Consequently, the spectral region of the metabolites is disturbed by a strong tail of the spectrum of water or fat. In order to produce a good metabolite image or a reliable table of concentrations in the case of spectroscopy, the disturbing water/fat contribution is to be removed from the signal.

Research area. Quantification of metabolite signals concerns estimation of the amounts of various metabolites in a selected volume of interest. The model function often used to represent the free induction decay signal is:

$$y_n = \bar{y}_n + e_n = \sum_{k=1}^{K} a_k e^{j\phi_k} e^{(-d_k + j2\pi f_k)t_n} + e_n \qquad n = 0, 1, \ldots, N-1, \qquad (3.1)$$

where K represents the number of different resonance frequencies and $j = \sqrt{-1}$. The amplitude a_k is proportional to the number of nuclei contributing to the spectral component with frequency f_k. The damping d_k provides, among others, information about the mobility and macromolecular environment of the nucleus. Often the acquisition of the signal is started after a time delay t_0, the receiver dead time. Therefore, the sampling time points t_n can be written as $t_n = n\Delta t + t_0$ where Δt denotes the sampling interval. The quantity $2\pi f_k t_0$ is also called the first-order phase. The phase ϕ_k consists of a so-called zero-order phase ϕ_0 and an additional phase factor ϕ'_k which represents extra degrees of freedom that may be required under certain experimental conditions (but usually all ϕ'_k are zero). The noise term e_n is assumed to be circular complex white Gaussian noise. Even though individual metabolite signals can intrinsically be represented by a damped complex exponential, in practical situations a perfect homogeneous magnetic field cannot be obtained throughout the sample under investigation. This results in deviations from the ideal model function.

The research aims at extracting these parameters out of the measured MRS data which are of low quality (low signal-to-noise ratio). This involves the use of advanced computational methods from numerical linear algebra such as the SVD, the TLS method, canonical-angles and subspace-based methods [7]. The main focus lies on the development of numerically reliable so-called blackbox (since the parameters do not have a physical meaning) algorithms by making use of these tools. The basic concepts of the method are briefly outlined. The data y_n, $n = 0, \ldots, N-1$ are arranged in a Hankel matrix as follows:

$$H = \begin{bmatrix} y_0 & y_1 & \cdots & y_M \\ y_1 & y_2 & \cdots & y_{M+1} \\ \vdots & \vdots & \vdots & \vdots \\ y_{N-M-1} & y_{N-M} & \cdots & y_{N-1} \end{bmatrix}, \text{ with } M \geq K,\ N-M > K. \quad (3.2)$$

Kung *et al.*'s method [10], introduced to the field of MRS by Barkhuijsen *et al.* [1] and called HSVD, computes the SVD of $H = U\Sigma V^H$ where Σ is a diagonal matrix containing the singular values $\sigma_1, \ldots, \sigma_p$ with $\sigma_1 \geq \ldots \sigma_p$, p being the rank of the matrix. The dimensions of the matrix H are chosen such that H is as square as possible to maximize parameter accuracy. The matrix H is truncated to rank k, that is, $H_K = U_K \Sigma_K V_K^H$ where U_K, V_K are the first K columns of U, V respectively and Σ_K is the upper-left $K \times K$ submatrix of Σ. Using U_K, the following set of equations is formed:

$$U_K^{\uparrow} \approx U_{K\downarrow} Q, \quad (3.3)$$

where the up (down) arrow stands for deleting the top (bottom) row respectively. HSVD computes an estimate of the matrix Q by solving (3.3) in a LS sense. The eigenvalues of the latter matrix are the estimates $\hat{z}_k$, $k = 1, \ldots, K$, of the signal poles. In order to compute amplitudes and phases, these estimates are backsubstituted in the model (3.1) and fitted to the original data with the LS method.

Again variants of HSVD exist. An improved variant is the HTLS algorithm [22] which computes the TLS solution of (3.3), leading to more accurate parameter estimates. In particular, the damping factors benefit from the use of TLS especially when the peak signal-to-noise ratio is low thereby also improving the accuracy of the amplitudes and phases. Experiments based on simulated data and *in vivo* NMR signals, as well as NMR signals from *in vitro* perfused organs, confirm the analysis. The HLSVD algorithm [12] is a computationally efficient version that computes only part of the SVD by using the Lanczos algorithm. In addition, improved signal enhancement techniques based on the truncated SVD and minimum variance estimation [2, 21] have been developed for 'cleaning up' the data before estimating the signal parameters, thereby improving the resolution performance of the basic HSVD and HTLS algorithms at a comparable parameter accuracy [3].

The use of these advanced blackbox techniques has also been investigated in the quantification of proton spectra. These signals are strongly dominated by the signal of water (fat) which is first to be removed before quantification of the metabolites of interest. However, the shape of these unwanted contributions deviates from (3.1), and this occurs in an unknown way. Fortunately, these deviating shapes can be successfully parametrized by black-box methods based on state space modelling and SVD truncation, as confirmed by simulations and *in vivo* spectroscopic images [15, 17].

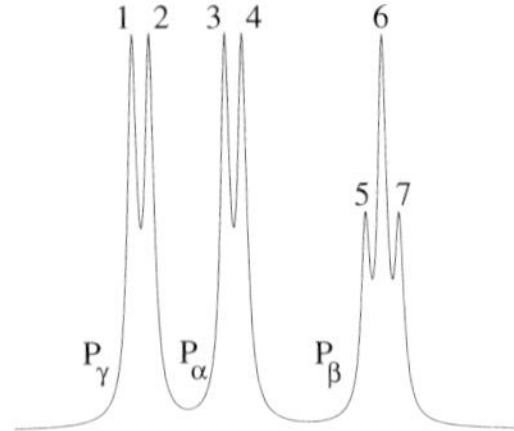

FIG. 1. Theoretical frequency spectrum of ATP.

Although these techniques are blackbox we were able to include several types of prior knowledge by translating this information, such as known signal poles, known frequency differences, and equality of damping factors, into linear algebraic concepts such as orthogonal projection onto subspaces, canonical angles, subspace intersections, thereby improving the quantified parameters [4, 5, 23].

Nevertheless, although these black-box techniques are very efficient and enable automation of the computations, they cannot make use of all available biochemical prior knowledge. Imposition of this prior knowledge is of paramount importance in order to further improve the accuracy. When the underlying assumptions concerning the model function and noise distribution are satisfied, obtaining the maximum likelihood (ML) estimates comes down to minimizing the squared difference between the data and the model function $\sum_{n=0}^{N-1} |y_n - \bar{y}_n|^2$ ($|\cdot|$ denotes the modulus of a complex entity).

This is possible by nonlinear least squares fitting using, for example, the variable projection method [18]. This prior knowledge can in some cases be derived from spin properties as is here illustrated using adenosine triphosphate (ATP) as an example. The three phosphorus atoms of ATP, P_α, P_β and P_γ, experience a different chemical/electrical environment and, as a result of shielding, they will resonate at slightly different frequencies, giving rise to three groups of spectral lines in the spectrum. As a result of the spin–spin coupling between the neighbouring phosphorus nuclei, the groups corresponding to the α and γ phosphates of ATP are split into two peaks (a doublet) and the β phosphate group is decomposed into three peaks (a triplet), as shown in Fig. 1. The improved performance of NLLS combined with all available prior knowledge is illustrated using an ^{31}P signal obtained from an *ex vivo* perfused rat liver and recorded at 81.1 MHz (4.7 T Bruker Biospec). The signal consists of 128 complex data points and the sampling interval is 0.2 ms. Peaks from an external standard (ES), phosphomonoesters (PME), inorganic phosphate (P_i), phosphodiesters (PDE) and adenosine triphosphate (α-, β-, and γ-ATP) can be observed. The broad resonance underlying the mentioned metabolites originates from phosphorous nuclei in less mobile molecules. In order to reduce its influence, the first five data points are excluded in the time-domain fit. The following prior knowledge was imposed: equal dampings within each of the multiplets of ATP, 16 Hz as frequency splitting within peaks belonging to the same multiplet, amplitude ratios

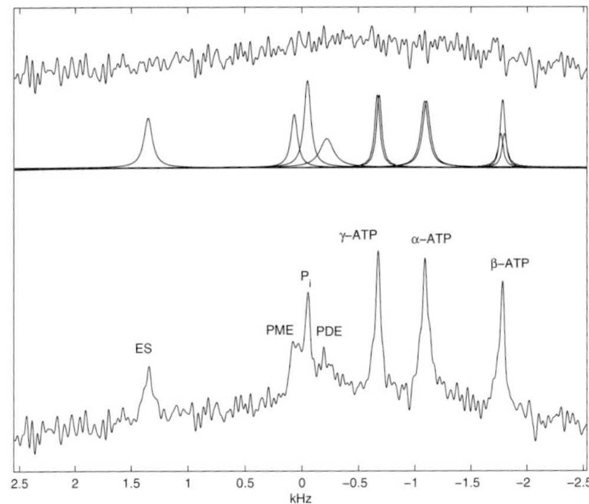

FIG. 2. *ex vivo* ^{31}P signal from a perfused rat liver. From bottom to top: the FT spectrum, the individual peaks fitted with AMARES and the residual.

of 1:2:1, 1:1, 1:1 between the β-, α- and γ-ATP respectively. The dampings of PME and P_i are constrained to be equal. The begin time was estimated and the phases ϕ_k of all peaks were constrained to be equal. The model function used is given by (3.1). The resulting fit with AMARES [16], a sophisticated NLLS fitting algorithm, is shown in Fig. 2.

A disadvantage of this technique is that suitable starting values of the model parameters to be fitted are to be provided. One way to alleviate this task is to filter out signals in frequency regions currently not of interest which can be done by means of a truncated SVD. The challenge is to combine the ease-of-use and efficiency of the black-box methods with the improved accuracy of nonlinear least-squares fitting. In summary, the main computational issues here are:

- Truncated SVD and variants enhance the signal and reject unwanted parts;
- Rank-revealing complete orthogonal decompositions [17] and Lanczos methods [15] outperform truncated SVD in speed for fast removal of residual water in proton spectra;
- Subspace-based tools (orthogonal complement, intersection, canonical angles) enable automated MRS data quantification using prior knowledge
- Fast structured TLS algorithms improve blackbox MRS data quantification;
- Methods using complex Principal Component Analysis are useful for alignment and fast quantification of large MRS images and data sets [25];
- Nonlinear least squares fitting and variable projection optimization methods for Time Domain MRS data quantification exploit all available prior knowledge, thereby maximizing parameter accuracy.

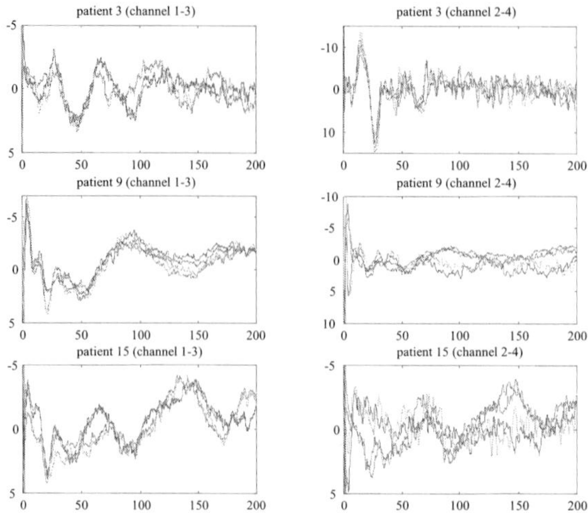

FIG. 3. Results of conventional EA TSEP signals of three different patients. Each channel consists of two runs. Post-stimulus analysis time is 200 ms.

4 Osseoperception

The very weak and noisy trigeminal somatosensory evoked potentials (TSEPs) are considered, which are evoked by electrical stimulation of an endosseous oral implant [24]. The aim is to make the tactile sensation through endosseous implants objective. Although a lot of nerves are damaged by extraction of teeth, it has been reported from psychophysical tests that patients rehabilitated with an osseointegrated implant seem to have a subjectively improved ability to feel through their prosthesis and the anchoring implant in the bone. We want to find an objective and noninvasive measurement method for the investigation of the trigeminal nerve and try to explain this feedback mechanism. The TSEPs are applied for this purpose. These are voltage changes evoked by peripheral stimulation of the trigeminal nerve. The potentials in the first 50 ms are believed to reflect cortical activity and can be recorded noninvasively in human subjects using scalp electrodes [6]. Since the amplitudes of the EP response are often smaller than those of the ongoing EEG signal, it is necessary to repeat the stimulus. By ensemble averaging (EA) of the responses to repetitive stimulation, the signal-to-noise ratio (SNR) is improved and the EP can be estimated. Figure 3 shows some EA TSEPs for three different subjects. Per channel, two signals, acquired during two consecutive runs, are displayed. Channels 1 and 3, and 2 and 4 are depicted in one graph, as their active recording electrodes are (nearly) the same. The recordings are very noisy and, as a consequence, an appropriate noise reduction algorithm is needed for enhancing the estimation of the TSEP waveform. By means of the normalized multichannel correlation coefficient [8], applied to the recordings with and without stimulation, a reliable statistical detection of the TSEP in the noise is performed. This coefficient is defined as follows:

$$D = \sum_{k=1}^{M} \Pi_{i=1}^{K} y_i(k) \;/\; \Pi_{i=1}^{K} \left(\sum_{k=1}^{M} (y_i(k))^K \right)^{1/K} \tag{4.1}$$

where M is the length of the analysis interval which is 1000 data points, K is the number of recorded signals per patient (normally $K = 8$), i is the recording index and k is the time index. This coefficient D can take values between -1 and $+1$ and reveals the presence of a transient signal (EP) in the noisy recordings whenever the logarithm of $|D|$, calculated from the recorded evoked potentials (or post-stimulus data), is larger than the logarithm of $|D|$, calculated from the EEG data (which are in fact the zero-stimulus or pre-stimulus data), as a function of M. For nine out of 15 patients, we are able to detect the transient signal in the background EEG activity. Since the TSEP and the background EEG are severely overlapping in time as well as in frequency, separation of these signals cannot be realised by simple windowing or classical (bandpass) filtering. More involved approaches are needed, such as the multichannel SVD-based filtering method applied here. The SVD-based filtering method for one channel was introduced by Cadzow [2] and decomposes the signal based on energetic criteria. As the TSEP is described by the most energetic component, it is extracted easily by selecting the largest singular values where p is the filter order to be estimated. This is done by arranging the data into a Hankel matrix of which the first column and last row contains the recorded signal. This matrix is reduced to lower rank p via a truncated SVD after which the Hankel structure is restored by averaging the elements along the anti-diagonals. Since the latter operation destroys the rank property, the procedure is repeated iteratively in order to converge to a Hankel matrix of desired rank p. The filtered signal is then retrieved from the first column and last row of the resulting matrix. This method has been successfully applied in diverse signal enhancement applications [21] (see also Section 3). Constructing a Hankel matrix for each signal of the different channels and applying this procedure to a block-Hankel matrix, which contains all the concatenated Hankel matrices, yield the multichannel SVD-based filtering method [14]. As this filter investigates all channels simultaneously the correlation between them can be fully exploited, thereby improving the SNR and enhancing the TSEP waveforms. The multichannel SVD-based filtered TSEP signals of Fig. 3 are shown in Fig. 4 (restricted to the first 60 ms.). Five iterations are used. The filter order is set to 9 and found to be a good compromise between maintaining sufficient relevant signal information and removing the high-frequency (low-energy) noise. In summary, the computational issues are:

- Thresholding the generalized correlation coefficient enables detection of evoked potentials;

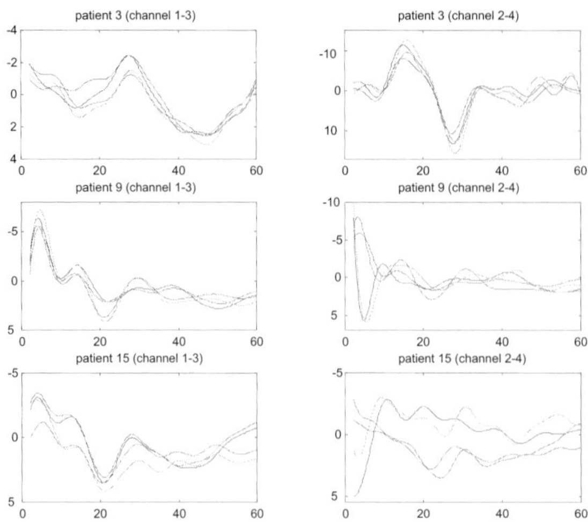

FIG. 4. Multichannel SVD based filtering of the TSEPs in Fig. 3.

- Multichannel filtering, based on truncated SVD and Hankelization of block matrices repeated iteratively, extracts the common information in the data (TSEPs) from the background EEG, thereby improving extraction of characteristic features, such as peak latencies and amplitudes.

5 Conclusions

Via a selected number of case studies, it is shown that advanced signal processing tools, strongly based on numerical linear algebra, are the key building blocks in biomedical signal processing software for improving accuracy, speed and automation. (Additional information, as well as other case studies in biomedical signal processing, are provided on the website http://www.esat.kuleuven.ac.be/sista/members/biomed/)

6 Acknowledgements

This paper presents research results of the Belgian Programme on Interuniversity Poles of Attraction (IUAP P4-02 and P4-24), initiated by the Belgian State, Prime Minister's Office—Federal Office for Scientific, Technical and Cultural Affairs, of the European Community TMR Programme, Networks,project CHRX-CT97-0160, of the Brite/Euram Programme, Thematic Network BRRT-CT97-5040 'Niconet', of the Concerted Research Action (GOA) projects of the Flemish Government MEFISTO-666 (Mathematical Engineering for Information and Communication Systems Technology), and of the IDO/99/03 project (K.U.Leuven).

Bibliography

[1] Barkhuysen, H., de Beer, R. and van Ormondt, D. (1987). Improved algorithm for noniterative time-domain model fitting to exponentially damped magnetic resonance signals. *J. Magn. Reson.*, **73**, 553–557.

[2] Cadzow, J.A. (1988). Signal Enhancement - A composite property mapping algorithm. *IEEE Trans. Acoust. Speech Signal Proc.*, **ASSP-36**, 49–62.

[3] Chen, H., Van Huffel, S., Decanniere, C. and Van Hecke, P. (1994). A signal-enhancement algorithm for the quantification of NMR data in the time domain. *J. Magn. Reson.*, **A 109**, 46–55.

[4] Chen, H., Van Huffel, S., van Ormondt, D. and de Beer, R. (1996). Parameter estimation with prior knowledge of known signal poles for the quantification of NMR spectroscopy data in the time domain. *J. Magn. Reson.*, **A 119**, 225–234.

[5] Chen, H., Van Huffel, S., van den Boom, A. and P. van den Bosch, P. (1997). Subspace-based parameter estimation of exponentially damped sinusoids using prior knowledge of frequency and phase. *Signal Process.*, **59**, 129–136.

[6] Fagade, O.O. and Wastell, D.G. (1990). Trigeminal somatosensory evoked potentials: technical parameters, reliability and potential in clinical dentistry. *J.Dent.*, **18**, 137–141.

[7] Golub, G.H. and Van Loan, C.F. (1996). *Matrix computations* (3rd edn). Johns Hopkins University Press, Baltimore, MD.

[8] Henning, G. and Husar, P. (1995). Statistical detection of visually evoked potentials. *IEEE Eng. Med. Biol. Magazine*, **July/August**, 386–389.

[9] Kailath, T. and Sayed, A.H. (1995). Displacement structure: Theory and Applications. *SIAM Rev.*, **37**, 297–386.

[10] Kung, S.Y., Arun, K.S. and Rao, D.V.B. (1983). State-space and singular-value decomposition-based approximation methods for the harmonic retrieval problem. *J. Opt. Soc. Am.*, **73**, 1799–1811.

[11] Mastronardi, N., Lemmerling, P. and Van Huffel, S. (2000). Fast structured total least squares algorithm for solving the basic deconvolution problem. *SIAM J. Matrix Anal. Appl.*, **22**, 533–553.

[12] Pijnappel, W.W.F., van den Boogaart, A., de Beer, R. and van Ormondt, D. (1992). SVD-based quantification of magnetic resonance signals. *J. Magn. Reson.*, **97**, 122–34.

[13] Rosen, J.B., Park, H. and Glick, J. (1996). Total least norm formulation and solution for structured problems. *SIAM J. Matrix Anal. Appl.*, **17**, 110–126.

[14] Swinnen, A., Van Huffel, S., Van Loven, K. and Jacobs, R. (2000). Detection and multichannel SVD based filtering of trigeminal somatosensory evoked potentials. *Med. Biol. Eng. Comput.*, **38**, 297–305.

[15] van den Boogaart, A., van Ormondt, D., Pijnappel, W.W.F., de Beer, R. and Ala-Korpela, M. (1994). Removal of the water resonance from ^{1}H mag-

netic resonance spectra. In *Mathematics in Signal Processing III*, (ed. J. G. McWhirter), Clarendon Press, Oxford, 175–195.

[16] Vanhamme, L., van den Boogaart, A. and Van Huffel S. (1997). Improved method for accurate and efficient quantification of MRS data with use of prior knowledge. *J. Magn. Reson.*, **129**, 35–43.

[17] Vanhamme, L., Fierro, R., Van Huffel, S. and de Beer, R. (1998). Fast removal of residual water in proton spectra. *J. Magn. Reson.*, **132**, 197-203.

[18] Vanhamme, L., Sundin, T., Van Hecke, P. and Van Huffel, S. (2001). MR Spectroscopy quantitation: a review of time-domain methods. *NMR in Biomedicine*, **14**, 233–246.

[19] Van Huffel, S., Vandewalle, J., De Roo, M.Ch. and Willems, J.L. (1987). Reliable and efficient deconvolution technique based on total linear least squares for calculating the renal retention function. *Med. Biol. Eng. Comput.*, **25**, 26–33.

[20] Van Huffel, S. and Vandewalle, J. (1991). *The Total Least Squares Problem: Computational Aspects and Analysis*. SIAM, Philadelphia.

[21] Van Huffel, S. (1993). Enhanced Resolution Based on Minimum Variance Estimation and Exponential Data Modeling. *Signal Process.*, **33**, 333–355.

[22] Van Huffel, S., Chen, H.,Decanniere, C. and Van Hecke, P. (1994). Algorithm for Time-Domain NMR Data Fitting Based on Total Least Squares, *J. Mag. Res.*, A **110**, 228–237

[23] Van Huffel, S., Lagae, L., Vanhamme, L., Van Hecke, P. and van Ormondt, D. (1998). Improving blackbox MRS data quantification by means of prior knowledge. Abstracts 15th Annual Meeting European Society for Magnetic Resonance in Medicine and Biology (ESMRMB 98, Geneva, Switzerland, Sept. 17–20), 1998, p. 15.

[24] Van Loven, K., Jacobs, R., Swinnen, A., Van Huffel, S., Van Hees, J., van Steenberghe, D. (2000). Sensations and trigeminal somatosensory-evoked potentials elicited by electrical stimulation of endosseous oral implants in humans. *Arch. Oral Biol.*, **45**, 1083–1090.

[25] Wang, Y., Van Huffel, S., Heyvaert, E., Vanhamme, L. and Van Hecke, P. (2001). Automatic frequency correction for quantitation of Magnetic resonance Spectroscopic Images. Proc. Int. IMA Workshop on Mathematics in Signal Processing, this volume.

Automatic Frequency Correction for Quantification of Magnetic Resonance Spectroscopic Images

Yu Wang, Sabine Van Huffel, Els Heyvaert and Leentje Vanhamme
Department of Electrical Engineering, Katholieke Universiteit Leuven Kasteelpark Arenberg 10, 3001 Leuven-Heverlee, Belgium

Paul Van Hecke
Biomedical NMR Unit, Faculty of Medicine, Katholieke Universiteit Leuven, Belgium

Abstract

We present two new methods based on Principal Component Analysis (PCA) using the complex Singular Value Decomposition (SVD) and cross-correlation of the magnitude spectra for correcting frequency shifts of resonances in sets of Magnetic Resonance (MR) spectra and spectroscopic images. After a description of the methods, we compare their performance on simulated data sets of Magnetic Resonance Spectroscopic Images and discuss their advantages and limitations.

1 Introduction

Magnetic Resonance Spectroscopic Imaging (MRSI) has been shown to be a valuable tool in medical diagnostics, in particular for tumour classification [5, 7, 8]. Recent improvements in MR technology enable the simultaneous acquisition of an MR spectrum in each voxel of a slice in the human body, called MRSI [6]. However, quantification of MRSI data, yielding information about metabolite concentrations, is always hampered by low signal-to-noise ratios (SNRs). S. Van Huffel *et al.* [11] applied the total least squares method to MRSI data quantification, assuming that the data are modelled by a sum of exponentially damped sinusoids. Their algorithm, called HTLS, is shown to improve the accuracy of all MRS parameter estimates as compared to HSVD. Recently, HTLS was generalized [3] to the quantification of MRSI data, resulting into the HTLSstack and HTLSsum algorithms.

As a widely used statistical technique, PCA was introduced in MRS data quantification by R. Stoyanova *et al.* [9]. Provided all spectra possess the same lineshape, PCA is able to quantify simultaneously all the spectra in one data set. In this way, the accuracy of the quantified data improves because PCA captures the common information in the spectral data set while the noise is significantly reduced. In [9] only the real part of the MRS data is considered and transformed to the frequency domain. An interval surrounding the peak of interest is selected and the points within this interval are stored row by row in a matrix, to which PCA is applied. In order to improve the accuracy and eliminate the influence of phase differences, PCA is performed on complex data using a complex SVD [4]. Mathematically, PCA decomposes the data set as a linear combination of basic features, the so-called principal components (PCs). It has been shown that under some conditions PCA successfully extracts quantitative metabolite information from data sets without any prior knowledge about the line shape [9, 4, 12].

The PCA method discussed above only applies if all spectra have the same frequency, line shape (and phase in case of real-valued data sets). Therefore, Brown and Stoyanova [2] developed a PCA-based method for aligning the spectra in frequency and phase using the real-valued part of the data set. This method was further improved in [12] by combining PCA with linear regression. However, working with the entire complex-valued data set has the big advantage that the spectra need only be aligned in frequency before quantifying the peak areas under each resonance using PCA. Phasing is no longer needed. This paper shows how these complex PCA based methods outperform the existing methods (HTLS, HTLSsum, HTLSstack, real PCA-based) and how to improve them further.

2 Methods

2.1 MRS data model function

It is assumed that, basically, each MR signal $y_n, n = 0, 1, \ldots, N-1$, acquired in the time domain in one particular voxel of a MRS image, can be modelled as a sum of exponentially damped complex-valued sinusoids (Also called Lorentzian. Note that other model functions, such as Gauss or Voigt line shapes, are sometimes used to model MRSI data.):

$$y_n = \breve{y}_n + e_n = \sum_{k=1}^{K} a_k e^{j\phi_k} e^{(-d_k + j2\pi f_k)t_n} + e_n \quad n = 0, 1, \ldots, N-1$$

where y_n is the nth measured data point of one MRS signal, $\breve{y}_n$ represents the nth value of the model function, $j = \sqrt{-1}$, a_k is the amplitude, ϕ_k the phase, d_k the damping factor, and f_k the frequency of the kth sinusoid ($k = 1, 2, \ldots, K$). K is the number of the sinusoids, $t_n = n\Delta t + t_0$ with Δt the sampling interval, t_0 the time between the effective time origin and the first data point to be included in the analysis, and e_n is complex white Gaussian noise. The amplitude a_k is directly related to the concentration of a certain metabolite, and should be estimated as precisely as possible.

2.2 Principal component analysis

Principal Component Analysis (PCA) is usually performed by means of an eigenvalue decomposition. Here we use the complex SVD. Although PCA can be applied to the time-domain data, the corresponding MR spectra are generally considered and are obtained from the original time-domain signal in each voxel by applying the discrete Fourier transformation (DFT). Assume P voxels and denote the MR spectrum in the ith voxel by $\mathbf{y_i}$. Arrange the P spectra $\mathbf{y}_1, \mathbf{y}_2, \ldots, \mathbf{y_P}$ as rows in a matrix D:

$$D = \begin{bmatrix} \mathbf{y}_1^T \\ \mathbf{y}_2^T \\ \vdots \\ \mathbf{y}_\mathbf{P}^T \end{bmatrix}. \tag{2.1}$$

Compute the complex SVD of the matrix D:

$$D_{P\times N} = U_{P\times P}\Sigma_{P\times N}V^H_{N\times N}$$

where $\Sigma = \mathrm{diag}(\sigma_1, \sigma_2, \ldots, \sigma_q)$, $q = \min(P, N)$. The columns of V contain the basic shapes of the signals, called PCs. The elements in $S_{P\times N} = U_{P\times P}\Sigma_{P\times N}$ are called the scores of each PC.

In fact, (2.1) can be rewritten as follows:

$$D = \begin{bmatrix} \mathbf{y}_1^T \\ \mathbf{y}_2^T \\ \vdots \\ \mathbf{y}_\mathbf{P}^T \end{bmatrix} = S_{P\times N}V^H_{N\times N}. \tag{2.2}$$

From (2.2), we can clearly see that PCA orthogonally transforms the original coordinate system into a new one, given by the PCs—the columns of V.

Now, suppose our set of P MRS signals is composed of a single resonance (peak), which means that the matrix D has rank one. Therefore, only $\sigma_1 \neq 0$, $\sigma_2 = \sigma_3 = \cdots = \sigma_q = 0$. Hence, the score matrix S contains only one column $S = \begin{bmatrix} s_1 & s_2 & \cdots & s_P \end{bmatrix}^T$. The first PC, that is, the first column $\mathbf{v}_1$ of V, represents the basic line shape, which in our case contains only one resonance.

Each signal $\mathbf{y}_p$ in the set can be expressed in the following way:

$$\mathbf{y}_p = s_p\mathbf{v}_1 \qquad p = 1, \ldots, P \tag{2.3}$$

Using frequency domain data, we need to normalize $\mathbf{v}_1$ so that it has unit area, that is,

$$\sum_{i=1}^{N} v_{i1} = 1. \tag{2.4}$$

In this case, (2.3) becomes (in matrix notation)

$$D = \tilde{S}\tilde{\mathbf{v}}_1^H = \left(S \sum_{i=1}^{N} v_{i1} \right) \frac{\mathbf{v}_1^H}{\sum_{i=1}^{N} v_{i1}}. \tag{2.5}$$

The modified scores $\tilde{S} = S \sum_{i=1}^{N} v_{i1}$ then represent the areas under the spectra, from which the corresponding metabolite concentrations can be obtained.

Using the properties of the DFT, it is easy to prove that PCA performed in the frequency domain is equivalent to its counterpart in the time domain except for the normalisation, which agrees with our experimental results.

Using the original time domain data as rows in the matrix D, the first PC $\mathbf{v}_1$ needs to be normalised in such a way that it represents a unit amplitude signal. To normalise $\mathbf{v}_1$, we need to divide $\mathbf{v}_1$ by its first element v_{11} so that (2.3) becomes (in matrix notation)

$$D = \tilde{S}\tilde{\mathbf{v}}_1^H = (S v_{11}) \frac{\mathbf{v}_1^H}{v_{11}}. \tag{2.6}$$

The modified scores $\tilde{S} = S v_{11}$ then represent the amplitudes of the original signals, from which the corresponding metabolite concentrations can be obtained.

Quantification of MRS signals containing more than one resonance or basic line shape using PCA is much more complex. Since PCA does not enable automated quantification of each resonance or line shape separately (each PC always contains a mixture of both resonances), we need to filter out each basic line shape before applying PCA. The best results are obtained using a maximum phase time domain FIR filter [10].

2.3 Frequency shift correction using complex PCA

The PCA method only performs well when there is only one common line shape in the whole data set. However in reality, due to fluctuations in instrumental parameters and other experimental variations, the resonances in an *in vivo* MRSI data set are always subject to small frequency shifts, phase misadjustment, and line shape distortions. This can mathematically be described as follows:

$$S_k(\tau) = A_k f(\tau), \qquad \text{with } \tau = \tau_0 + \Delta\tau_k, \tag{2.7}$$

where $S_k(\tau)$ is the kth spectrum $\mathbf{y}_k$ and function of τ, A_k is the complex amplitude of the kth spectrum, $f(\tau)$ is the line shape function, and $\Delta\tau_k$ is the complex shift (mixture of frequency shift and line shape distortion) of the kth spectrum. Expanding $f(\tau)$ in a Taylor series:

$$S_k(\tau) = A_k \{ f(\tau_0) + \frac{\partial f}{\partial \tau}\bigg|_{\tau_0} \Delta\tau_k + \frac{1}{2}\frac{\partial^2 f}{\partial \tau^2}\bigg|_{\tau_0} \Delta\tau_k^2 + \cdots \}.$$

If $\Delta\tau_k$ is small enough, we can neglect the second and higher order terms:

$$S_k(\tau) \approx A_k \left\{ f(\tau_0) + \frac{\partial f}{\partial \tau}\bigg|_{\tau_0} \Delta\tau_k \right\}. \tag{2.8}$$

We can also apply PCA to the same data set to obtain

$$S_k(\tau) = s_{1k} \cdot \mathbf{v}_1 + s_{2k} \cdot \mathbf{v}_2 + \cdots .$$

The first PC $\mathbf{v}_1$ somehow represents the basic line shape $f(\tau_0)$ in the data set, $\mathbf{v}_1 \approx af(\tau_0)$, a is an arbitrary positive real number. Therefore we can approximate $\left.\frac{\partial f}{\partial \tau}\right|_{\tau_0}$ by using the derivative of the first PC, $\mathbf{v}_1^{'} \approx a\left.\frac{\partial f}{\partial \tau}\right|_{\tau_0}$. By applying Linear Regression (LR), each spectrum $S_k(\tau)$ is approximated by a linear combination of $\mathbf{v}_1$ and $\mathbf{v}_1^{'}$:

$$S_k(\tau) \approx c_1 \cdot \mathbf{v}_1 + c_2 \cdot \mathbf{v}_1^{'}. \tag{2.9}$$

Equating the coefficient ratios of the expressions (2.8) and (2.9) yields the complex shift, that is,

$$\frac{c_2}{c_1} = \frac{aA_k \Delta\tau_k}{aA_k} = \Delta\tau_k.$$

We then transform $S_k(\tau)$ to the time domain, correct the frequency by multiplying the nth point by $e^{-j\Delta\tau_k 2\pi t_n}$ and retransform back to the frequency domain. Note that the calculation of $\left.\frac{\partial f}{\partial \tau}\right|_{\tau_0}$ is not trivial if the complex shift is a mixture of frequency shifts and line shape distortion. But if the resonances are only subject to frequency shifts, we only take the real part of $\Delta\tau_k$ and can simply use numerical differentiation to calculate $\left.\frac{\partial f}{\partial \tau}\right|_{\tau_0}$, where τ represents frequency.

The above-described algorithm, denoted by cPCA-LR(f), is applied iteratively to an MRSI data set until frequency shifts become negligible compared to the line width.

cPCA-LR(f) can be further improved by adding higher-order terms of the Taylor expansion. For example, the second-order derivative can be added to the linear regression equation as follows:

$$S_k(\tau) \approx c_1 \cdot \mathbf{v}_1 + c_2 \cdot \mathbf{v}_1^{'} + c_3 \cdot \mathbf{v}_1^{''} \tag{2.10}$$

where $\mathbf{v}_1^{''} = \left.\frac{\partial^2 f}{\partial \tau^2}\right|_{\tau_0}$ is approximated by using numerical differentiation of $\mathbf{v}_1^{'}$. This algorithm is called cPCA2-LR(f).

2.4 Numerical differentiation of the first PC

In their PCA research papers, Stoyanova and Witjes used first-order finite differences in order to approximate the derivative of the first PC. Given a set of grid points x_i, with $x_{i+1} - x_i = h$ and corresponding first PC values $\mathbf{v}(x_i)$, the derivative can be approximated by

$$\mathbf{v}^{'}(x_i) \approx \frac{\mathbf{v}(x_{i+1}) - \mathbf{v}(x_i)}{h}. \tag{2.11}$$

There are more accurate ways to calculate the derivatives [1]. For instance, taking higher-order finite differences such as fourth-order finite differences yields better approximations, which clearly improves the performance of cPCA-LR(f), as shown in Section 3:

$$\mathbf{v}'(x_i) \approx \frac{1}{h}\left[-\frac{1}{12}\mathbf{v}(x_{i-2}) + \frac{2}{3}\mathbf{v}(x_{i-1}) - \frac{2}{3}\mathbf{v}(x_{i+1}) + \frac{1}{12}\mathbf{v}(x_{i+2})\right]. \qquad (2.12)$$

The discrete differentiation process can be described as a matrix-vector multiplication:

$$\mathbf{v}' \approx h^{-1} \begin{bmatrix} & & \ddots & & & \frac{1}{12} & -\frac{2}{3} \\ & & \ddots & -\frac{1}{12} & & & \frac{1}{12} \\ & & \ddots & \frac{2}{3} & \ddots & & \\ & & \ddots & 0 & \ddots & & \\ & & \ddots & -\frac{2}{3} & \ddots & & \\ -\frac{1}{12} & & & \frac{1}{12} & \ddots & & \\ \frac{2}{3} & -\frac{1}{12} & & & \ddots & & \end{bmatrix}^T \mathbf{v} \qquad (2.13)$$

2.5 Cross-correlation

Cross-correlation is a well-known tool for measuring similarity between two different data sets. It can also be used to align the spectra in the set in frequency by computing the corresponding frequency shift for each spectrum in the given data set. This method, denoted here by cCross(f), is described as follows:

1. Calculate the magnitude spectrum of each complex MR spectrum;
2. Compute the cross-correlation between the first spectrum and each remaining spectrum in the data set. The maximal value points out the frequency shift to be applied to each spectrum;
3. Apply the frequency shifts to the original data set (as described in Section 2.3).

Unlike PCA-based methods, which are limited to small frequency shifts because of the use of the first-order Taylor expansion, cross-correlation is more robust and can correct large frequency shifts as shown in the next section.

3 Results

We have compared Witjes *et al.*'s [12] real-valued PCA method plus linear regression, abbreviated as rPCA-LR(f,p), complex PCA plus linear regression (cPCA-LR(f) and cPCA2-LR(f)), and cross-correlation (cCross(f)) on simulated MRSI data sets.

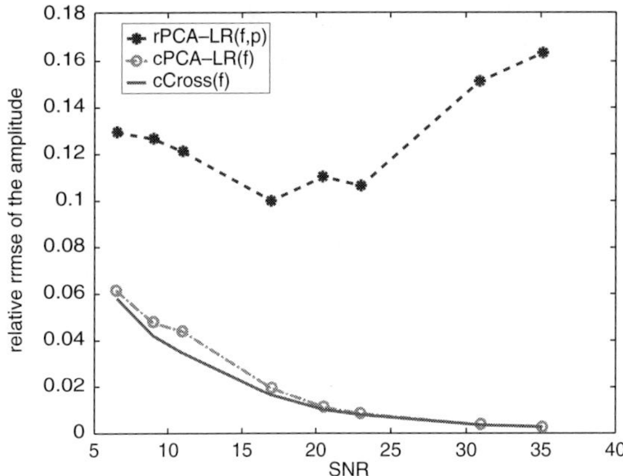

FIG. 1. rrmse of amplitude estimation versus SNR.

3.1 Comparison between different frequency shift alignment algorithms

First, we would like to show the advantages of using complex-valued methods. In Fig. 1, we use an MRSI data set with 100 spectra of 512 data points. The amplitudes are all set to the same value. Each spectrum contains a single Lorentzian line shape centred at position 256 with line width = 30 data points. Gaussian white noise ($\mu = 0$, $\sigma = 1$) is added to the data set. We vary the amplitudes of the signals in order to test the performance of each algorithm versus different SNRs. We add uniformly distributed phase (between $-90°$ and $90°$) and frequency shifts (the central position of each spectrum falls between $256 - line\,width/2$ and $256 + line\,width/2$) to the data set. The experiment is repeated 10 times. cPCA-LR(f) is better than rPCA-LR(f,p), which is unstable due to the large phase shift. The cross-correlation algorithm demonstrates the lowest relative root-mean-squared error (rrmse) although the differences with cPCA-LR(f) are not significant, in particular for the higher SNRs.

In Fig. 2, the simulated data sets contain 30 spectra of 512 data points. The remaining features are the same as those described in Fig. 1 except that we vary the frequency shifts from 0 up to 10 times the line width and the height of the Lorentzian peak is equal to 50. At small frequency shifts, cPCA-LR(f) and cCross(f) have comparable performance which is clearly much better than that of rPCA-LR(f,p). However, at large frequency shifts the performance of cPCA-LR(f) and rPCA-LR(f,p) decreases (partly due to the use of first-order Taylor expansions) while the performance of cross-correlation remains very good no matter how big the frequency shift is.

3.2 Two ways to improve the performance of PCA-based algorithms

One way to improve the performance of PCA-based algorithms is to add the second-order derivative of the Taylor expansion of each spectrum in the set to

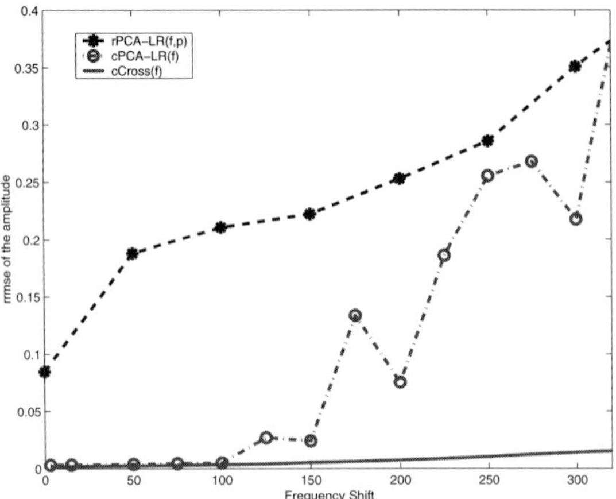

FIG. 2. rrmse of amplitude estimation versus frequency shift.

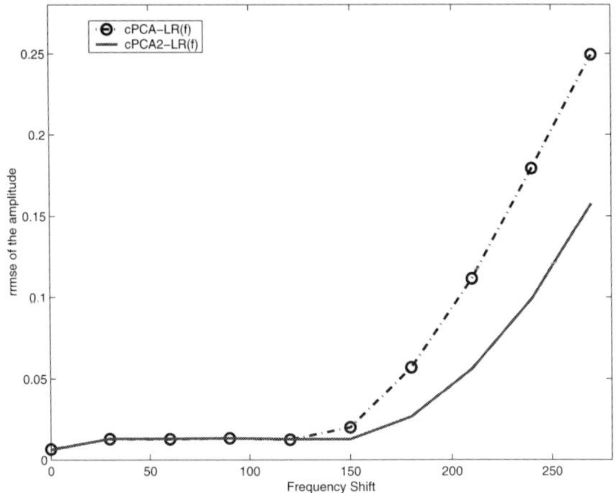

FIG. 3. rrmse of amplitude estimation versus frequency shift.

the LR equation (see (2.10)). In Figs 3 and 4, where both algorithms cPCA-LR(f) and cPCA2-LR(f) are tested for different frequency shifts and SNR, the advantage of cPCA2-LR(f) is obvious. The test data set is produced in the same way as described in the previous section, except that it contains 30 spectra of 512 data points and the test is repeated 100 times. cPCA2-LR(f) yields better results when the frequency shift is large and the SNR is low.

The advantages of using fourth-order finite differences instead of first-order finite differences for calculating the derivatives of the first PC are shown in Figs 5 and 6.

At low SNR and large frequency shifts, we clearly observe the improvement

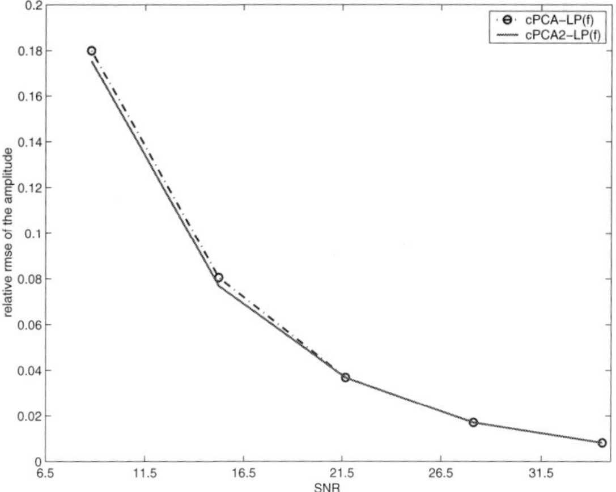

FIG. 4. rrmse of amplitude estimation versus SNR.

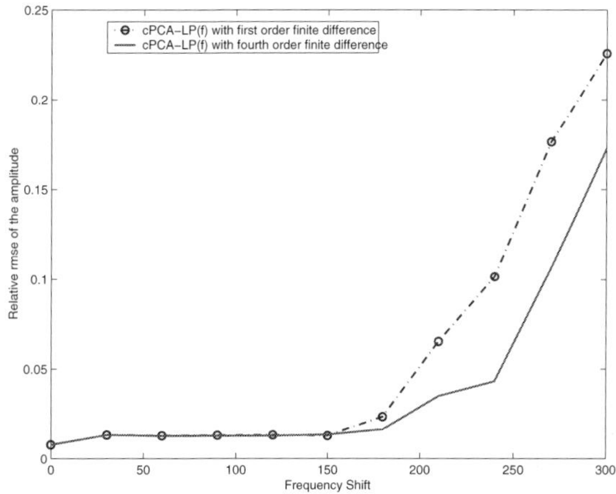

FIG. 5. rrmse of amplitude estimation versus frequency shift.

of cPCA-LR(f) by using fourth-order finite differences instead of first-order finite differences.

4 Conclusions

Frequency shift alignment of MRSI data, based on the complex SVD, is shown to improve the accuracy of the peak area estimation. Furthermore, this method is insensitive to phase shifts so that phasing of the spectra is no longer needed. The use of higher-order terms in Taylor expansions and higher-order finite differences can further improve the performance of PCA-based frequency shift algorithms. In our simulation experiments, this method has comparable performance at small frequency shifts as cross-correlation. However, at large frequency shifts,

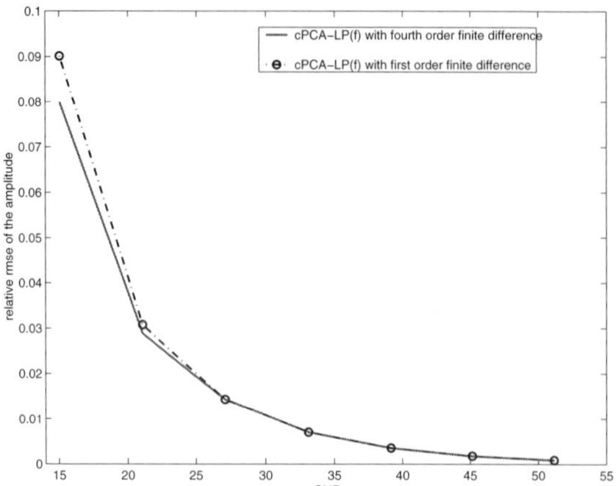

FIG. 6. rrmse of amplitude estimation versus SNR.

the cross-correlation algorithm demonstrates the best performance in terms of accuracy, robustness, and computational efficiency.

Acknowledgements

This paper presents research results of the Belgian Programme on Interuniversity Poles of Attraction (IUAP P4-02 and P4-24), initiated by the Belgian State, Prime Minister's Office—Federal Office for Scientific, Technical and Cultural Affairs, of the European Community TMR Programme, Networks, project CHRX-CT97-0160, of the Brite-Euram Programme, Thematic Network BRRT-CT97-5040 'Niconet', of the Concerted Research Action (GOA) projects of the Flemish Government MEFISTO-666 (Mathematical Engineering for Information and Communication Systems Technology), of the IDO/99/03 project (K.U.Leuven).

Bibliography

[1] Trefethen, L. N. (2000). Spectral Methods in Matlab. SIAM, Philadelphia, USA.

[2] Brown, T. R. and Stoyanova, R. (1996) NMR spectral quantitation by principal-component analysis II. Determination of frequency and phase shifts. *J. Magn. Res.*, B **112**, 32–43.

[3] Vanhamme, L. and Van Huffel, S. (1998), Multichannel quantitation of biomedical magnetic resonance spectroscopy signals. Advanced signal processing algorithms, architectures, and implementations VIII, (ed. F. T. Luk) *Proc. SPIE*, **3461**, pp. 237–248.

[4] Elliott, M. A., Walter, G. A., Swift, A., Vandenborne, K., Schotland, J. D.and Leigh, J. S. (1999). Spectral quantitation by principal component analysis using complex singular value decomposition. *Mag. Res. Med.*, **41**, 450–455.

[5] McKnight, T. R., Noworolski, S. M., Vigneron, D. B., Nelson, S. J. and Nat, D. R. (2001). An automated technique for the quantitative assessment of 3D-MRSI data from patients with glioma. *J. Mag. Res. Imag.*, **13**, 167–177.

[6] Nelson, S. J., Vigneron, D. B., Star-Lack, J. and Kurhanewicz, J. (1997). High spatial resolution and speed in MRSI. *NMR in Biomedicine*, **10**, 411–422.

[7] Preul, M. C., Caramanos, Z., Leblanc, R., Villemure, J. G. and Arnold, D. L. (1998). Using pattern analysis of in vivo proton MRSI data to improve the diagnosis and surgical management of patients with brain tumors. *NMR in Biomedicine*, **11**, 192–200.

[8] Tran, T. C., Vigneron, D. B., Sailasuta, N., Tropp, J., Le Roux, P., Kurhanewicz, J., Nelson, S. and Hurd, R. (2000). Very selective suppression pulses for clinical MRSI studies of brain and prostate cancer. *Mag. Res. Med.*, **43**, 23–33

[9] Stoyanova, R., Kuesel, A. C. and Brown, T. R. (1995) Application of Principal-Component Analysis for NMR Spectral Quantitation. *J. Mag. Res.*, A **115**, 265–269.

[10] Vanhamme, L., Sundin, T., Van Hecke, P., Van Huffel, S. and Pintelon, R. (2000). Frequency selective quantitation of biomedical magnetic resonance spectroscopy data. *J. Mag. Res.*, **143**, 1–16.

[11] Van Huffel, S., Chen, H., Decanniere, C. and Van Hecke, P. (1994). Algorithm for Time-Domain NMR Data Fitting Based on Total Least Squares. *J. Mag. Res.*, A **110**, 228–237

[12] Witjes, H., Melssen, W. J., in 't Zandt, H. J. A., van der Graaf, M., Heerschap, A. and Buydens, L. M. C., (2000) Automatic Correction for Phase Shifts, Frequency shifts and Lineshape Distortions across a Series of Single Resonance Lines in Large Spectral Datasets. *J. Mag. Res.*, **144**, 35–44.

Time–Frequency and Time–Scale Analysis of Embolic Signals

Nizamettin Aydin
Department of Electronics and Electrical Engineering, the University of Edinburgh, Edinburgh, UK

Abstract

Early and accurate detection of emboli is important for monitoring of preventive therapy in stroke-prone patients. Emboli can be detected by using Doppler ultrasound. Most Doppler ultrasound systems employ quadrature demodulation techniques at the detection stage. The windowed Fourier transform has been widely used for the analysis of quadrature signals. The Fourier transform maps directional information in the frequency domain. However, it is not ideally suited to analysis of short duration embolic signals due to an inherent trade-off between time and frequency resolution. Alternatively, the wavelet transform ideally suits to the analysis of nonstationary signals. Mapping directional information in the scale domain is also desirable. A method based on the utilization of complex wavelets and negative scales has been proposed. The results reveal that the wavelet transform provides an optimized time–frequency resolution for analysis and detection of low-intensity embolic signals and the utilization of complex wavelets and negative scales eliminates the intermediate processing stages by mapping directional information in the scale domain.

1 Introduction

Early and accurate detection of emboli, which are particles larger than red blood cells, may be important in the identification of patients at high risk of stroke. Embolic signals reflected by emboli have some distinctive characteristics when compared to the Doppler signals from normal blood flow and artifacts [1, 2]. They usually have larger amplitude than the signals from normal blood flow and show a transient characteristic because of their reflectivity and size compared to the red blood cells. They are also frequency focused. Figure 1 shows some examples of embolic signals seen *in vivo*. Unlike many artefacts, embolic signals are unidirectional and usually contained within the flow spectrum. Therefore any analysis method must be capable of processing directional Doppler signals. Asymptomatic circulating emboli can be detected by transcranial Doppler ultrasound (TCD) [3], which is based on the pulsed wave Doppler ultrasound detection principle operating at 1 or 2 MHz transducer frequency. Because the quadrature demodulation method is used to preserve directional information, the outputs of TCD systems are quadrature audio Doppler ultrasound signals. The TCD systems in common use were designed for determining flow velocity rather than for

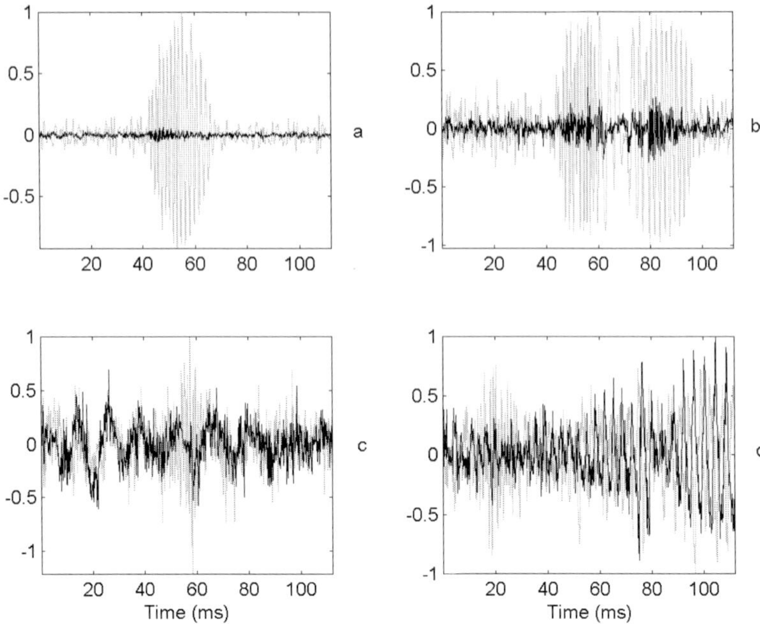

FIG. 1. Examples of embolic signals seen *in vivo*. For clarity, forward (light) and reverse (dark) flow components are shown. Note that the signals in (c), and (d) are corrupted by large artifacts.

embolic signal detection [4]. This has resulted in problems using these systems for embolic signal detection.

A quadrature Doppler signal pair can be defined mathematically as a complex expression

$$s(t) = s_i(t) + js_q(t) = \{s_f(t) + jH[s_f(t)]\} + j\{s_r(t) - jH[s_r(t)]\} \tag{1.1}$$

where $s_i(t)$ and $s_q(t)$ are real in-phase and quadrature-phase signals and contain information concerning the forward and reverse channel signals ($s_f(t)$ and $s_r(t)$) and their Hilbert transforms ($H[s_f(t)]$ and $H[s_r(t)]$). Using the frequency response of the Hilbert transform, it is easy to obtain the Fourier transform (FT) of $s(t)$:

$$F\{s(t)\} = S(\omega) = \begin{cases} 2S_f(\omega) & \text{for } \omega > 0 \\ j2S_r(\omega) & \text{for } \omega < 0 \end{cases} . \tag{1.2}$$

It is obvious that the positive frequencies of $S(w)$ contain only the spectrum of the forward channel signal $s_f(t)$, and its negative frequencies contain only 90^0 phase shifted spectrum of the reverse channel signal $s_r(t)$. In fact, the complex signal $s(t)$ is the base-band equivalent of a modulated narrow-band signal having upper and lower side-bands with different information contents, which is in our case forward and reverse flow information. Considering a quadrature Doppler signal as complex expression has a very useful practical result. The FT of such signal results in mapping the directional information (upper and lower side-bands) in the frequency domain. This is discussed in detail in Section 2.1.

2 Analysis and detection of embolic signals

Commercial Doppler ultrasound systems for emboli detection use the fast Fourier transform (FFT) [5]. The assumption is made that the Doppler signals obtained from the blood flow are stationary within the analysis window. However, the presence of short duration embolic signals invalidates this assumption [4]. Embolic signals are highly non-stationary and last only for a short time. Therefore time–frequency (TF) analysis of embolic signals is necessary. The FFT is not ideally suited to the analysis of embolic signals due to an inherent trade-off between frequency resolution (Δf) and time resolution (Δt) [6], which can be expressed as

$$\Delta t = \frac{W}{f_s}, \qquad \Delta f = \frac{f_s}{W} \tag{2.1}$$

where W is window length and f_s is sampling frequency. Alternatively, the wavelet transform (WT) appears to be ideal for analysis of embolic signals [7].

2.1 Windowed Fourier transform and time–frequency analysis

The windowed Fourier transform (WFT) introduces time dependency in the FT by pre-windowing the signal $s(t)$ around a particular time t, and calculating its FFT. This is repeated for each time instant t. The WFT of $s(t)$ is given by

$$F_s(t, f) = \int_{-\infty}^{+\infty} s(\tau) g^*(\tau - t) e^{-j2\pi f \tau} d\tau \tag{2.2}$$

where $g(t)$ is a short time analysis window function. Because multiplication by relatively short window suppresses the signal outside a neighbourhood around analysis time point $\tau = t$, the WFT is a local spectrum of the signal $s(t)$ around a particular time t. The WFT results of quadrature Doppler signals are presented as a sonogram, which is a form of TF representation and defined as $|F_s(t, f)|^2$. The information concerning flow direction, which is encoded in the phase relationship between in-phase and quadrature-phase channels, is decoded in frequency domain by complex FFT and mapped over the TF plane [8]. In the context of the quadrature Doppler signals an analytic signal is a subset of the complex quadrature signals, in which both positive- and negative-frequency components have a physical meaning. The positive-frequency range on a TF plane represents the signals resulting from the forward flow and the negative-frequency range represents the signals resulting from the reverse flow. Then the forward and reverse signals in frequency domain can be given as

$$S_f(\omega) = \begin{cases} 2S(\omega) & \text{for } \omega \geq 0 \\ 0 & \text{for } \omega < 0 \end{cases} \qquad S_r(\omega) = \begin{cases} 0 & \text{for } \omega \geq 0 \\ 2S(\omega) & \text{for } \omega < 0 \end{cases}. \tag{2.3}$$

In view of the potential difficulties in applying the FFT to processing short embolic signals, it is important that the processing parameters such as window size, window type, and overlap ratio are optimized. When using FFT analysis to study the frequency spectrum of signals, there are two basic problems. We can only sample the signal for a limited time, and the FFT only calculates results for certain discrete frequency values. The FFT makes an implicit assumption that the signal within the processing frame is repetitive. Most real signals, however, will have discontinuities at the ends of the processing frame so that when the FFT assumes the signal repeats it also incorporates discontinuities that are not there. This will cause spectral leakage. Discontinuities can be minimized by multiplying the signal with a suitable window function, which forces the signal to fall smoothly to zero at each end of the processing frame [9, 10].

When the frames are formed from adjacent non-overlapping groups of samples, processing data of length N_s by using an analysis window of length W will result in a TF distribution having a dimension of $M = [N_s/W] \times [W+1]$. In overlapped WFT, shifting the analysis window by less than the window length results in overlap of the frames. The data frames of length W are processed sequentially by sliding the window $W - O_L$ times at each processing stage, where O_L is the number of overlapped samples. Consequently, overlapping FFT windows produces higher-dimensional WFT. Some of the information in an overlapped WFT is redundant, and some of it is novel. The overlapping process also introduces a predictable time shift on the actual location of a transient event on the TF plane of the FFT. The duration of the time shift depends on the overlap ratio used, while the direction of the time shift is dictated by the way that the data are arranged prior to the FFT. The duration of the time shift can be estimated as $(O_L/2) \times (1/f_s)$. The time shift can be adjusted by adding zeros equally at both ends of the original data array. In this case the dimension of the overlapped WFT is $M = [N_s/(W - O_L)] \times [W + 1]$. It is obvious that a degree of overlap is required for embolic signal detection. An inadequate overlap may result in an underestimation of embolic signal intensity, or even in embolic signals not appearing on the spectral display at all. On the other hand, a higher overlap ratio than required will result in redundant data and imposes a computational burden on the system. Since the FFT analysis depends critically on the choice of the window, its application requires *a priori* information concerning the time evolution of the signal properties in order to make an *a priori* choice of the window. Once a window is chosen the resolution in both time and frequency is fixed in the entire TF plane. The best combination of Δt and Δf is not immediately obvious. This depends on the signal being processed and the optimal TF resolution trade-off may be determined empirically or analytically.

2.2 Wavelet transform and time scale analysis

The WT analysis is a relatively recent mathematical development, with many powerful applications in the analysis of real world signals and images [11–13]. A complete WT process creates a two-dimensional time scale (TS) representation of a one-dimensional signal, typically with the horizontal axis as time and vertical axis corresponding to the wavelet scale. The third dimension (colour in a two-dimensional display) is the amplitude of the WT coefficients. This representation allows exact localization of any abrupt change, or an exact time and duration to be attributed to a short signal, which may not be evidenced by conventional signal processing techniques. The continuous WT is performed by projecting a signal $s(t)$ onto a family of zero-mean functions (the wavelets) derived from an elementary function $\psi(t)$ (the mother wavelet) by translations and dilations, and given by

$$W_s(a,b) = |a|^{-1/2} \int_{-\infty}^{+\infty} s(t)\psi^* \left(\frac{t-b}{a} \right) dt \tag{2.4}$$

where * denotes the complex conjugate and $\psi^*(t)$ is the analysing wavelet. The variable $a(\neq 0)$ controls the scale of the wavelet, such that taking $a > 1$ dilates the wavelet ψ and taking $a < 1$ compresses ψ. The variable b is the time translation and controls the position of the wavelet. The wavelet transform is characterized by the following properties:

(1) It is a linear transformation;

(2) It is covariant under translations

$$s(t) \rightarrow s(t-u), \;\; W_s(a,b) \rightarrow W_s(a, b-u); \tag{2.5}$$

(3) It is covariant under dilations

$$s(t) \rightarrow s(kt), \;\; W_s(a,b) \rightarrow k^{-1/2} W_s(ka, kb). \tag{2.6}$$

Unlike the WFT, when the scale factor a is changed, the duration and bandwidth of the wavelet are both changed but its shape remains the same. The WT uses short windows at high frequencies and long windows at low frequencies. The bandwidth B is proportional to the frequency ω. The WT results of quadrature Doppler signals are presented as a scalogram, which is a form of TS representation and defined as $|W_s(a,b)|^2$. A signal $s(t)$ can be restored from its WT $W_s(a,b)$ using the following formula:

$$s(t) = \frac{1}{C_\psi} \int_{-\infty}^{+\infty} \int_{-\infty}^{+\infty} W_s(a,b)\psi \left(\frac{t-b}{a} \right) \frac{da db}{a^2} \tag{2.7}$$

providing that the FT of the wavelet $\psi(t)$, denoted $\Psi(w)$ satisfies the following admissibility condition:

$$C_\psi = \int_{-\infty}^{+\infty} \frac{|\Psi(\omega)|^2}{\omega} < \infty \tag{2.8}$$

which shows that the wavelet $\psi(t)$ has to oscillate and decay.

2.3 Complex wavelets

If a signal is real, its FT coefficients are symmetric about the origin and no information is lost by considering only the positive-frequency part. However, in many applications it is useful to work with complex signals, as this allows representing a modulated band-limited signal as its base-band equivalent. It was shown in Section 2.1 that the FT coefficients are no longer symmetric and both the positive- and negative-frequency parts of the spectrum should be considered independently. Likewise, it is convenient to work with the analytic wavelets in this case. The real wavelets are often used to detect sharp signal transitions while the complex wavelet, which can separate amplitude and phase components, can be used to measure the time evolution of frequency transients [14]. The existence and usefulness of a negative scale for the WT analogous to those of a negative-frequency spectrum in the FT for the complex quadrature signals bears scrutiny. The latter is ensured by use of the complex exponential. A comparable property for the WT can be attained by the use of complex wavelets.

Theorem 2.1 *The WT for processing complex quadrature Doppler signals can be implemented in such a way that wavelet coefficients resulting from the forward flow components are obtained when the scale is positive and the wavelet coefficients resulting from the reverse flow components are obtained when the scale is negative. If the FT of a mother wavelet $\psi(t)$ is defined as the sum of its positive- and negative-frequency components ($\Psi(\omega) = \Psi^{+}(\omega) + \Psi^{-}(\omega)$), in addition to the required standard properties, such a wavelet must also satisfy the following property:*

$$\Psi(\omega) = \begin{cases} \Psi^{+}(\omega) & \text{for } a > 0 \\ \Psi^{-}(\omega) & \text{for } a < 0 \end{cases} . \tag{2.9}$$

This is attained by the sine–cosine formulation, which naturally exists in some common wavelets such as Morlet and Cauchy wavelets [15–17].

2.3.1 *Morlet wavelet*

The Morlet wavelet, which is a locally periodic wavetrain, is obtained by localizing a complex sine wave with a Gaussian envelope as given by

$$\psi(t) = e^{i\omega_0 t} e^{-t^2/2} \tag{2.10}$$

where ω_0 is nondimensional frequency and usually assumed to be 5 to 6 to satisfy the admissibility condition. Since the directional coefficients are directly defined by the scale parameter a, ignoring the translation parameter b, the scale-dependent Morlet wavelet is

$$\psi(t/a) = e^{i\omega_0 t/a} e^{-t^2/2a^2} . \tag{2.11}$$

The FT of the scale-dependent Morlet wavelet is given by

$$\Psi(\omega) = \begin{cases} |a|\sqrt{2\pi} e^{\frac{-(\omega_0 - a\omega)^2}{2}} \mathrm{H}(\omega) & \text{for } a > 0 \\ |a|\sqrt{2\pi} e^{\frac{-(\omega_0 + a\omega)^2}{2}} \mathrm{H}(-\omega) & \text{for } a < 0 \end{cases} \tag{2.12}$$

where H stands for the Heaviside step function and is defined as

$$\mathrm{H}(\omega) = \begin{cases} 1 & \text{for } \omega > 0 \\ 0 & \text{for } \omega < 0. \end{cases} \tag{2.13}$$

From (2.12), we can observe that a frequency spectrum of a progressive wavelet is obtained for $a > 0$ and a frequency spectrum of a regressive wavelet is obtained for $a < 0$.

2.3.2 *Cauchy wavelet*

The Cauchy wavelet is also a complex exponential function and given by

$$\psi(t) = \frac{\Gamma(m+1)}{2\pi(1-it)^{m+1}} \tag{2.14}$$

where $m\,(> 0)$ is the order of the wavelet and the gamma function is defined by

$$\Gamma(m) = \int_0^{+\infty} t^{m-1} e^{-t}\, dt. \tag{2.15}$$

Again ignoring the translation parameter b, the scale-dependent Cauchy wavelet is given by

$$\psi(t/a) = \frac{\Gamma(m+1)}{2\pi} \left(1 - i\frac{t}{a}\right)^{-(m+1)}. \tag{2.16}$$

Taking the FT of (2.16) reveals the directionality property of this wavelet:

$$\Psi(\omega) = \begin{cases} a^{m+1}\omega^m e^{-a\omega}\mathrm{H}(\omega) & \text{for } a > 0 \\ a^{m+1}(-\omega)^m e^{a\omega}\mathrm{H}(-\omega) & \text{for } a < 0. \end{cases} \tag{2.17}$$

The Morlet and Cauchy wavelets with corresponding frequency spectra for a positive scale are shown in Fig. 2.

3 Method and simulation

In order to see how negative scales have been utilized for obtaining directional wavelet coefficients, let us evaluate the WT equation for the Morlet wavelet. In this case, the specific form of the WT of a signal $s(t)$ is given by

$$W_s(a,b) = |a|^{-1/2} \int_{-\infty}^{+\infty} s(t) e^{\frac{i\omega_0(t-b)}{a} - \frac{(t-b)^2}{2a^2}}\, dt. \tag{3.1}$$

This is simply a convolution of the signal $s(t)$ with the Morlet wavelet. A complete set of directional wavelet coefficients can be mapped over the scales from $a = -J$ to $a = J$, excluding $a = 0$, by evaluating (3.1). If the translation parameter $b = 0$ for simplicity, then

$$W_s(a,0) = |a|^{-1/2} \int_{-\infty}^{+\infty} s(t) e^{\frac{i\omega_0 t}{a} - \frac{t^2}{2a^2}}\, dt. \tag{3.2}$$

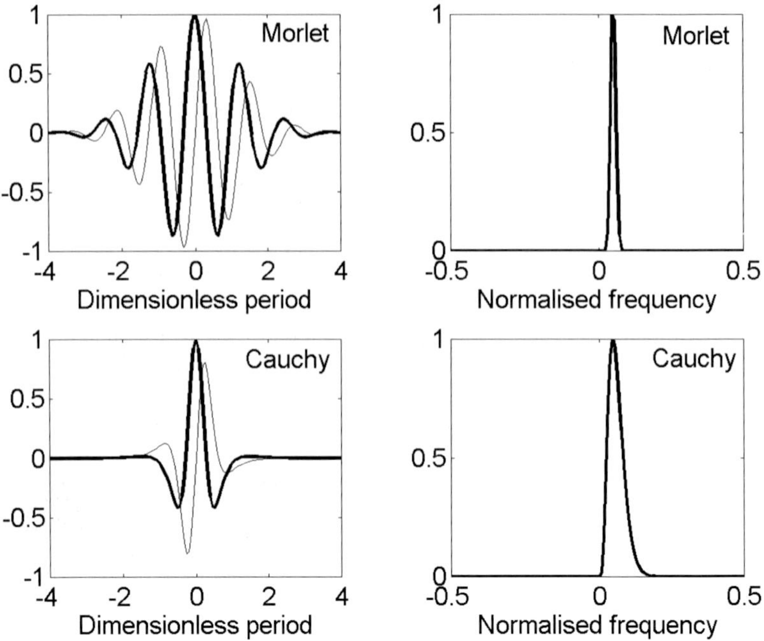

FIG. 2. Real (thick curve) and imaginary (thin curve) components of two complex wavelets and respective frequency spectra ($m = 4$ for the Cauchy wavelet).

For $a = \pm 1$, the WT of the signal $s(t)$ is given by

$$W_s(\pm 1, 0) = \int_{-\infty}^{+\infty} s(t) e^{\pm i\omega_0 t - 0.5t^2} dt. \tag{3.3}$$

Since the WT is equivalent to the convolution of the wavelet function and the signal under investigation, it can be easily implemented using the fast convolution, which corresponds to taking FTs of the wavelet function and the signal independently and multiplying them in frequency domain [18]. Taking the FT of (3.3) yields

$$F\{W_s(\pm 1, 0)\} = CS(\omega) e^{-0.5(\omega_0 \mp \omega)} \mathrm{H}(\pm\omega) \tag{3.4}$$

where $S(w)$ is the FT of the signal $s(t)$ and C is a constant. For positive (negative) scales, the inverse FT of (3.4) yields complex wavelet coefficients resulting from only the forward (reverse) flow components. In the case of the Cauchy wavelet with $m = 4$, assuming $b = 0$, for $a = \pm 1$, directional wavelet coefficients in the frequency domain can be calculated as

$$F\{W_s(\pm 1, 0)\} = CS(\omega)(\pm\omega)^4 e^{\mp\omega} \mathrm{H}(\pm\omega) \tag{3.5}$$

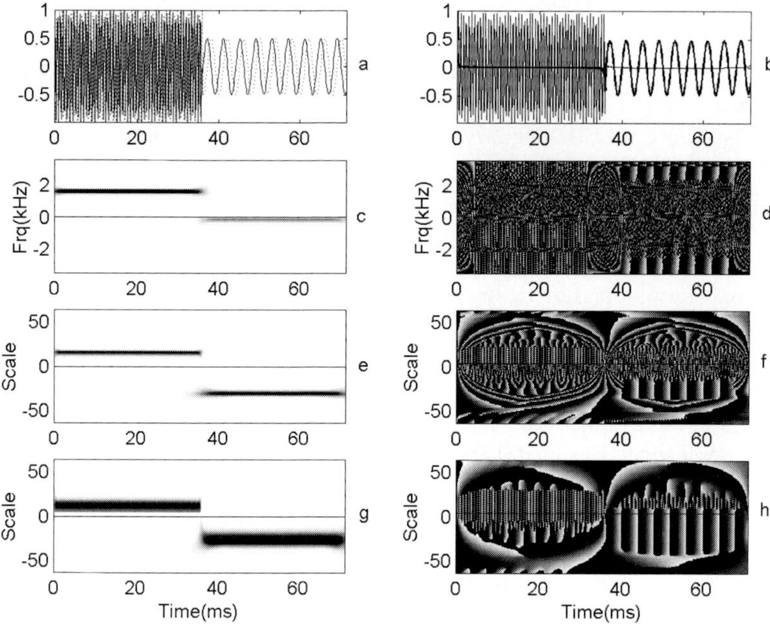

FIG. 3. (a) Simulated complex quadrature signal having two monotonic signals at each direction, (b) corresponding directional signals, (c) TF, (d) associated phase plots, (e, g) TS, and (f, h) associated phase plots.

where $S(w)$ is the FT of the signal $s(t)$ and C is a constant.

A simple simulated quadrature signal can be used to test the algorithm:

$$s(t) = e^{2\pi f_f t}[u(t) - u(t - t_1)] + 0.5e^{-2\pi f_r t}[u(t - t_1) - u(t - t_2)] \tag{3.6}$$

where u is the unit step function. It consists of two complex signals representing constant forward ($f_f = 1500$ Hz) and reverse ($f_r = 250$ Hz) flow signal components with a sampling frequency of 7150 Hz, length of 512. The forward flow signal stops and the reverse flow signal starts at time t_1 (35.8 ms). The TF plot, the TS plots with the Morlet and Cauchy wavelets, and related phase plots are shown in Fig. 3. The complex WT algorithm maps directional information in the scale domain as the complex WFT does in the frequency domain.

4 Comparison of TF and TS representations

For a quantitative comparison of TF and TS representations of embolic signals, 25 consecutive embolic signals from patients with symptomatic carotid stenosis were used for analysis. Recordings were made from a middle cerebral artery using an axial sample volume of 5 mm. The quadrature ultrasonic Doppler signals had been recorded using a TCD system. The sampling frequency was 7150 Hz and the data length was 2048 point (286 ms). The embolic signals were included in the first half of the data. The data were analysed using both the WFT and the WT. For the WT analysis, a 64 scale WT using complex Morlet wavelet was applied, producing a total of 128×2048 data array representing the TS distribution of the signals. For investigation of the effect of window size, the WFT was evaluated for each sample point. Prior to the FFT, the Hanning window was applied. The FFT size was fixed to 512 and the parameters were evaluated for six different window sizes (16, 32, 64, 128, 256, and 512). The TF representations of embolic signals using the FFT and the WT were compared by calculating embolic signal to background blood signal ratio (EBR), half-width maximum for both time and frequency (HWMT, HWMF), and embolic signal onset (ESO) as described by Aydin *et al.* [7].

Mean and standard deviations of the EBR, HWMF, HWMT, and ESO for 25 embolic signals are presented in Table 1. Although the differences are insignificant the WT gives the best EBR ratio for this data set. The HWMT for the WFT increases and the HWMF for the WFT decreases with increasing window size. The TF description of a signal using the WFT is mainly a function of the window length, as shown in Fig. 4. In contrast, the TS description of a signal using the WT is mainly a function of the wavelet used, as illustrated in Fig.3. The HWMT for the WT was as good as the WFT with short window sizes. The HWMF for the WT was as good or better then the WFT with window sizes up to 128 point. Time localization using the WT was also much better than the WFT and shows very close agreement with the measurement from the time domain signal.

Table 1 Mean and standard deviations of the EBR, HWMT, HWMF, and ESO for the 25 embolic signals. (ESOT: ESO measured from time domain signals)

	EBR (dB)		HWMT (ms)		HWMF (ms)		ESO (ms)	
	Mean	SD	Mean	SD	Mean	SD	Mean	SD
FFT(16p)	12.2	1.7	6.8	4.5	1018	245	65.9	6.2
FFT(32p)	12.9	1.5	7.2	4.8	769	307	65.5	6.4
FFT(64p)	13.4	1.2	8.3	4.6	623	332	64.5	6.5
FFT(128p)	13.1	1.3	10.9	4.3	547	368	60.7	7.4
FFT(256p)	12.5	1.4	17.7	3.9	498	340	52.6	10.7
FFT(512p)	11.6	1.4	38.8	17.6	497	342	34.2	16.9
WT	13.9	1.8	7.1	5.2	627	298	65.9	6.7
ESOT							66.2	6.6

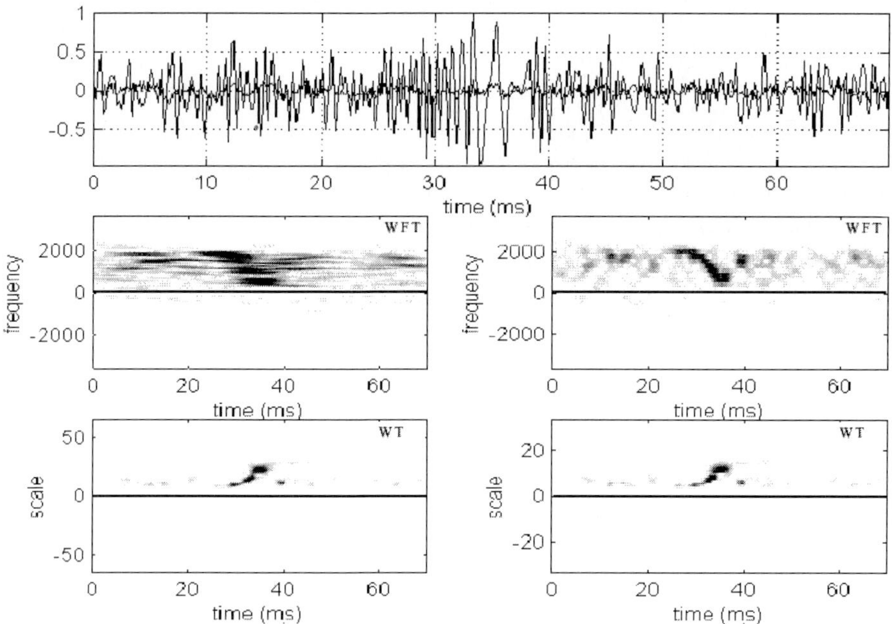

FIG. 4. A low-intensity embolic signal and corresponding TF and TS distributions (left: 128 point WFT and 64 scales WT, right: 32 point WFT and 32 scale WT).

5 Concluding remarks

The wavelet transform appears well suited to the analysis of embolic signals offering better time localization compared to the WFT. As the wavelet transform is local both in time and frequency domains, the TF representation of an embolic signal using the WT is optimized in terms of the TF resolution. This is apparent from Fig. 4, where the CWT results appear to be clearer, confirming the results presented in Table 1. The WT had as high temporal resolution as the WFT with short window and the best EBR as given in Table 1. The WT provides good frequency resolution for the low-frequency signals and good time resolution for the high-frequency signals and hence an optimized TF localization. This optimization allows the WT to capture an abrupt change and a slowly changing trend in a signal at the same time. In the case of the WFT, three important parameters must be defined prior to analyses: window size, window type, and overlap ratio. If the properties of the signal of interest do not change within the analysis window, the WFT analysis is likely to yield a desired TF representation. However, in practice, statistical properties of a non-stationary signal cannot be known prior to the analysis. The time resolution is better and frequency resolution is poorer for shorter windows, while the time resolution is poorer and the frequency resolution is better for longer windows, as depicted in Fig. 4. Here

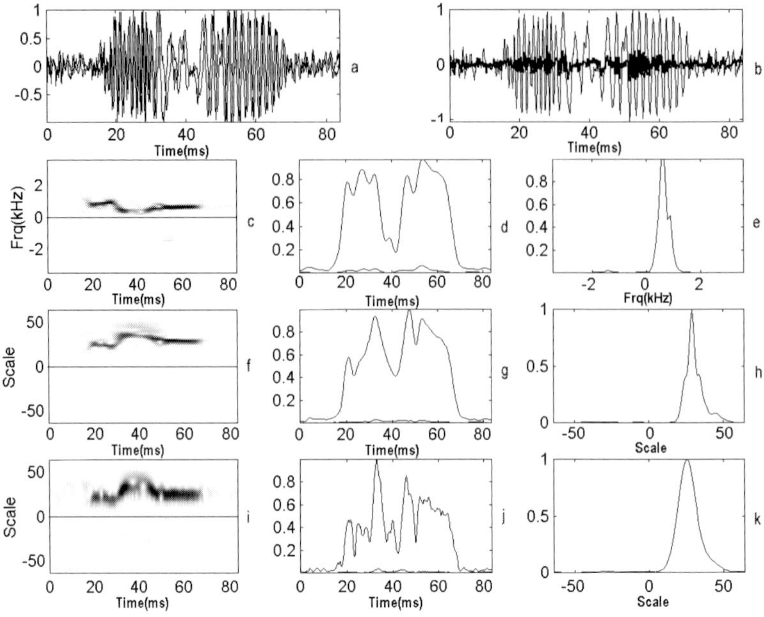

Fig. 5. (a) An embolic quadrature Doppler signal with (b) corresponding directional signals. (c) TF plot and TS plots for (f) Morlet and (i) Cauchy wavelets. Associated instantaneous envelopes (d, g, j) and averaged Fourier and wavelet energy spectra (e, h, k).

the description of the low-intensity embolic signal is clearly a function of analysis window size. In the WT case, this limitation is eliminated by projecting the signal of interest on multiple scales. However, its results are affected by the type of the wavelet chosen for analysis. This is clearly shown in Fig. 5. The Morlet wavelet appears to give better frequency resolution as indicated by the averaged spectral energy, which is calculated by averaging the TS distribution over the time (Fig. 5(h)). On the other hand, the Cauchy wavelet has better temporal resolution, as indicated by the instantaneous envelope calculated by averaging the TS distribution over the scales (Fig. 5(j)). The choice of the wavelet is mainly dictated by the goals of the analysis. If one knows the characteristics of the signal or pattern being sought, the wavelet should be chosen to have a similar pattern. The phase plot of the complex WT is also more informative than the WFT. The transition point from one frequency to another one is defined very well in Figs 3(f) and (h).

A method to obtain directional wavelet coefficients from quadrature Doppler signals has been described. The use of complex wavelets allows negative and positive scales to be utilized in order to map directional wavelet coefficients. From the results we can conclude that the complex WT described above yields the wavelet coefficients resulting from the signal components having only positive-frequency spectrum for the positive scales. On the other hand, negative scales will produce the wavelet coefficients resulting from the signal components having only negative-frequency spectrum. By the method proposed above, intermediate processing stages for obtaining directional Doppler signals for wavelet analysis have been eliminated. This also eliminates the computational errors introduced by the FIR Hilbert transform and associated delay filters, which is the standard technique for obtaining directional time domain signals from quadrature Doppler signals, and reduces the computational requirements for the wavelet analysis of complex quadrature Doppler signals. Furthermore, directional time domain signals can be reconstructed by simply evaluating the reconstruction formula. If the inverse WT is evaluated over the positive scales, the forward signal and associated Hilbert-transformed forward signal are obtained. If the inverse WT is evaluated over the negative scales, the reverse signal and associated Hilbert transformed reverse signal are obtained. The proposed method can also be extended to the discrete WT case and associated fast implementations.

Acknowldegements

The author would like to thank to the British Heart Foundation for financial support.

Bibliography

[1] Markus, H.S. and Tegeler, C. (1995). Experimental aspects of high-intensity transient signals in detection of emboli. *J. Clin. Ultrasound*, **23**, 81–87.

[2] Ringelstein, E.B., Droste, D.W., Babikian, V.L., Evans, D.H., Grosset, D.G., Kaps, M., Markus, H.S., Russell, D. and Siebler, M. (1998). Consensus on microembolus detection by TCD. *Stroke*, **29**, 725–729.

[3] Spencer M.P., Thomas, G.I., Nicholls, S.C. and Sauvage, L.R. (1990). Detection of middle cerebral emboli during carotid endarterectomy using transcranial Doppler ultrasonograpy. *Stroke*, **21**, 415–423.

[4] Aydin, N. and Markus, H.S. (2000). Optimisation of processing parameters for the analysis and detection of embolic signals. *Eur. J. Ultrasound*, **12**, 69–79.

[5] Aaslid, R., Markwalder, T. and Nornes, H. (1982). Noninvasive transcranial Doppler ultrasound recording of flow-velocity in basal cerebral arteries. *J. Neurosurg.*, **57**, 769–774.

[6] Cohen, L. (1989). Time frequency distributions—a review. *Proc. IEEE*, **77**, 941–981.

[7] Aydin, N., Padayachee, S. and Markus, H.S. (1999). The use of the wavelet transform to describe embolic signals. *Ultrasound Med. Biol.*, **25**, 953–958.

[8] Aydin, N. and Evans, D.H. (1994). Implementation of directional Doppler techniques using a digital signal processor. *Med. Biol. Eng. Comput.*, **32**, S157–S164.

[9] Harris, F.J. (1978). On the use of windows for harmonic analysis with discrete Fourier transform. *Proc. IEEE*, **66**, 51–83.

[10] Kay, S.M. (1988). *Modern Spectral Estimation*. Prentice Hall, Englewood Cliffs, NJ.

[11] Akay, M (1995). Wavelets in biomedical engineering. *Annals of Biomedical Engineering*, **23**, 531–542.

[12] Unser, M. and Aldroubi, A. (1996). A review of wavelets in biomedical applications. *Proc. IEEE*, **84**, 626–638.

[13] Farge, M., Kevlahan, N., Perrier, V., and Goirand, E. (1996). Wavelets and turbulence. *Proc. IEEE*, **84**, 639–669.

[14] Mallat, S. (1998). *A Wavelet Tour of Signal Processing*. Academic, San Diego.

[15] Martinet, R.K., Morlet, J. and Grossmann, A. (1986). Analysis of sound patterns through wavelet transforms. *Int. J. Pattern Recogn. and Artif. Intell.*, **1**, 273–302.

[16] Daubechies, I. (1990). The wavelet transform time–frequency localization and signal analysis. *IEEE Trans. Inform. Theory*, **36**, 961–1004.

[17] Aydin, N. and Markus, H.S. (2000). Directional wavelet transform in the context of complex quadrature Doppler signals. *IEEE Signal Process. Lett.*, **10**, 278–280.

[18] Holschneider, M. (1995). *Wavelets: an Analysis Tool*. Clarendon, Oxford.

Instantaneous Frequency Estimation of Quadratic FM Signals Corrupted by Multiplicative and Additive Noise

Braham Barkat

Nanyang Technological University, School of Electrical and Electronic Engineering, Block S2, Nanyang Avenue, Singapore 639798, Singapore

Abstract

In this paper, we propose the peak of the sixth-order polynomial Wigner–Ville distribution as an instantaneous frequency estimator of quadratic frequency-modulated signals in the presence of multiplicative and additive complex Gaussian processes. We show that this estimator is unbiased and we derive an analytic expression of its asymptotic variance. Simulation results, based on Monte Carlo realizations, are presented in order to show the validity of the theoretical derivations.

1 Introduction

In many engineering applications such as radar, sonar, and telecommunications the instantaneous frequency (IF) characterizes important physical parameters of the signals [1]. Therefore, in many situations, it is of great importance to have effective techniques to estimate the instantaneous frequency of a given signal.

Diverse IF estimation techniques have been developed for *constant*, or slowly varying, amplitude frequency-modulated (FM) signals embedded in additive noise [2, 3]

In practice, in addition to the additive noise, the signal under consideration may be subjected to a *random* amplitude modulation which behaves as multiplicative noise [4, 5]. Thus, it is important to have suitable analysis methods for such signals.

We can distinguish two major approaches to estimate the IF of FM signals: a parametric and a non-parametric approach. In general, parametric methods use a signal model and the goal, in this case, is to estimate some parameters in order to obtain the IF [6, 7, 8]. Non-parametric methods, on the other hand, use methods that do not require a parametrization of the signal and can, therefore, be applied to a larger class of signals. A well-known class of non-parametric methods, for IF estimation, is based on time–frequency distributions of the signal. Time-frequency distributions (TFDs) are natural extensions of the Fourier transform. They map a one dimensional signal, function of time only, to a two dimensional quantity, function of time and frequency [9].

In this paper, we use the peak of a particular time–frequency distribution, called the sixth-order polynomial Wigner–Ville distribution [10], to estimate the IF of a quadratic FM signal in the presence of both multiplicative and additive complex Gaussian noises. The choice of this TFD stems from the fact that it yields maximum energy concentration, about the signal IF, for quadratic FM signals. A statistical performance evaluation shows that this estimator is unbiased for such noisy signals. We also derive an analytic expression for the asymptotic variance of the estimator. An example, using Monte Carlo simulations, is presented to show the validity of the derived theoretical results. The presented technique can be extended to other nonlinear, not necessarily polynomial, FM signals.

2 Problem formulation

Let us consider the following model:

$$y(t) = m(t) \cdot z(t) + w(t) \tag{2.1}$$

where the stationary processes $m(t)$ and $w(t)$ are assumed to be circular complex Gaussian and independent, with means and variances given by (μ_m, σ_m^2) and $(0, \sigma_w^2)$ respectively. The noiseless quadratic FM signal $z(t)$ is given by $z(t) = e^{j\phi(t)} = \exp\left\{j\sum_{i=0}^{3} a_i t^i\right\}$ with a_i being real coefficients.

Our primary interest is in estimating the instantaneous frequency (IF) of the signal $z(t)$ defined as

$$f_i(t) = \frac{1}{2\pi}\frac{d\phi(t)}{dt} = \exp\left\{j\sum_{i=1}^{3} i a_i t^{i-1}\right\}$$

Expressing the non-zero mean multiplicative noise as $m(t) = \mu_m + m_o(t)$, with $m_o(t)$ being a zero-mean complex Gaussian noise process with variance σ_m^2, we can re-write the expression in (2.1) as

$$\begin{aligned} y(t) &= \mu_m \cdot z(t) + m_o(t)z(t) + w(t) \\ &= |\mu_m| \cdot e^{j(\Theta_{\mu_m} + \phi(t))} + w_1(t). \end{aligned} \tag{2.2}$$

In (2.2), $\Theta_{\mu_m} = \tan^{-1}\left[\frac{\mathrm{Im}(\mu_m)}{\mathrm{Re}(\mu_m)}\right]$ and $w_1(t) = m_o(t)z(t) + w(t)$ is a zero-mean circular complex Gaussian noise with variance equal to $\sigma_{w_1}^2 = (\sigma_m^2 + \sigma_w^2)$. Note that the instantaneous frequencies of the signals $e^{j(\Theta_{\mu_m}+\phi(t))}$ and $e^{j\phi(t)}$ are exactly the same, that is.,

$$f_i(t) = \frac{1}{2\pi} \cdot \frac{d\phi(t)}{dt} = \frac{1}{2\pi} \cdot \frac{d(\Theta_{\mu_m} + \phi(t))}{dt}.$$

We, therefore, see that the problem of estimating the IF of the quadratic FM signal, $z(t) = e^{j\phi(t)}$, affected by multiplicative and additive noises is equivalent to that of estimating the IF of a constant amplitude quadratic FM signal ($z_1(t) = |\mu_m| \cdot e^{j(\Theta_{\mu_m}+\phi(t))}$) having the same IF as $z(t)$ but affected by additive noise only.

3 Bias and variance of the IF estimator

To evaluate the statistical performance of the estimator, let us consider the discrete-time version of the model in (2.2), namely

$$y(nT) = z_1(nT) + w_1(nT), \qquad n = 0, \ldots, N-1. \tag{3.1}$$

where $z_1(nT)$ is the quadratic FM signal whose IF is to be estimated (and whose amplitude is equal to $|\mu_m|$), N the number of observations and T the sampling period. Without loss of generality, we will assume $T = 1$.

Since the sixth-order polynomial Wigner–Ville distribution (PWVD6) yields maximum energy concentration for quadratic FM signals (that is, a row of delta functions about the signal IF) [10], we propose its peak as an IF estimator. This TFD is given by [10]

$$\begin{aligned} W_z^{(6)}(n,f) = & \sum_{m=-M}^{M} [z(n+0.62m)\, z^*(n-0.62m)][z(n+0.75m)\, z^*(n-0.75m)] \\ & \times \; [z(n-0.87m)\, z^*(n+0.87m)] \cdot e^{-j2\pi fm} \\ = & \sum_{m=-M}^{M} K_z(n,m) \cdot e^{-j2\pi fm}. \end{aligned} \tag{3.2}$$

From the above expression, we observe that the kernel of the noisy signal $y(n)$ is equal to

$$\begin{aligned} K_y(n,m) &= [z_1(n+0.62m) + w_1(n+0.62m)][z_1^*(n-0.62m) + w_1^*(n-0.62m)] \\ &\times [z_1(n+0.75m) + w_1(n+0.75m)][z_1^*(n-0.75m) + w_1^*(n-0.75m)] \\ &\times [z_1(n-0.87m) + w_1(n-0.87m)][z_1^*(n+0.87m) + w_1^*(n+0.87m)]. \end{aligned}$$

Evaluating the above expression, we obtain

$$K_y(n,m) = K_{z_1}(n,m) + C_T(n,m) + \cdots \tag{3.3}$$

where

$$\begin{aligned} K_{z_1}(n,m) &= [z_1(n+0.62m)\, z_1^*(n-0.62m)][z_1(n+0.75m)\, z_1^*(n-0.75m)] \\ &\times [z_1(n-0.87m)\, z_1^*(n+0.87m)] \end{aligned}$$

and

$$\begin{aligned} C_T(n,m) &= [z_1(n+0.62m)z_1^*(n-0.62m)z_1(n+0.75m)z_1^*(n-0.75m) \\ &\quad \times z_1(n-0.87m)w_1^*(n+0.87m)] \\ &+ [z_1(n+0.62m)z_1^*(n-0.62m)z_1(n+0.75m)z_1^*(n-0.75m) \\ &\quad \times z_1^*(n+0.87m)w_1(n-0.87m)] \\ &+ [z_1(n+0.62m)z_1^*(n-0.62m)z_1(n+0.75m)z_1(n-0.87m) \\ &\quad \times z_1^*(n+0.87m)w_1^*(n-0.75m)] \\ &+ [z_1(n+0.62m)z_1^*(n-0.62m)z_1^*(n-0.75m)z_1(n-0.87m) \\ &\quad \times z_1^*(n+0.87m)w_1(n+0.75m)] \\ &+ [z_1(n+0.62m)z_1(n+0.75m)z_1^*(n-0.75m)z_1(n-0.87m) \\ &\quad \times z_1^*(n+0.87m)w_1^*(n-0.62m)] \\ &+ [z_1^*(n-0.62m)z_1(n+0.75m)z_1^*(n-0.75m) \\ &\quad \times z_1(n-0.87m)z_1^*(n+0.87m)w_1(n+0.62m)]. \end{aligned}$$

The terms indicated by dots in (3.3), correspond to quantities which involve products of more than one noise term. If we assume the power in $K_{z_1}(n,m)$ and $C_T(n,m)$ to be much larger than that in the discarded terms, we can write

$$K_y(n,m) \approx K_{z_1}(n,m) + C_T(n,m).$$

Using the linearity of the expectation operator and the i.i.d. property of the complex Gaussian noise $w_1(n)$, we can easily show that the power in $C_T(n,m)$ is equal to

$$P = \mathrm{E}\left[C_T(n,m) \cdot C_T^*(n,m)\right] = 6\,|\mu_m|^{10}\sigma_{w_1}^2.$$

The PWVD-based estimate $\hat{f}_i$ of the true IF f_i is defined as the frequency at which the function $|W_y^{(6)}(n,f)|^2$ is maximum, that is,

$$\left[\frac{\partial}{\partial f}|W_y^{(6)}(n,f)|^2\right]_{f=\hat{f}_i} = 0. \tag{3.4}$$

To determine $\hat{f}_i$, consider Taylor's expansion of $|W_y^{(6)}(n,f)|^2$ up to second order around $(f-f_i)$, that is,

$$\begin{aligned} |W_y^{(6)}(n,f)|^2 &\approx |W_y^{(6)}(f_i)|^2 + (f-f_i)\left[\frac{\partial|W_y^{(6)}(n,f)|^2}{\partial f}\right]_{f=f_i} \\ &\quad + \frac{1}{2}(f-f_i)^2\left[\frac{\partial^2|W_y^{(6)}(n,f)|^2}{\partial f^2}\right]_{f=f_i}. \end{aligned} \tag{3.5}$$

Differentiating (3.5) with respect to f results in

$$\frac{\partial |W_y^{(6)}(n,f)|^2}{\partial f} \approx \left[\frac{\partial |W_y^{(6)}(n,f)|^2}{\partial f}\right]_{f=f_i} + (f - f_i)\left[\frac{\partial^2 |W_y^{(6)}(n,f)|^2}{\partial f^2}\right]_{f=f_i} \tag{3.6}$$

and using equation (3.4) we obtain:

$$\hat{f}_i \approx f_i - \left\{\left[\frac{\partial |W_y^{(6)}(n,f)|^2}{\partial f}\right]_{f=f_i} \Big/ \left[\frac{\partial^2 |W_y^{(6)}(n,f)|^2}{\partial f^2}\right]_{f=f_i}\right\}. \tag{3.7}$$

Taking the expected value of (3.7), the expressions for the bias and variance in the estimate of f_i are found to be

$$\mathbf{Bias}(\hat{f}_i(n)) \approx -\mathrm{E}\left[\left[\frac{\partial |W_y^{(6)}(n,f)|^2}{\partial f}\right]_{f=f_i} \Big/ \left[\frac{\partial^2 |W_y^{(6)}(n,f)|^2}{\partial f^2}\right]_{f=f_i}\right] \tag{3.8}$$

$$\mathbf{Var}(\hat{f}_i(n)) \approx \mathrm{E}\left[\left\{\left[\frac{\partial |W_y^{(6)}(n,f)|^2}{\partial f}\right]_{f=f_i} \Big/ \left[\frac{\partial^2 |W_y^{(6)}(n,f)|^2}{\partial f^2}\right]_{f=f_i}\right\}^2\right]. \tag{3.9}$$

Using similar derivations as those in [3], the expressions in the previous equations result in

$$\mathbf{Bias}(\hat{f}_i(n)) = 0 \tag{3.10}$$

$$\begin{aligned} \mathbf{Var}(\hat{f}_i(n)) &\approx \frac{18\,\sigma_{w_1}^2}{(2\pi)^2|\mu_m|^2 M(M+1)(2M+1)} \\ &\approx \frac{18\,(\sigma_m^2 + \sigma_w^2)}{(2\pi)^2|\mu_m|^2 M(M+1)(2M+1)} \end{aligned} \tag{3.11}$$

where in the last expression we used the analysis result of the previous section, namely $\sigma_{w_1}^2 = (\sigma_m^2 + \sigma_w^2)$.

Note that the variance decreases with increasing values of M or increasing values of the window length in the PWVD6 (see (3.2)). In the middle of the time interval (assuming N odd) where the window length can be chosen equal to the signal length (that is, $2M+1 = N$), the variance is minimum and equal to

$$\mathbf{Var}(\hat{f}_i) = \frac{72\,(\sigma_m^2 + \sigma_w^2)}{(2\pi)^2|\mu_m|^2 N(N^2-1)}. \tag{3.12}$$

At other time instants n, the implementation of the PWVD necessitates smaller window lengths ($2M + 1 < N$) resulting in a decrease in the statistical performance of the estimator.

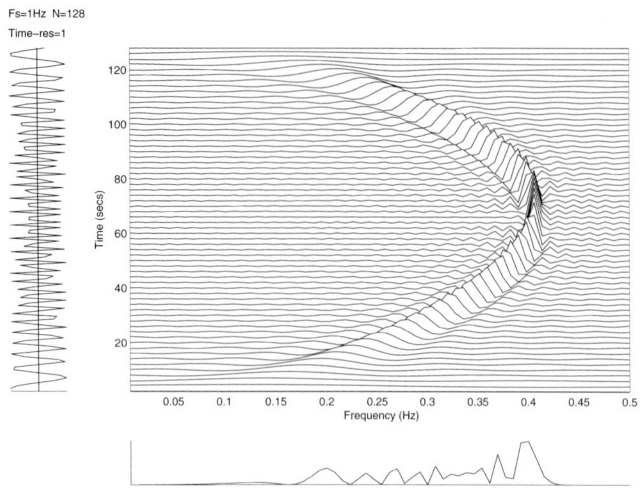

FIG. 1. The PWVD6 of a noiseless quadratic FM signal.

4 Example

In this section, we use Monte Carlo simulations to validate the theoretical results. For that, let us consider the IF estimation of a quadratic FM signal at the middle of the signal interval.

The noiseless quadratic FM signal $z(t)$ is modulated by a complex circular Gaussian noise $m(t)$ (μ_m, σ_m^2) and added to an independent circular complex Gaussian noise $w(t)$ $(0, \sigma_w^2)$, as suggested in (2.1). The noisy signal $y(t)$ is sampled at $T = 1$ and the number of observations N is chosen equal to 129.

In Fig. 1, we display the PWVD6 for the noiseless quadratic FM signal $z(t)$; and in Fig. 2, we display the same TFD for a noisy realization of the signal once affected by a multiplicative $(|\mu_m| = 3, \sigma_m^2 = 1)$ and additive noise $(0, \sigma_w^2 = 1)$.

In order to have a statistical performance evaluation for the IF estimator we fix the value of $|\mu_m|$ and vary the values of the variances σ_m^2 and σ_w^2. We define the overall signal-to-noise ratio (SNR) as $\mathrm{SNR}_{w_1} = 10\log_{10}(|\mu_m|^2/(\sigma_m^2 + \sigma_w^2))$. The SNR is varied in a 1 dB step from 0 to 15 dB. Monte Carlo simulations for 1000 realizations are run for each value of SNR_{w_1}. The results of two different experiments, one conducted for $|\mu_m| = 0.01$ and the other conducted for $|\mu_m| = 1$, are displayed in Fig. 3. We observe that, above a certain threshold, the estimated variances represented by '+' (for $|\mu_m| = 0.01$) and 'o' (for $|\mu_m| = 1$) are in agreement with the derived theoretical ones given by (3.11) and represented by the continuous lines (superimposed). In this example, we fixed $\sigma_m^2 = \sigma_w^2$. We used other values for $|\mu_m|$ and considered the case $\sigma_m^2 \neq \sigma_w^2$ and found similar results, that is, the estimated results confirm the theoretical ones.

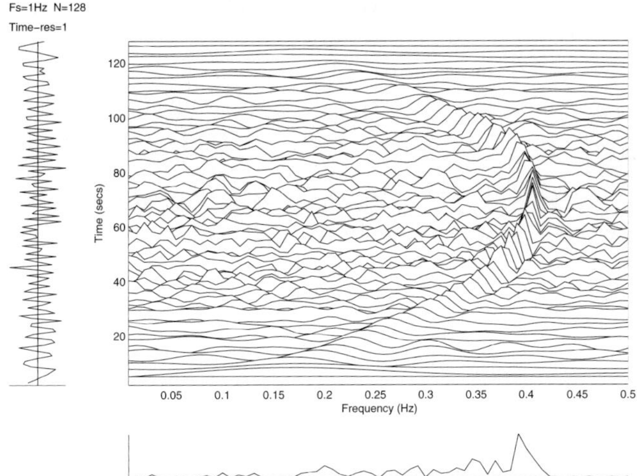

FIG. 2. The PWVD6 of a quadratic FM signal corrupted by multiplicative and additive noises.

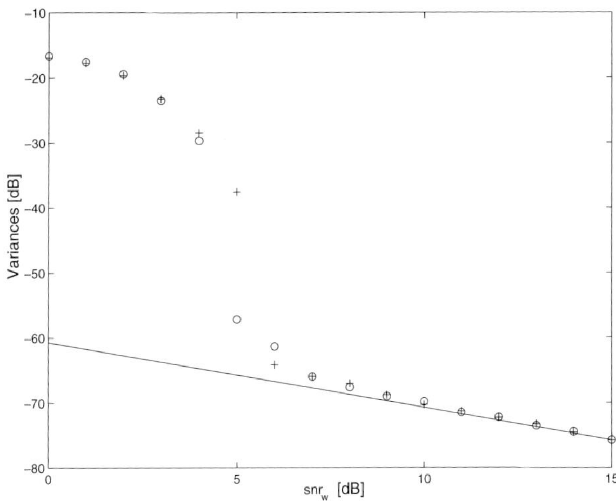

FIG. 3. Peak of the sixth-order PWVD is used as an IF estimator, at the middle of the observation interval, for a quadratic FM signal corrupted by complex Gaussian multiplicative and additive noise processes. The window length is equal to the signal length ($2M + 1 = N = 129$). The continuous lines (superimposed) represent the theoretical variances while '+' and 'o' correspond to the estimated variances for $|\mu_m| = 0.01$ and $|\mu_m| = 1$ respectively.

5 Conclusion

In this paper, we proposed the peak of the sixth-order polynomial Wigner–Ville distribution to estimate the instantaneous frequency of quadratic FM signals corrupted by multiplicative and additive complex Gaussian noises. We showed that this estimator is unbiased and derived an analytic expression for its asymptotic variance. We presented an example, using Monte Carlo simulations, to validate the theoretical derived results.

Bibliography

[1] Boashash, B. Interpreting and estimating the instantaneous frequency of a signal—part I: Fundamentals. *Proc. IEEE*, **80**, 520–538, 1992.

[2] Rao, P. and Taylor, F.J. Estimation of IF using the discrete Wigner-Ville Distribution. *Electron. Lett.*, **26**, 246–248, 1990.

[3] Barkat, B. and Boashash, B. Instantaneous frequency estimation of polynomial FM signals using the peak of the PWVD: statistical performance in the presence of additive Gaussian noise. *IEEE Trans. on Signal Processing*, **47**, 2480–2490, 1999.

[4] Stein, S. Fading channel issues in system engineering. *IEEE J. on Selected Areas in Communications*, **5**, 68–89, 1987.

[5] Kennedy, R. *Fading Dispersive Communication Channels.* Wiley, New York, 1969.

[6] Shamsunder, S., Giannakis, G.B. and Friedlander, B. Estimating random amplitude polynomial phase signals: a cyclostationary approach. *IEEE Trans. on Signal Process.*, **43**, 492–505, 1995.

[7] Besson, O., Ghogho, M. and Swami, A. On Estimating Random Amplitude Chirp Signals. In *Proc. ICASSP*, Vol. III, pp. 1561–1564, (Arizona), 1999.

[8] Swami, A. Polyphase signals in additive and multiplicative noise: CRLB and HOS. In *Proc. 6th IEEE Signal Processing Workshop DSP*, pp. 109–112, (California), 1994.

[9] Cohen, L. *Time-Frequency Analysis.* Englewood Cliffs, NJ, Prentice-Hall, 1995.

[10] Barkat, B. and Boashash, B. Design of higher-order polynomial Wigner–Ville distributions. *IEEE Trans. on Signal Processing*, **47**, 2608–2611, 1999.

Analysis and Fast Implementation of Oversampled Modulated Filter Banks

Stephan Weiss
Department of Electronics and Computer Science, University of Southampton, UK

Abstract

Oversampled modulated filter banks (OSFBs) are popularly employed for a number of applications such as acoustic echo cancellation in order to reduce the processing complexity of a signal processing algorithm. Hence, an efficient implementation of OSFBs themselves is mandatory. In this paper, a polyphase description is used to remove redundancies in the filter operations and to factorize the OSFB into filter components depending on the prototype filter, and the modulating transform. Based on a state-space representation of this derived polyphase factorization, signal flow graphs can be obtained which permit a very simple and efficient OSFB implementation. The analysis is performed for a number of different classes of OSFBs, and a comparison to existing methods is drawn.

1 Introduction

Oversampled filter banks (OSFBs) find a wide range of applications, where computational reductions for resource-demanding signal processing algorithms are sought by means of subband approaches. Examples include subband adaptive filters used in acoustic echo control [1, 2], line enhancement [3], or beamforming [4]. Another use of OSFBs is, for example, transmultiplexers for the transmission of several users over a single channel [5]. Particularly in the light of computational efficiency, simple low-complexity realizations for the filter banks themselves are therefore desirable. Despite this motivation and in contrast to their critically decimated counterparts [6, 7], numerically efficient implementations of non-critically sampled (or 'oversampled') filter banks have received little attention.

A simple filter bank system giving K subband signals decimated by $N \leq K$ is shown in Fig. 1. An efficient implementation is based on modulation of all K analysis filters $H_k(z)$ and all synthesis filter $G_k(z)$ from one prototype filter [8]. For OSFBs with non-critical decimation $N < K$, an efficient implementation has been reported by Wackersreuther [9], where a time domain approach leads to a factorization of the analysis filter bank operation into a filtering operation linked to the prototype filter coefficients, a cyclical shift, and the applications of the appropriate modulating transform (e.g. a DFT). A similar approach resulting in a different sequence of execution is presented in [10, 5], with a time-

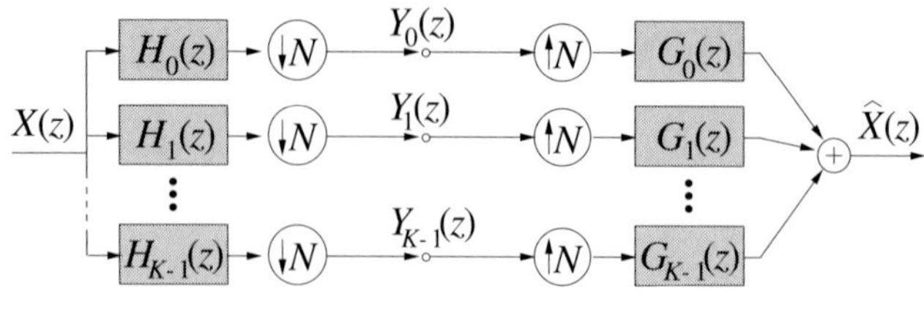

FIG. 1. Analysis and synthesis filter bank for subband decomposition of $X(z)$.

varying excitation of different components of the prototype filter followed by the modulating transform. More recently, polyphase factorizations in the z-domain have been presented [11, 12], which also permit a separation into filtering operations based on the prototype filter, and the modulating transform. For all cases [9, 10, 11, 12, 5], a dual implementation can be found for the synthesis filter bank operation.

The polyphase approach [11, 12] can be utilized as a starting point to derive a filter bank factorization yielding a very simple and low cost implementation [13]. Here, this factorization is generalized to arbitrary length prototype filters and arbitrary modulations, for which the analysis and factorization is presented in Section 2. Based on a state-space representation of the OSFB operations in Section 3, signal flow graphs for analysis and synthesis filter bank are derived in Section 4. Further, in the latter the implementation and computational complexity of the resulting circuits is discussed and compared to existing methods.

In our notation, boldface uppercase variables are matrix-valued, boldface lowercase or uppercase underlined quantities refer to vector-valued variables. An $N \times N$ identity matrix is denoted by $\mathbf{I}_N$, an $N \times M$ matrix with zero elements by $\mathbf{0}_{N \times M}$.

2 Filter bank analysis

2.1 Polyphase notation

Let us consider the analysis filter bank of Fig. 1 producing K subband signals. To exploit computational redundancies arising from the decimation by N, a polyphase description is utilized [7]. A polyphase notation for the kth analysis filters

$$H_k(z) = \sum_{n=0}^{N-1} z^{-n} H_{k|n}(z^N) \tag{2.1}$$

produces a decomposition into N type-I polyphase components $H_{k|n}(z)$ [7]. Similarly, the input signal $X(z)$ is decomposed into N type-II polyphase components

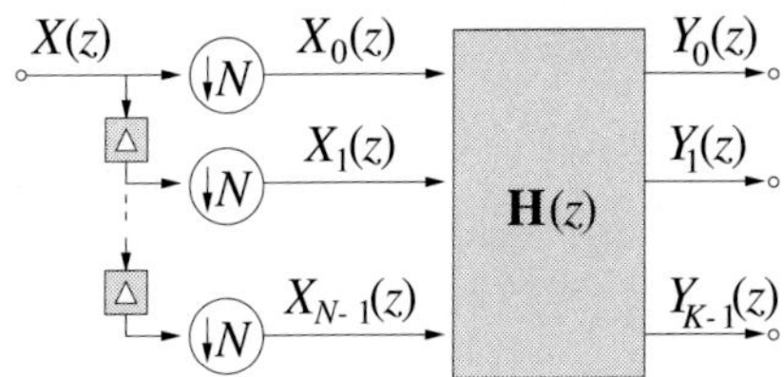

FIG. 2. Analysis filter bank with demultiplexer and $\mathbf{H}(z)$ describing an $N \times K$ MIMO system.

$X_n(z)$,

$$X(z) = \sum_{n=0}^{N-1} z^{-N+n-1} X_n(z^N). \tag{2.2}$$

If the polyphase components are organized in vector form,

$$\underline{H}_k(z) = \left[H_{k|0}(z)\ H_{k|1}(z)\ \cdots\ H_{k|N-1}(z)\right]^{\mathrm{T}} \tag{2.3}$$

$$\underline{X}(z) = [X_0(z)\ X_1(z)\ \cdots\ X_{N-1}(z)]^{\mathrm{T}} \tag{2.4}$$

a subband signal $Y_k(z)$ can be denoted as

$$Y_k(z) = \underline{H}_k^{\mathrm{T}}(z) \cdot \underline{X}(z). \tag{2.5}$$

For compatibility, in the following we assume that filters are always subjected to type-I and signals to type-II polyphase decompositions.

2.2 Analysis filter bank

For a compact notation of the analysis filter bank operations, the K subband signals are collected in a vector $\underline{Y}(z) = [Y_0(z)\ Y_1(z)\ \cdots\ Y_{K-1}(z)]^{\mathrm{T}}$. Inserting (2.5) gives

$$\underline{Y}(z) = \left[\underline{H}_0(z)\ \underline{H}_1(z)\ \cdots\ \underline{H}_{K-1}(z)\right]^{\mathrm{T}} \cdot \underline{X}(z) \tag{2.6}$$

$$= \mathbf{H}(z) \cdot \underline{X}(z) \tag{2.7}$$

where $\mathbf{H}(z) \in \mathbb{C}^{K \times N}(z)$ is the polyphase analysis matrix [11]. With the description (2.7), the analysis filter bank in Fig. 1 can be implemented by a demultiplexer followed by a linear time-invariant multi-input multi-output (MIMO) system $\mathbf{H}(z)$ as shown in Fig. 2. We are now interested in a particular factorization of $\mathbf{H}(z)$.

It is assumed that the analysis filters are FIR with L_{p} coefficients, and derived from a prototype filter $P(z)$ by modulation. With the coefficients of the kth analysis filter organized in a vector $\mathbf{h}_k \in \mathbb{C}^{L_{\mathrm{p}}}$,

$$\mathbf{h}_k = \left[h_k[0]\ h_k[1]\ \cdots\ h_k[L_{\mathrm{p}}-1]\right]^{\mathrm{T}}, \tag{2.8}$$

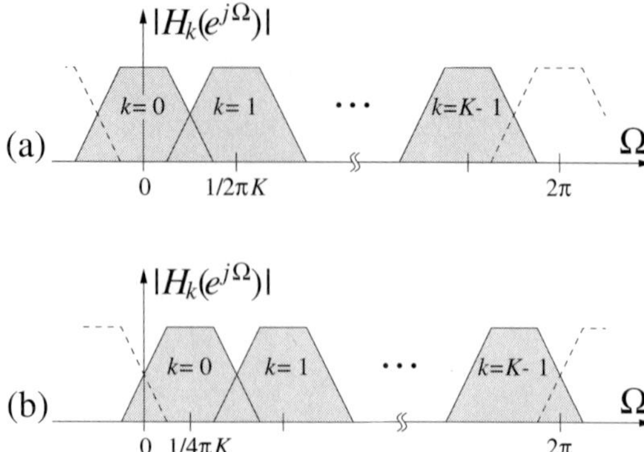

FIG. 3. (a) Even-stacked and (b) odd-stacked K-channel filter bank.

the polyphase components in (2.3) can be written as

$$\underline{H}_k(z) = \underbrace{\begin{bmatrix} \mathbf{I}_N & z^{-1}\mathbf{I}_N & \cdots & z^{-\lfloor L_\mathrm{p}/N \rfloor + 1}\mathbf{I}_N & \begin{matrix} z^{-\lfloor L_\mathrm{p}/N \rfloor}\mathbf{I}_R \\ \mathbf{0}_{N-R\times R} \end{matrix} \end{bmatrix}}_{\mathbf{L}_1^\mathrm{T}(z)} \cdot \mathbf{h}_k \tag{2.9}$$

where $\mathbf{I}_N$ is an $N \times N$ identity matrix, $\lfloor \cdot \rfloor$ the floor operator, and $R = \mathrm{mod}_N L_\mathrm{p}$ (the remainder of the division of the filter length L_p by the decimation factor N). The dependency of $H_k(z)$ on the underlying prototype filter with coefficients $p[i]$, $i = 0 \ldots L_\mathrm{p}-1$ is incorporated as

$$\mathbf{h}_k = \underbrace{\begin{bmatrix} p[0] & & & \mathbf{0} \\ & p[1] & & \\ & & \ddots & \\ \mathbf{0} & & & p[L_\mathrm{p}-1] \end{bmatrix}}_{\mathbf{P}} \cdot \underbrace{\begin{bmatrix} t_k[0] \\ t_k[1] \\ \vdots \\ t_k[L_\mathrm{p}-1] \end{bmatrix}}_{\mathbf{t}_k} , \tag{2.10}$$

based on the (generally complex) modulation sequence contained in $\mathbf{t}_k \in \mathbb{C}^{L_\mathrm{p}}$. The matrix $\mathbf{P} \in \mathbb{R}^{L_\mathrm{p} \times L_\mathrm{p}}$ is a diagonal matrix holding the prototype filter coefficients.

Depending on which modulation is invoked for the filter bank, different periodicities of the sequence $t_k[i]$, $i = 0 \ldots L_\mathrm{p}-1$, result. If the filter bank is even-stacked as shown in Fig. 3(a), the periodicity is K. In this case, a compact notation of the modulation sequence can be found

$$\mathbf{t}_k = \underbrace{\begin{bmatrix} \mathbf{I}_K & \mathbf{I}_K & \cdots & \mathbf{I}_K & \begin{matrix} \mathbf{I}_S \\ \mathbf{0}_{K-S\times S} \end{matrix} \end{bmatrix}^\mathrm{T}}_{\mathbf{L}_2^\mathrm{T}} \cdot \tilde{\mathbf{t}}_k, \tag{2.11}$$

where $\tilde{\mathbf{t}}_k \in \mathbb{C}^K$, $\mathbf{L}_2 \in \mathbb{N}^{L_\text{p} \times K}$, and $S = \text{mod}_K L_\text{p}$. Odd-stacked filter banks as in Fig. 3(b) exhibit a $2K$ periodicity of $t_k[i]$. Additional symmetries in the modulation sequence can, however, be exploited by alternating sign changes on blocks of K coefficients,

$$\mathbf{t}_k = \begin{bmatrix}\mathbf{I}_K & -\mathbf{I}_K & \mathbf{I}_K & -\mathbf{I}_K & \cdots\end{bmatrix}^\text{T} \cdot \tilde{\mathbf{t}}_k. \tag{2.12}$$

Instead of modifying the matrix $\mathbf{L}_2$ of odd-stacked filter banks, this sign change can be incorporated into $\mathbf{P}$ in (2.10) and hence the prototype filter $P(z)$, a trick that is also known from discrete cosine transform implementations [7]. It is assumed that for odd-stacked filter banks such a modification of the prototype filter is performed by negating the coefficients' signs in every second K-block of coefficients. With this assumptions, in the following even- and odd-stacked filter banks can be treated alike using (2.11).

The modulation sequences $\tilde{\mathbf{t}}_k$, $k = 0 \ldots K-1$, are collected in a matrix

$$\mathbf{T} = \begin{bmatrix}\tilde{\mathbf{t}}_0 & \cdots & \tilde{\mathbf{t}}_{K-1}\end{bmatrix}^\text{T} \quad \in \mathbb{C}^{K \times K} \tag{2.13}$$

which for example for a DFT-modulated filter bank would be a K-point DFT matrix. Applying (2.13) to the substitution of (2.11) and (2.10) into (2.6) gives

$$\mathbf{H}(z) = \mathbf{T} \cdot \mathbf{L}_2 \cdot \mathbf{P} \cdot \mathbf{L}_1(z) \tag{2.14}$$

as notation for the polyphase analysis matrix. With (2.14) a factorization into prototype filter components and a rotation by a transform matrix $\mathbf{T}$ has been established similar to [11, 12]. The difference is that the diagonal matrix $\mathbf{P}$ contains no sparse filters but only the prototype filter coefficients, which will be exploited for the implementation in Section 4.

2.3 Synthesis filter bank

Dual to the analysis filter bank, the synthesis filter bank as shown in Fig. 1 upsamples the subband signals by a factor N and applies interpolation filters $G_k(z)$. The condition that all filters $G_k(z)$ and $H_k(z)$ are derived from the same prototype lowpass filter and that the filter bank is perfectly reconstructing is guaranteed by $\mathbf{H}(z)$ being paraunitary [11]. Reconstruction is then given by the polyphase synthesis matrix $\mathbf{G}(z) \in \mathbb{C}^{N \times K}(z)$, which is the parahermitian of $\mathbf{H}(z)$, and relates the subband samples back to the polyphase components of the fullband signal,

$$\underline{\hat{X}}(z) = \mathbf{G}(z) \cdot \underline{Y}(z) \quad . \tag{2.15}$$

For causality, a delay has to be introduced such that $\hat{X}(z) = z^{-L_\text{p}+1} \cdot X(z)$ in Fig. 1. In the N-polyphase domain, this is expressed as [7]

$$\underline{\hat{X}}(z) = \underbrace{\begin{bmatrix} \mathbf{0}_{N-R \times R} & z^{-\lfloor L_\text{p}/N \rfloor + 1} \cdot \mathbf{I}_{N-R} \\ z^{-\lfloor L_\text{p}/N \rfloor} \cdot \mathbf{I}_R & \mathbf{0}_{R \times N-R} \end{bmatrix}}_{\Delta(z)} \cdot \underline{X}(z). \tag{2.16}$$

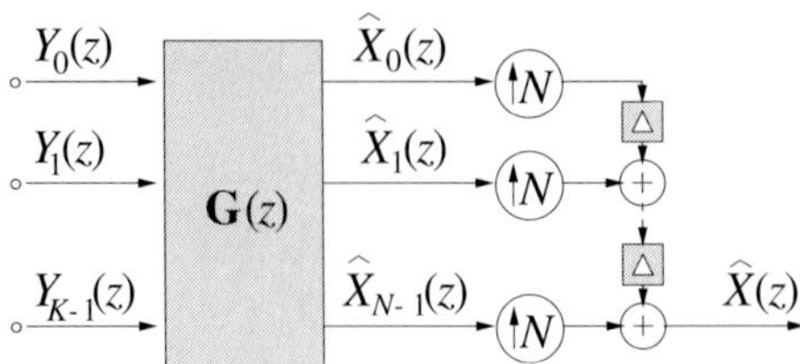

FIG. 4. Synthesis filter bank with $\mathbf{G}(z)$ describing a $K \times N$ MIMO system followed by a multiplexer.

Incorporating this delay into the perfect reconstruction condition, $\mathbf{G}(z) \cdot \mathbf{H}(z) = \Delta(z)$, the polyphase synthesis matrix is given by

$$\begin{aligned} \mathbf{G}(z) &= \Delta(z) \cdot \mathbf{H}^{\mathrm{H}}(z^{-1}) && (2.17) \\ &= \Delta(z) \cdot \mathbf{L}_1^{\mathrm{T}}(z^{-1}) \cdot \mathbf{P} \cdot \mathbf{L}_2^{\mathrm{T}} \cdot \mathbf{T}^{\mathrm{H}}. && (2.18) \end{aligned}$$

Evaluating $\Delta(z) \cdot \mathbf{L}_1^{\mathrm{T}}(z^{-1})$ in (2.18) gives

$$\hat{\mathbf{L}}_1^{\mathrm{T}}(z) = \Delta(z) \cdot \mathbf{L}_1^{\mathrm{T}}(z^{-1}) = \mathbf{L}_1^{\mathrm{T}}(z) \cdot \mathbf{J}_{L_{\mathrm{p}}}, \tag{2.19}$$

where $\mathbf{J}_{L_{\mathrm{p}}}$ is an $L_{\mathrm{p}} \times L_{\mathrm{p}}$ reverse identity matrix left–right flipping the causal matrix $\mathbf{L}_1^{\mathrm{T}}(z)$. The polyphase synthesis matrix in (2.18) leads to the signal flow graph in Fig. 4 with the MIMO system $\mathbf{G}(z)$ followed by a multiplexer. Its functionality is identical to the original synthesis filter bank in Fig. 1, but multiplications with expanding zeros in the interpolations filters are avoided.

Paraunitarity of $\mathbf{H}(z)$ is equivalent to the filter bank implementing a tight frame decomposition, which offers useful properties such as a fixed energy relation between the fullband signal $X(z)$ and the subband samples in $\underline{Y}(z)$. In a more general case, $H_k(z)$ and $G_k(z)$ can be based on different prototype lowpass filters (with different coefficients or even filter lengths) and therefore the filter bank system is not necessarily perfectly reconstructing. In this case the factorization of the polyphase analysis matrix remains as in Section 2.2, while the matrix factors of $\mathbf{G}(z)$ in (2.18) are built accordingly, whereby the parameters of the synthesis prototype filter have to be applied for $\mathbf{L}_1(z)$, $\mathbf{P}$, and $\mathbf{L}_2$.

3 Canonical state-space representations

Before trying to find suitable implementations for the previously factorized filter bank operations, canonical state-space representations for Figs 2 and 4 are derived in this section. These representations take the form

$$\begin{bmatrix} z\underline{W}(z) \\ \underline{V}(z) \end{bmatrix} = \begin{bmatrix} \mathbf{A} & \mathbf{B} \\ \mathbf{C} & \mathbf{D} \end{bmatrix} \cdot \begin{bmatrix} \underline{W}(z) \\ \underline{U}(z) \end{bmatrix}. \tag{3.1}$$

A flow graph of (3.1) is given in Fig. 5. Appropriate system matrices $\mathbf{A}$, $\mathbf{B}$, $\mathbf{C}$, $\mathbf{D}$, a state vector $\underline{W}(z)$, input $\underline{U}(z)$, and output $\underline{V}(z)$ need to be defined in the following.

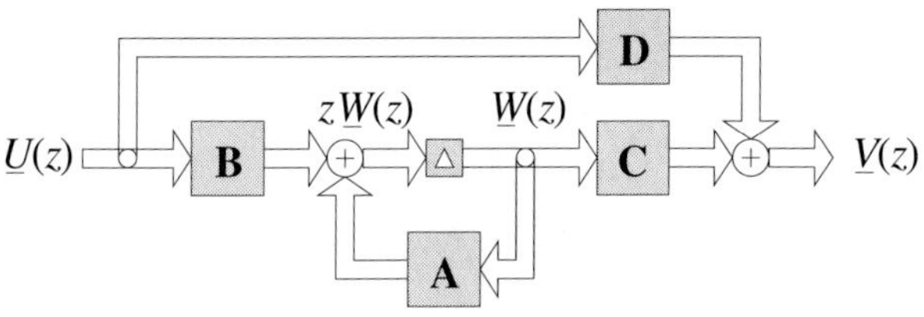

FIG. 5. Flow graph of state-space representation with states $\underline{W}(z)$.

3.1 Analysis filter bank representation

For the analysis filter bank, $\underline{U}(z)$ contains N demultiplexed samples of the input signal (that is $\underline{U}(z) = \underline{X}(z)$), and the output $\underline{V}(z) = \underline{Y}(z)$ holds the K subband signals. For a canonical form, the state vector $\underline{W}(z)$ must not have more than $L_\mathrm{p} - N$ elements and can be found by reformulating the analysis operation by introduction of an intermediate variable $\underline{Q}(z)$:

$$\underline{Y}(z) = \mathbf{T} \cdot \mathbf{L}_2 \cdot \mathbf{P} \cdot \underline{Q}(z) \qquad \text{with} \quad \underline{Q}(z) = \mathbf{L}_1(z) \cdot \underline{X}(z) \tag{3.2}$$

$$\underline{Q}(z) = \begin{bmatrix} \mathbf{0}_{N \times L_\mathrm{p}-N} & \mathbf{0}_{N \times N} \\ \mathbf{I}_{L_\mathrm{p}-N} & \mathbf{0}_{L_\mathrm{p}-N \times N} \end{bmatrix} z^{-1} \cdot \underline{Q}(z) + \begin{bmatrix} \mathbf{I}_N \\ \mathbf{0}_{L_\mathrm{p}-N \times N} \end{bmatrix} \underline{X}(z). \tag{3.3}$$

With the recursive update in (3.3), the lower portion with $L_\mathrm{p} - N$ elements of $\underline{Q}(z)$ now forms the state vector $\underline{W}(z)$, and the state-space system matrices can be identified as

$$\mathbf{A} = \begin{bmatrix} \mathbf{0}_{N \times L_\mathrm{p}-2N} & \mathbf{0}_{N \times N} \\ \mathbf{I}_{L_\mathrm{p}-2N} & \mathbf{0}_{L_\mathrm{p}-2N \times N} \end{bmatrix} \qquad \mathbf{B} = \begin{bmatrix} \mathbf{I}_N \\ \mathbf{0}_{L_\mathrm{p}-2N \times N} \end{bmatrix} \tag{3.4}$$

$$\mathbf{C} = \mathbf{T} \cdot \mathbf{L}_2 \cdot \mathbf{P} \cdot \begin{bmatrix} \mathbf{0}_{N \times L_\mathrm{p}-N} \\ \mathbf{I}_{L_\mathrm{p}-N} \end{bmatrix} \qquad \mathbf{D} = \mathbf{T} \cdot \mathbf{L}_2 \cdot \mathbf{P} \cdot \begin{bmatrix} \mathbf{I}_N \\ \mathbf{0}_{L_\mathrm{p}-N \times N} \end{bmatrix}. \tag{3.5}$$

Note that all memory exhibiting matrices in (2.14) have been replaced by memoryless operations due to the recursive updating in (3.1).

3.2 Synthesis filter bank representation

For the synthesis, the input in Fig. 5 now contains the K subband samples, $\underline{U}(z) = \underline{Y}(z)$, which are used for the reconstruction of N polyphase components of the fullband signal $\hat{X}(z)$, held in the output $\underline{V}(z) = \underline{\hat{X}}(z)$. A suitable state vector $\underline{W}(z)$ is sought from (2.18) with (2.19) inserted:

$$\underline{\hat{X}}(z) = \hat{\mathbf{L}}_1^\mathrm{T}(z) \cdot \mathbf{P} \cdot \mathbf{L}_2^\mathrm{T} \cdot \mathbf{T}^\mathrm{H} \cdot \underline{Y}(z) = \begin{bmatrix} \mathbf{0}_{N \times L_\mathrm{p}-N} & \mathbf{I}_N \end{bmatrix} \cdot \underline{Q}(z) \tag{3.6}$$

$$\underline{Q}(z) = \begin{bmatrix} \mathbf{0}_{N \times L_\mathrm{p}-N} & \mathbf{0}_{N \times N} \\ \mathbf{I}_{L_\mathrm{p}-N} & \mathbf{0}_{L_\mathrm{p}-N \times N} \end{bmatrix} z^{-1} \cdot \underline{Q}(z) + \mathbf{P} \cdot \mathbf{L}_2^\mathrm{T} \cdot \mathbf{T}^\mathrm{H} \cdot \underline{Y}(z). \tag{3.7}$$

The upper $L_\mathrm{p} - N$ elements of the intermediate variable $\underline{Q}(z)$ form the state vector $\underline{W}(z)$, which together with the recursive formulation (3.7) gives rise to the following state-space matrices:

$$\mathbf{A} = \begin{bmatrix} \mathbf{0}_{N\times L_\mathrm{p}-2N} & \mathbf{0}_{N\times N} \\ \mathbf{I}_{L_\mathrm{p}-2N} & \mathbf{0}_{L_\mathrm{p}-2N\times N} \end{bmatrix} \qquad \mathbf{B} = \left[\mathbf{I}_{L_\mathrm{p}-N} \quad \mathbf{0}_{L_\mathrm{p}-N\times N}\right]\cdot\mathbf{P}\cdot\mathbf{L}_2^\mathrm{T}\cdot\mathbf{T}^\mathrm{H} \tag{3.8}$$

$$\mathbf{C} = \left[\; \mathbf{0}_{N\times L_\mathrm{p}-2N} \quad \mathbf{I}_N \;\right] \qquad \mathbf{D} = \left[\; \mathbf{0}_{N\times L_\mathrm{p}-N} \quad \mathbf{I}_N \;\right]\cdot\mathbf{P}\cdot\mathbf{L}_2^\mathrm{T}\cdot\mathbf{T}^\mathrm{H}. \tag{3.9}$$

The system matrices $\mathbf{A}$ in both (3.4) and (3.8) represent tapped delay lines (TDLs), in which the state values are shifted by N states for every update iteration.

4 Filter bank implementation

Based on the factorizations in Section 2 and the state-space representations in Section 3, we now aim to find signal flow graphs that provide simple and efficient OSFB implementations.

4.1 Analysis filter bank implementation

Inspecting (3.4) and (3.5), the analysis filter bank operation in (2.14) can be executed in two steps. As mentioned above, the state values form a TDL, which is shifted by $\mathbf{A}$ and updated with N fresh samples in every subband sampling period by $\mathbf{B}$. Hence, $\mathbf{C}$ and $\mathbf{D}$ are excited by $L_\mathrm{p}-N$ old and N current input samples in $\underline{W}(z)$ and $\underline{U}(z)$, respectively. Due the similiarity of $\mathbf{C}$ and $\mathbf{D}$, a single TDL $[\underline{U}^\mathrm{T}(z)\ \underline{W}^\mathrm{T}(z)]^\mathrm{T}$ holding a total of L_p current and past input samples can be assembled. The result is the flow graph in Fig. 6. There, the demultiplexing of the scalar OSFB input into N parallel samples in $\underline{U}(z)$ as shown in Fig. 2 is already incorporated into the TDL.

According to (3.5), in Fig. 6 the TDL vector $[\underline{U}^\mathrm{T}(z)\ \underline{W}^\mathrm{T}(z)]^\mathrm{T}$ is passed into the block $\mathbf{P}$, multiplying each value by a prototype filter coefficients $p[i]$. From the results, $\mathbf{L}_2$ creates K sub-summations. After rotation of these subsums by the modulation matrix $\mathbf{T}$, finally a set of K subband samples has been calculated.

Note that the only memory-exhibiting operation in this analysis filter bank realization is the TDL holding L_p samples of the input signal. The number of states has increased by N over the canonical state-space representation in (3.4) and (3.5) due to the inclusion of the demultiplexer of Fig. 2. Therefore, with respect to the overall circuit running at the fullband rate, the circuit in Fig. 6 is canonical.

4.2 Synthesis filter bank implementation

Let us consider the system matrices $\mathbf{B}$ and $\mathbf{D}$ defined for the synthesis OSFB in (3.8) and (3.9), respectively, and the state-space representation in Fig. 5. It is obvious that the subband samples in $\underline{U}(z) = \underline{Y}(z)$ are derotated by $\mathbf{T}^\mathrm{H}$ and duplicated to L_p values by $\mathbf{L}_2^\mathrm{T}$ to finally excite the L_p prototype filter coefficients in $\mathbf{P}$. Of these L_p products, the upper $L_\mathrm{p}-N$ values are latched by $\mathbf{B}$ onto the TDL implemented by $\mathbf{A}$ in (3.8). The lower N product values are added with the lower N elements of the state vector to form the output $\underline{V}(z) = \underline{\hat{X}}(z)$. Incorporating the multiplexing of the output $\underline{\hat{X}}(z)$ to $\hat{X}(z)$, the TDL structure

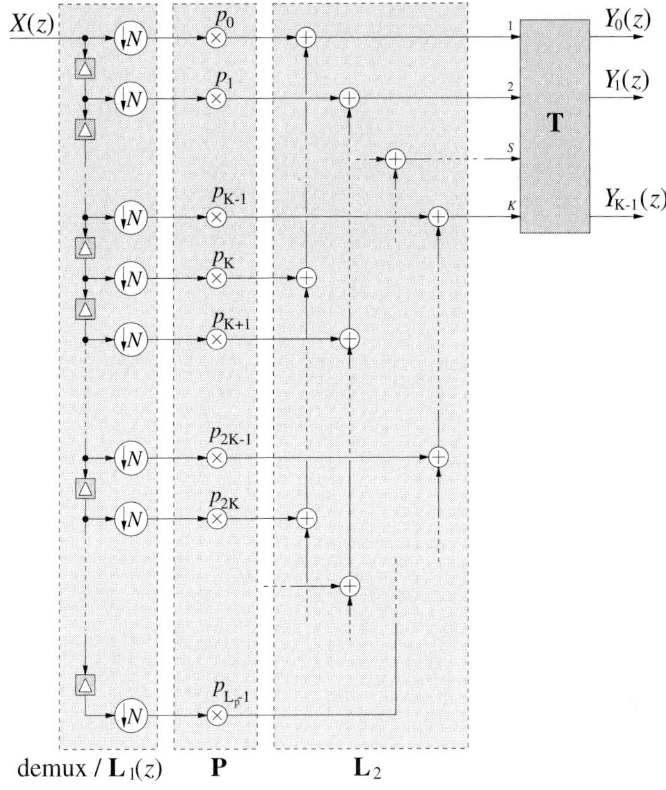

FIG. 6. Analysis filter bank signal flow graph.

can be realized as shown in Fig. 7. This can also be motivated by merging the multiplexer of Fig. 4 with $\hat{\mathbf{L}}_1^{\mathrm{T}}(z)$.

Note that similar to the analysis OSFB in Fig. 6 this circuit only requires a single TDL, onto which the results of the operation $\mathbf{P} \cdot \mathbf{L}_2^{\mathrm{T}} \cdot \mathbf{T}^{\mathrm{H}} \cdot \underline{\hat{Y}}(z)$ are accumulated. This operation is performed once in every subband sampling period. Afterwards, the values in the TDL can be shifted N times, clocked at the fullband rate, to form the fullband output samples. Similar arguments as in Section 4.1 hold for the canonical property of the flow graph in Fig. 7.

4.3 Special filter bank cases

A number of special filter bank implementations arise from different choices of the transform matrix $\mathbf{T}$, which in the following is to be refined in its implementation. In the case of a DFT-modulated filter bank, resulting in an even-stacked design as shown in Fig. 3(a), $\mathbf{T}$ is a $K \times K$ DFT matrix, $\mathbf{T} = \mathbf{T}_{\mathrm{DFT}}$. In the case of a generalized DFT (GDFT, [14]) filter bank, $\mathbf{T}$ is a GDFT matrix. This GDFT matrix is comprised of elements $t_{k,n} = e^{j\frac{2\pi}{K}(k+k_0)(n+n_0)}$, with $n = 0 \ldots L_{\mathrm{p}}-1$ and $k = 0 \ldots L_{\mathrm{p}}-1$. Note that for $k_0 = \frac{1}{2}$, the odd-stacked filter bank characteristic

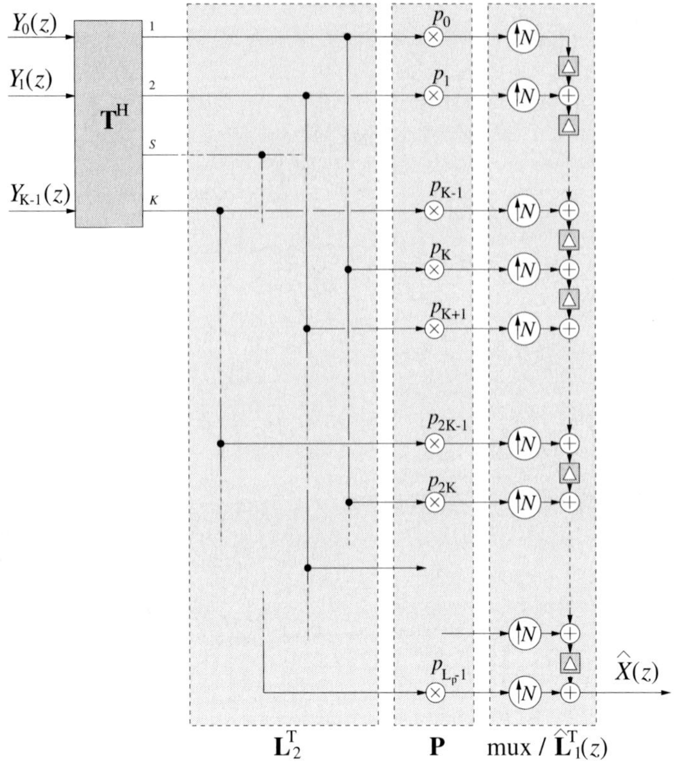

FIG. 7. Synthesis filter bank signal flow graph.

of Fig. 3(b) arises. A GDFT matrix generally permits a factorization

$$\mathbf{T} = \mathbf{D}_1 \cdot \mathbf{T}_{\mathrm{DFT}} \cdot \mathbf{D}_2. \tag{4.1}$$

For an odd-stacked filter bank, all matrices are $K \times K$ and $\mathbf{D}_1$ and $\mathbf{D}_2$ have diagonal form if inversion of the signs of prototype coefficients are performed as discussed in Section 2.2 [12].

Efficient implementations make use of an FFT routine instead of performing $\mathbf{T}_{\mathrm{DFT}}$ as a matrix operation. Further, if the input signal $X(z)$ is real-valued, (i) all operations except the transform matrix can be performed with real arithmetic, and (ii) redundancies arise due to approximately one-half of the subband signals being complex conjugate copies of others. Hence about one-half (depending on the DFT/GDFT transform and K being odd or even) of the subbands do not need to be calculated nor processed.

Single sideband (SSB) modulated OSFBs for real-valued subband signals can be obtained by modification of a GDFT-modulated filter bank decimated by only $N/2$ [8]. An additional complex modulation is performed on the subband signals $\underline{Y}(z)$ followed by a real operation. On the synthesis side, this is compensated by a matching demodulation prior to feeding into a GDFT synthesis filter bank.

Table 1 Computational complexity of filter banks implementations

	C_{real} / [MACs]	C_{complex} / [MACs]
DFT	$\frac{1}{N}(L_{\text{p}} + 4K\log_2 K)$	$\frac{1}{N}(2L_{\text{p}} + 4K\log_2 K)$
GDFT	$\frac{1}{N}(L_{\text{p}} + 4K\log_2 K + 4K)$	$\frac{1}{N}(2L_{\text{p}} + 4K\log_2 K + 8K)$
SSB	$\frac{2}{N}(L_{\text{p}} + 4K\log_2 K + 5K)$	$\frac{2}{N}(2L_{\text{p}} + 4K\log_2 K + 10K)$

4.4 Computational complexity

From the signal flow graphs for analysis and synthesis in Figs 6 and 7, the computational complexities for both operations can be evaluated in terms of multiply-accumulates (MACs) per sampling period. The latter is the period of the fullband signals prior to analysis or after synthesis. The complexities are given in Table 1 and are identical for analysis and synthesis, but differ for the choice of T and depend on whether the input signal is real- or complex-valued. Note, that the multiplication of the complex samples with the real-valued prototype filter coefficients accrues to $2L_p$ MACs. The modulation matrix $\mathbf{T}_{\text{DFT}}$ is assumed to be implemented by a K-point FFT requiring $4K\log_2 K$ real-valued MACs, which is invariably applied for all types of filter banks.

Although a number of methods reported in the literature give identical complexities in terms of MACs, the realizations in Figs 6 and 7 do not require any additional time-varying circular shifts [9] or switching [5], the indexing of time-varying filters [10], or filters with sparse coefficients [11, 12]. Further, the signal flow graphs in Figs 6 and 7 only require a single circular buffer and hence a minimum amount of pointers for addressing, and permit an arbitrary prototype filter length L_{p} independent of both the decimation ratio N and the channel number K.

5 Conclusion

Oversampled filter banks where all filters are derived from a prototype filter by modulation have been analysed using the well-known polyphase decomposition. Similar to previous analyses in the literature, this decomposition was factorized to reduce all filter operations to operations on the prototype filter coefficients, and a multiplication by the modulation matrix. This ensured a minimum amount of multiply-accumulate operations. However, the factorization was exploited via a state-space representation to locate all memory-requiring operations next to the multiplexers and demultiplexers of the circuit. The benefit is an implementation with a only single TDL that can be conveniently updated.

The presented implementation can be applied to a variety of modulated filter banks, such as shown for DFT, GDFT, or SSB filter banks, and with a large flexibility for the length of the prototype filter. Although not explicitly derived here, the analysis and synthesis filter bank implementations can be similarly applied to filter banks where filters $H_k(z)$ and $G_k(z)$ originate from more than one prototype filter [15].

Bibliography

[1] Kellermann, W. Analysis and design of multirate systems for cancellation of acoustical echoes. In *Proc. IEEE ICASSP*, **V**, pp. 2570–2573, (New York) 1988.

[2] Breining, C., Dreiseitel, P., Hänsler, E., Mader, A., Nitsch, B., Puder, H., Schertler, T., Schmidt, G. and Tilp, J. Acoustic echo control. *IEEE Signal Process. Mag.*, **16**, 42–69, 1999.

[3] Somayazulu, V.S., Mitra, S.K. and Shynk, J.J. Adaptive line enhancement using multirate techniques. In *Proc. IEEE ICASSP*, **2**, 928–931, 1989.

[4] Kellermann, W. Strategies for combining acoustic echo cancellation and adaptive beamforming microphone arrays. In *Proc. IEEE ICASSP*, **I**, 219–222, 1997.

[5] Cherubini, G., Eleftheriou, E. Ölcer, S. and Cioffi, J.M. Filter bank modulation techniques for very high-speed digital subscriber lines. *IEEE Commun. Mag.*, **38**, 98–104, 2000.

[6] Bellanger, M., Bonnerot, G. and Coudreuse, M. Digital filtering by polyphase network: application to sample rate alteration and filter banks. *IEEE Trans. on Acoustics, Speech, and Signal Process.*, **24**, 109–114, 1978.

[7] Vaidyanathan, P.P. *Multirate Systems and Filter Banks.* Prentice-Hall, Englewood Cliffs, NJ, 1993.

[8] Wackersreuther, G. Some new aspects of filters for filter banks. *IEEE Trans. on Acoustics, Speech, and Signal Process.*, **34**, 1182–1200, 1986.

[9] Wackersreuther, G. *Ein beitrag zum entwurf digitaler filterbänke.* Ph.D. Thesis, Institut für Nachrichtentechnik, Universität Erlangen-Nürnberg, 1987.

[10] Stasinski, R. Efficient Implementation of Uniform Filter Banks in the Absence of Critical Sampling. *IEE Electr. Lett.*, **30**, 118–120, 1994.

[11] Cvetković, Z. and Vetterli, M. Tight Weyl-Heisenberg frames in $l^2(\mathbb{Z})$. *IEEE Trans. Signal Process.*, **46**, 1256–1259, 1998.

[12] Weiss, S. *On adaptive filtering in oversampled subbands.* Ph.D. Thesis, Signal Processing Division, University of Strathclyde, Glasgow, 1998.

[13] Weiss, S. and Stewart, R.W. Fast implementation of oversampled modulated filter banks. *IEE Electr. Lett.*, **36**, 1502–1503, 2000.

[14] Crochiere, R.E. and Rabiner, L.R. *Multirate Digital Signal Processing.* Prentice-Hall, Englewood Cliffs, NJ, 1983.

[15] Harteneck, M., Weiss, S. and Stewart, R.W. Design of Near Perfect Reconstruction Oversampled Filter Banks for Subband Adaptive Filters. *IEEE Trans. on Circuits and Systems II*, **46**, 1081–1086, 1999.

Exponential Decomposition and Hankel Matrix

Franklin T. Luk
Department of Computer Science and Engineering, Chinese University of Hong Kong, Shatin, NT, Hong Kong

Sanzheng Qiao
Department of Computing and Software, McMaster University, Hamilton, Ontario L8S 4L7 Canada

David Vandevoorde
Edison Design Group, Belmont, CA 94002-3112 USA

1 Introduction

We consider an important problem in mathematical modeling.

Exponential decomposition. Given a finite sequence of signal values $\{s_1, s_2, \ldots, s_n\}$, determine a minimal positive integer r, complex coefficients $\{c_1, c_2, \ldots, c_r\}$, and distinct complex knots $\{z_1, z_2, \ldots, z_r\}$, so that the following exponential decomposition signal model is satisfied:

$$s_k = \sum_{i=1}^{r} c_i z_i^{k-1}, \quad \text{for } k = 1, \ldots, n. \tag{1.1}$$

We usually solve the above problem in three steps. First, determine the rank r of the signal sequence. Second, find the knots z_i. Third, calculate the coefficients $\{c_i\}$ by solving an $r \times r$ Vandermonde system:

$$W^{(r)}\mathbf{c} = \mathbf{s}^{(r)}, \tag{1.2}$$

where

$$W^{(r)} = \begin{bmatrix} 1 & 1 & \cdots & 1 \\ z_1 & z_2 & \cdots & z_r \\ \vdots & \vdots & \ddots & \vdots \\ z_1^{r-1} & z_2^{r-1} & \cdots & z_r^{r-1} \end{bmatrix}, \quad \mathbf{c} = \begin{bmatrix} c_1 \\ c_2 \\ \vdots \\ c_r \end{bmatrix} \quad \text{and } \mathbf{s}^{(r)} = \begin{bmatrix} s_1 \\ s_2 \\ \vdots \\ s_r \end{bmatrix}.$$

A square Vandermonde system (1.2) can be solved efficiently and stably [1].

There are two well known methods for solving our problem: namely, due to Prony [8] and Kung [6]. Vandevoorde [9] presented a single-matrix pencil-based framework that includes [8] and [6] as special cases. By adapting an implicit

Lanczos process in our matrix pencil framework, we present an exponential decomposition algorithm which requires only $O(n^2)$ operations and $O(n)$ storage, compared to $O(n^3)$ operations and $O(n^2)$ storage required by previous methods.

The relation between this problem and a large subset of the class of Hankel matrices was fairly widely known, but a generalization was unavailable [3]. We propose a new variant of the Hankel–Vandermonde decomposition and prove that it exists for any given Hankel matrix. We also observe that this Hankel–Vandermonde decomposition of a Hankel matrix is rank-revealing. As with many other rank-revealing factorizations, the Hankel–Vandermonde decomposition can be used to find a low-rank approximation to the given perturbed matrix. In addition to the computational savings, our algorithm has an advantage over many other popular methods in that it produces an approximation that exhibits the exact desired Hankel structure.

This paper is organized as follows. After introducing notations in Section 2, we discuss the techniques for determining rank and separating signals from noise in Section 3. In Section 4, we describe the matrix pencil framework and show how it unifies two previously known exponential decomposition methods. We devote Section 5 to our new fast algorithm based on the framework and Section 6 to our new Hankel–Vandermonde decomposition.

2 Notation

We adopt the Matlab notation. For example, the symbol $A_{i:j,k:l}$ denotes a submatrix formed by intersecting rows i through j and columns k through l of A, and $A_{:,k:l}$ is a submatrix consisting of columns k through l of A. We define $E^{(m)}$ as an $m \times m$ exchange matrix:

$$E^{(m)} = \begin{bmatrix} 0 & \cdots & 0 & 1 \\ \vdots & \iddots & 1 & 0 \\ 0 & \iddots & \iddots & \vdots \\ 1 & 0 & \cdots & 0 \end{bmatrix}, \quad \text{for} \quad m = 1, 2, \ldots .$$

We also define a sequence of n-element vectors $\{\mathbf{w}^{(0)}(z), \mathbf{w}^{(1)}(z), \ldots, \mathbf{w}^{(n-1)}(z)\}$ as follows:

$$\mathbf{w}^{(0)}(z) = \begin{bmatrix} 1 \\ \vdots \\ z^{k-1} \\ z^k \\ z^{k+1} \\ \vdots \\ z^{n-1} \end{bmatrix} \quad \text{and} \quad \mathbf{w}^{(k)}(z) = \begin{bmatrix} 0 \\ \vdots \\ 0 \\ 1 \\ (k+1)z \\ \vdots \\ \frac{(n-1)!}{k!\,(n-1-k)!} z^{n-1-k} \end{bmatrix},$$

for $k = 1, 2, \ldots, n-1$. We see that $\mathbf{w}^{(k)}(z)$ is a *scaled* kth derivative of $\mathbf{w}^{(0)}(z)$, normalized so that the first nonzero element equals unity. We introduce the concept of a *confluent* Vandermonde slice of order $n \times m$:

$$C^{(m)}(z) = \left[\mathbf{w}^{(0)}(z), \mathbf{w}^{(1)}(z), \ldots, \mathbf{w}^{(m-1)}(z)\right], \quad \text{for} \quad m = 1, 2, \ldots.$$

A confluent Vandermonde matrix [5] is a matrix of the form

$$\left[C^{(m_1)}(z_1), C^{(m_2)}(z_2), \ldots, C^{(m_j)}(z_j)\right].$$

The positive integer m_i is called the multiplicity of z_i in the confluent Vandermonde matrix. For our purposes, we define a new and more general structure that we call a *biconfluent* Vandermonde matrix:

$$\left[B^{(m_1)}(z_1), B^{(m_2)}(z_2), \ldots, B^{(m_j)}(z_j)\right],$$

where $B^{(m_i)}(z_i)$ equals either $C^{(m_i)}(z_i)$ or $E^{(n)}C^{(m_i)}(z_i)$.

3 Finding rank and denoising data

Recall our sequence of signal values. We want to find r, the rank of the data. Since the values contain errors, finding the rank means separating the signals from the noise. We modify our notations and use $\{\hat{s}_j\}$ to represent the sequence of 'polluted' signal values. Although we do not know the exact value of r, we would know a range in which r must lie. So we find two positive integers k and l such that

$$k + 1 \geq r, \quad l \geq r \quad \text{and} \quad k + l \leq n.$$

Since n is usually large and r quite small, choosing k and l to satisfy the above inequalities is not a problem. We construct a $(k+1) \times l$ Hankel matrix $\widehat{H}$ by

$$\widehat{H} = \begin{bmatrix} \hat{s}_1 & \hat{s}_2 & \cdots & \hat{s}_l \\ \hat{s}_2 & \hat{s}_3 & \cdots & \hat{s}_{l+1} \\ \vdots & \vdots & \ddots & \vdots \\ \hat{s}_k & \hat{s}_{k+1} & \cdots & \hat{s}_{k+l-1} \\ \hat{s}_{k+1} & \hat{s}_{k+2} & \cdots & \hat{s}_{k+l} \end{bmatrix}.$$

If there is no noise in the signal values, the rank of $\widehat{H}$ will be exactly r. Due to noise, the rank of $\widehat{H}$ will be greater than r. Compute a singular value decomposition (SVD) of $\widehat{H}$:

$$\widehat{H} = U\Sigma V^{\mathrm{H}},$$

where U and V are unitary matrices, and

$$\Sigma = \mathrm{diag}(\sigma_1, \sigma_2, \ldots, \sigma_{\min(k+1,l)}) \in R^{(k+1)\times l}.$$

We find r from looking for a 'gap' in the singular values:

$$\sigma_1 \geq \sigma_2 \geq \cdots \geq \sigma_r \gg \sigma_{r+1} \geq \sigma_{r+2} \geq \cdots \geq \sigma_{\min(k+1,l)} \geq 0. \tag{3.1}$$

There could be multiple gaps in the sequence of singular values, in which case we could either pick the first gap or use other information to choose r from the several possibilities. This is how we determine the numerical rank r of $\widehat{H}$, and thereby the rank of the signal data. A $(k+1) \times l$ matrix A of exact rank r is obtained from

$$A = U_{:,1:r}\Sigma_{1:r,1:r}(V_{:,1:r})^{\mathrm{H}}.$$

Unfortunately, the matrix A would have lost its Hankel structure. We want to find a Hankel matrix H that will be 'close' to A.

Hankel matrix approximation. Given a $(k+1) \times l$ matrix A of rank r, find a $(k+1) \times l$ Hankel matrix H of rank r such that $\|A - H\|_F = \min$.

A simple way to get a Hankel structure from A is to average along the antidiagonals; that is,

$$s_1 = a_{11},\ s_2 = (a_{12} + a_{21})/2,\ s_3 = (a_{13} + a_{22} + a_{31})/3,\ \ldots.$$

However, the rank of the resultant Hankel matrix would inevitably be greater than r. An SVD could be used to reduce the rank to r, but the decomposition would destroy the Hankel form. Cadzow [2] proposed that we iterate using a two-step cycle of an SVD followed by averaging along the antidiagonals. He proved that the iteration would converge under some mild assumptions.

Note that in the matrix approximation problem, the obvious selection of the Frobenius norm may not be appropriate, since this norm would give an unduly large weighting to the central elements of the Hankel structure. A weighted Frobenius matrix could be constructed, but such a norm is no longer unitarily invariant.

Nonetheless, let us assume that we have denoised the data and obtained a $(k+1) \times l$ Hankel matrix H of rank r:

$$H = \begin{bmatrix} s_1 & s_2 & \cdots & s_l \\ s_2 & s_3 & \cdots & s_{l+1} \\ \vdots & \vdots & \ddots & \vdots \\ s_k & s_{k+1} & \cdots & s_{k+l-1} \\ s_{k+1} & s_{k+2} & \cdots & s_{k+l} \end{bmatrix}. \tag{3.2}$$

4 Matrix pencil framework

For this section, we take

$$l = r$$

in (3.2) and we assume that $\text{rank}(H) = r$. We present a matrix pencil framework which includes the methods of Prony [8] and Kung [6] as special cases.

We begin by defining two k-by-r submatrices of H:

$$H_1 \equiv H_{1:k,:} \quad \text{and} \quad H_2 \equiv H_{2:k+1,:}. \tag{4.1}$$

So,

$$H_1 = \begin{bmatrix} s_1 & s_2 & \cdots & s_{r-1} & s_r \\ s_2 & s_3 & \cdots & s_r & s_{r+1} \\ \vdots & \vdots & \ddots & \vdots & \vdots \\ s_k & s_{k+1} & \cdots & s_{r+k-2} & s_{r+k-1} \end{bmatrix}$$

and

$$H_2 = \begin{bmatrix} s_2 & s_3 & \cdots & s_r & s_{r+1} \\ s_3 & s_4 & \cdots & s_{r+1} & s_{r+2} \\ \vdots & \vdots & \ddots & \vdots & \vdots \\ s_{k+1} & s_{k+2} & \cdots & s_{r+k-1} & s_{r+k} \end{bmatrix}.$$

Note that column i of H_2 equals column $(i+1)$ of H_1, for $i = 1, 2, \ldots, r-1$. What about the last column of H_2? Let $\{z_i\}$ denote the roots of the polynomial

$$q_r(z) = z^r + \gamma_{r-1} z^{r-1} + \cdots + \gamma_1 z + \gamma_0. \tag{4.2}$$

For any $m = 1, 2, \ldots, k$, we can verify that

$$s_{r+m} + \sum_{i=0}^{r-1} \gamma_i s_{i+m} = c_1 z_1^{m-1} q_r(z_1) + c_2 z_2^{m-1} q_r(z_2) + \cdots + c_r z_r^{m-1} q_r(z_r) = 0.$$

It follows that

$$s_{r+m} = -\sum_{i=0}^{r-1} \gamma_i s_{i+m}.$$

Thus,

$$H_1 X = H_2, \tag{4.3}$$

where

$$X = \begin{bmatrix} 0 & 0 & \cdots & 0 & -\gamma_0 \\ 1 & 0 & \cdots & 0 & -\gamma_1 \\ 0 & 1 & \cdots & 0 & -\gamma_2 \\ \vdots & \vdots & \ddots & \vdots & \vdots \\ 0 & 0 & \cdots & 1 & -\gamma_{r-1} \end{bmatrix}.$$

Hence X is the companion matrix of the polynomial (4.2), and the knots z_i are the eigenvalues of the X.

Let $\mathbf{v}_i$ denote an eigenvector of X corresponding to z_i; that is

$$X\mathbf{v}_i = z_i\mathbf{v}_i.$$

It follows from (4.3) that

$$z_i H_1 \mathbf{v}_i = H_2 \mathbf{v}_i.$$

In general, the knots z_i are the eigenvalues of matrix Y satisfying the matrix equation:

$$FH_1GY = FH_2G, \tag{4.4}$$

for any nonsingular F and G. In other words, let F and G be any nonsingular matrices of orders k and r respectively, then the knots z_i are the eigenvalues of the pencil equation

$$FH_2G\mathbf{v} - \lambda FH_1G\mathbf{v} = 0.$$

4.1 Prony's method

For Prony's method, we consider the case when $n = 2r$ and set $k = r$. Let $F = G = I$ in the framework (4.4), then the knots are the eigenvalues of the matrix Y satisfying $H_1Y = H_2$. Now, (4.3) shows that the companion matrix X of the polynomial (4.2) satisfies this matrix equation. Thus the roots of the polynomial (4.2) are the knots. Moreover, the last column of (4.3) indicates that the coefficient vector $\mathbf{g} = (\gamma_0, \gamma_1, \ldots, \gamma_{r-1})^{\mathrm{T}}$ of the polynomial (4.2) is the solution of the $r \times r$ Yule–Walker system:

$$H_1\mathbf{g} = -(s_{r+1}, s_{r+2}, \ldots, s_{2r})^{\mathrm{T}}. \tag{4.5}$$

This leads to Prony's method.

Prony's method

(1) Solve a square equation (4.5), that is, $n = 2r$, for the coefficients γ_i;
(2) Determine the roots z_i of the polynomial $q_r(z)$ of (4.2);
(3) Find the coefficients c_i by solving the square Vandermonde system: $W^{(r)}\mathbf{c} = \mathbf{s}^{(r)}$ of (1.2).

4.2 Kung's method

Now, we derive Kung's Hankel SVD method [6] using our framework. Let

$$H = U\Sigma V^{\mathrm{H}}$$

be the SVD of H in (3.2). Denote

$$U_1 = U_{1:k,:} \quad \text{and} \quad U_2 = U_{2:k+1,:}.$$

Then we have

$$H_1 = U_1 \Sigma V^{\mathrm{H}} \quad \text{and} \quad H_2 = U_2 \Sigma V^{\mathrm{H}}.$$

Setting $F = I$ and $G = V\Sigma^{-1}$ in the framework (4.4), we know that the eigenvalues of Y in

$$U_1 Y = U_2 \tag{4.6}$$

are the knots z_i. The matrix Y can be solved by an approximation method; for example, least squares.

The following procedure is sometimes referred to as the *Hankel SVD* (HSVD) algorithm.

Kung's HSVD method

(1) Solve (4.6) for Y in a least squares sense;
(2) Find the knots z_i as the eigenvalues of Y;
(3) Find the coefficients c_i by solving (1.2).

Kung's method computes better results than Prony's method. See [9] for numerical examples. However, it is based on the SVD, which is expensive to compute.

5 Lanczos process

In this section, we assume that $k = r$. First, we expand H into a $2r \times 2r$ Hankel matrix:

$$\widehat{H} = \begin{bmatrix} s_1 & \cdots & s_{2r} \\ \vdots & \iddots & \\ s_{2r} & & 0 \end{bmatrix}$$

and assume that it is strongly nonsingular, that is, all its leading principal submatrices are nonsingular. Suppose that we have a triangular decomposition $\widehat{H} = R^{\mathrm{T}} D R$ where R is upper triangular and D diagonal. Since the Hankel matrix H in (3.2) equals $\widehat{H}_{1:k+1,1:r}$, we have

$$H = (R_{1:r,1:k+1})^{\mathrm{T}} D_{1:r,1:r} R_{1:r,1:r}. \tag{5.1}$$

Thus,

$$H_1 \equiv H_{1:k,:} = (R_{1:r,1:k})^{\mathrm{T}} D R_{1:r,1:r}$$

and

$$H_2 \equiv H_{2:k+1,:} = (R_{1:r,2:k+1})^{\mathrm{T}} D R_{1:r,1:r}.$$

Setting $F = I$ and $G = (R_{1:r,1:r})^{-1}(D_{1:r,1:r})^{-1}$ in our framework (4.4), we know that the eigenvalues of a matrix Y satisfying

$$(R_{1:r,1:k})^{\mathrm{T}} Y = (R_{1:r,2:k+1})^{\mathrm{T}} \tag{5.2}$$

are the desired knots. Now, using the Lanczos process, we show that we can find a tridiagonal matrix satisfying (5.2).

Let $\mathbf{e} = (1, 1, \ldots, 1)^{\mathrm{T}}$ and $D(\mathbf{z}) = \mathrm{diag}(z_1, \ldots, z_r)$, then the transpose of the Vandermonde matrix

$$W^{\mathrm{T}} = [\mathbf{e}, D(\mathbf{z})\mathbf{e}, \ldots, D(\mathbf{z})^{k+1}\mathbf{e}]$$

is a Krylov matrix. Similar to the standard Lanczos method [5], we can find a tridiagonal matrix T such that

$$BT = D(\mathbf{z})B \quad \text{or} \quad T^{\mathrm{T}} B^{\mathrm{T}} = B^{\mathrm{T}} D(\mathbf{z}), \tag{5.3}$$

where B is orthogonal with respect to $D(\mathbf{c}) = \mathrm{diag}(c_1, \ldots, c_r)$, that is, $B^{\mathrm{T}} D(\mathbf{c}) B = D$, a diagonal matrix.

The Lanczos method also gives a QR decomposition of the Krylov matrix

$$W^{\mathrm{T}} = BR_{1:r,1:k+1}, \tag{5.4}$$

where R is an $r \times (k+1)$ upper trapezoidal matrix. It follows that

$$(W_{1:k,:})^{\mathrm{T}} = BR_{1:r,1:k} \quad \text{and} \quad (W_{2:k+1,:})^{\mathrm{T}} = BR_{1:r,2:k+1}.$$

It is obvious that $(W_{2:k+1,:})^{\mathrm{T}} = D(\mathbf{z})(W_{1:k,:})^{\mathrm{T}}$. Then we have

$$D(\mathbf{z})BR_{1:r,1:k} = BR_{1:r,2:k+1}.$$

From (5.3), substituting $D(\mathbf{z})B$ with BT in the above equation, we get

$$TR_{1:r,1:k} = R_{1:r,2:k+1} \quad \text{or} \quad (R_{1:r,1:k})^{\mathrm{T}} T^{\mathrm{T}} = (R_{1:r,2:k+1})^{\mathrm{T}}. \tag{5.5}$$

On the other hand, it can be easily verified from straightforward multiplication that

$$H = W_{1:k+1,:} D(\mathbf{c}) (W_{1:r,:})^{\mathrm{T}}. \tag{5.6}$$

Then, from (5.4), we have the decomposition $H = (R_{1:r,1:k+1})^{\mathrm{T}} DR_{1:r,1:r}$. This says that the upper trapezoidal matrix $R_{1:r,1:k+1}$ computed by the Lanczos method gives the decomposition (5.1) of H and we can find a tridiagonal T^{T} satisfying (5.2). Thus the eigenvalues of T^{T} are the knots. In fact, the equations in (5.3) also show that the eigenvalues of T^{T} are the knots z_i. Moreover, (5.5) gives a three-term recursion of the rows of R, which leads to an efficient method for generating the tridiagonal T and the rows of R. To start the process, we note that the first row of R, or the first column of R^{T}, is the scaled first column of $\widehat{H}$ since $\widehat{H} = R^{\mathrm{T}} DR$ and R is upper triangular and D is diagonal. The process is outlined as follows. For details, see [9].

Lanczos process

(1) Initialize $R_{1,1:2r} = (s_1, \ldots, s_{2r})/s_1$;
(2) Set $\alpha_1 = R_{1,2}$;
(3) Use the recurrence

$$R_{i-2,i:2r-i+1} = R_{i-1,i+1:2r-i+2} - \alpha_{i-1} R_{i-1,i:2r-i+1} - \beta_{i-1}^2 R_{i,i:2r-i+1}$$

to compute a new row $R_{i,i:2r-i+1}$;
(4) Compute new $\alpha_i = R_{i,i+1} - R_{i-1,i}$; $\beta_{i-1}^2 = R_{i,i+2} - \alpha_i R_{i,i+1} - R_{i-1,i+1}$.

The tridiagonal T^{T} is of the form

$$T^{\mathrm{T}} = \begin{bmatrix} \alpha_1 & \beta_1^2 & & & 0 \\ 1 & \alpha_2 & \beta_2^2 & & \\ & \ddots & \ddots & \ddots & \\ & & 1 & \alpha_{r-1} & \beta_{r-1}^2 \\ 0 & & & 1 & \alpha_r \end{bmatrix}.$$

Then we transform T into a complex symmetric tridiagonal matrix by scaling its rows and columns. Since the transformed T is complex symmetric, we apply the complex-orthogonal transformations in the QR method [4, 7] to compute its eigenvalues, that is, the knots.

6 A new Hankel–Vandermonde decomposition

The matrix product formulation (5.6) does not apply to an arbitrary Hankel matrix. Even when it exists, the exponential decomposition is not always unique. In this section, we will generalize this Hankel factorization so that the new decomposition will exist for any Hankel matrix and exhibit the rank of the given Hankel matrix. We start by establishing some useful properties of Hankel matrices which are closely related to Prony's method.

Observe that the signal model (1.1) describes the solution of a *linear difference equation* and that Prony's method computes the coefficients γ_i in its characteristic polynomial

$$y_{k+r} + \gamma_{r-1} y_{k+r-1} + \cdots + \gamma_0 y_k = 0,$$

whose solution starts with the sequence $(s_1, s_2, \ldots, s_{2r-1})$. Indeed, the coefficients γ_i form the solution of the Hankel system:

$$\begin{bmatrix} s_1 & s_2 & \cdots & s_r \\ s_2 & s_3 & \cdots & s_{r+1} \\ \vdots & \vdots & \ddots & \vdots \\ s_{r-1} & s_r & \cdots & s_{2r-2} \\ s_r & s_{r+1} & \cdots & s_{2r-1} \end{bmatrix} \begin{bmatrix} \gamma_0 \\ \gamma_1 \\ \vdots \\ \gamma_{r-1} \end{bmatrix} = - \begin{bmatrix} s_{r+1} \\ s_{r+2} \\ \vdots \\ s_{2r-1} \\ \eta \end{bmatrix}, \tag{6.1}$$

where η can be chosen arbitrarily. Thus γ_i are determined by the parameter η.

In preparation of a general result, we introduce a new concept of a *Prony η-continuation.*

Let H be an $r\times r$ nonsingular Hankel matrix. The $k\times m$ Prony η-continuation of H is the Hankel matrix

$$\operatorname{Pro}(H,\eta) = \begin{bmatrix} s_1(\eta) & s_2(\eta) & \cdots & s_m(\eta) \\ s_2(\eta) & s_3(\eta) & \cdots & s_{m+1}(\eta) \\ \vdots & \vdots & \ddots & \vdots \\ s_k(\eta) & s_{k+1}(\eta) & \cdots & s_{k+m-1}(\eta) \end{bmatrix},$$

whose entries $s_i(\eta)$ are determined by the solution $(\gamma_0,\ldots,\gamma_{r-1})^{\mathrm{T}}$ of (6.1).

Note that $s_i(\eta) = s_i$ for $i < 2r$ and $s_{2r}(\eta) = \eta$ and $\operatorname{rank}(\operatorname{Pro}(H,\eta)) = r$ if $k \geq r$ and $m \geq r$. Thus we embed a square and nonsingular H into a general Hankel matrix. Using the Prony continuation, we are able to prove that a Hankel matrix can be decomposed into a sum of two Prony continuations of two nonsingular Hankel matrices. For the proof see [9].

Let H be a $k \times k$ Hankel matrix of rank r, then there exist values η_1 and η_2, and nonsingular Hankel matrices H_1 and H_2, with respective sizes $r_1 \times r_1$ and $r_2 \times r_2$, such that $r = r_1 + r_2$ and

$$H = \operatorname{Pro}(H_1,\eta_1) + E_k \operatorname{Pro}(E_k H_2 E_k, \eta_2) E_k.$$

Then, using confluent Vandermonde matrix, we obtain the following theorem of Hankel–Vandermonde decomposition [9].

Let H be an arbitrary $k \times m$ real or complex Hankel matrix. There exist positive integers m_i $(i = 1,\ldots,j)$ and distinct complex numbers z_i $(i = 1,\ldots,j)$ such that

$$H = W_k \operatorname{diag}(C_1, C_2, \ldots, C_j) W_m^{\mathrm{T}}, \tag{6.2}$$

where W_l denotes an $l \times r$ biconfluent Vandermonde matrix:

$$W_l = \left[B_{m_1}(z_1), \ldots, B_{m_j}(z_j)\right],$$

with $r = m_1 + m_2 + \cdots + m_j$, and the C_i are nonsingular 'left' (upper) triangular $m_i \times m_i$ Hankel matrices:

$$C_i = \begin{bmatrix} c_1^{(i)} & c_2^{(i)} & \cdots & c_{m_i}^{(i)} \\ c_2^{(i)} & \cdot^{\cdot^{\cdot}} & \cdot^{\cdot^{\cdot}} & 0 \\ \vdots & \cdot^{\cdot^{\cdot}} & \cdot^{\cdot^{\cdot}} & \\ c_{m_i}^{(i)} & 0 & & 0 \end{bmatrix}.$$

Obviously, (6.2) is a generalization of (5.6) in that H can be any Hankel matrix. Also, we note that

- $\text{rank}(H) = \min(k, m, r)$;
- The values z_i can be chosen to lie on the unit disc, that is, $|z_i| \leq 1$.

Conclusion We have proposed a matrix-pencil framework, which unifies previously known exponential decomposition methods. We have developed a new fast Lanczos based exponential decomposition algorithm and have presented a general Hankel–Vandermonde decomposition. A low-rank Hankel approximation to a given Hankel matrix can be derived from our general Hankel–Vandermonde decomposition (see [9]).

Bibliography

[1] Björck, A. and Pereyra, V. Solution of Vandermonde systems of equations, *Math. Comput.*, **24**, (1970), 893–903.

[2] Cadzow, J.A. Signal enhancement – a composite property mapping algorithm, *IEEE Trans. Acoust., Speech, Sig. Proc.*, **36**, (1988), 49–62.

[3] Chen, H., Van Huffel, S. and Vandewalle, J. *Bandpass filtering for the HTLS estimation algorithm: Design, evaluation and SVD analysis*, In SVD and Signal Processing III, (eds, M. Moonen and B. De Moor), pp. 391–398, Amsterdam, Elsevier, 1995.

[4] Cullum, J.K. and Willoughby, R.A. A QL procedure for computing the eigenvalues of complex symmetric tridiagonal matrices, *SIAM J. Matrix Anal. Appl.*, **17**, (1996), 83–109.

[5] Golub, G.H. and Van Loan, C.F. Matrix Computations, 3rd edition, Johns Hopkins University Press, Baltimore, 1996.

[6] Kung, S.Y., Arun, K.S. and Rao, D.B. State-space and singular value decomposition-based approximation methods for the harmonic retrieval problem, *J. Opt. Soc. Am.*, **73**, (1983), 1799–1811.

[7] Luk, F.T. and Qiao, S. *Using complex-orthogonal transformations to diagonalize a complex symmetric matrix.* In Advanced Signal Processing: Algorithms, Architectures, and Implementations VII, (ed. F. T. Luk), Proc. SPIE 3162, pp. 418–425, 1997.

[8] Prony, R. *Essai expérimental et analytique sur les lois de la delatabilité et sur celles de la force expansive de la vapeur de l'eau et de la vapeur de l'alkool, à différentes température*, J. de l'Ecole Polytechnique, 1 (1795), pp. 24–76.

[9] Vandevoorde, D. *A fast exponential decomposition algorithm and its applications to structured matrices*, Ph.D. Thesis, Rensselaer Polytechnic Institute, Troy, New York, 1996.

A Robustness Property of Algorithms Using Second-Order Statistics

Jean-Pierre Delmas
Institut National des Télécommunications. 9 rue Charles Fourier, 91011 Evry Cedex, France

Abstract

This paper re-examines the asymptotic performance analysis of second-order methods for parameter estimation in a general context. It provides a unifying framework to investigate the asymptotic performance of second-order methods under the stochastic model assumption in which both the waveforms and noise signals are possibly temporally correlated, possibly non-Gaussian, real or complex (possibly noncircular) random processes. Thanks to a functional approach and a new central limit theorem about the sample covariance matrix, the conditions under which the asymptotic covariance of a parameter estimator are dependent or independent of the distribution of the signal involved are specified. In particular, some classic asymptotic robustness properties are questioned. Finally, we demonstrate the application of our general results to direction of arrival (DOA) estimation, identification of finite impulse response models, sinusoidal frequency estimation for mixed spectra time series and frequency estimation of sinusoidal signal with very lowpass envelope.

1 Introduction

The problem of estimating parameters of waveforms embedded in additive noise based on second-order algorithms has been intensively studied in the signal processing community, due to its wide applicability, mostly explicit physical interpretation, simplicity of implementation and often good performance. Performance analyses of such algorithms derive from several signal models. The deterministic (or conditional) and the stochastic (or unconditional) model are the main models that have appeared in the literature (see, for example [9, 12]. The noise is assumed to be a temporally uncorrelated Gaussian random process in these two models. But the waveforms are assumed to be fixed in all the realizations in the deterministic model and to be generally Gaussian temporally uncorrelated random processes in the stochastic model. Many authors have compared the asymptotic performance of parameter estimators with these two models and connected their performance to the deterministic or stochastic Cramer–Rao bound (see, for example, [9, 12] and the references therein). Among the performance studies carried out in the stochastic model, Cardoso and Moulines [5] have shown that the asymptotic performance of most high resolution covariance-based DOA estimators is independent of the distribution of the source signals for independent snapshots. This robustness property was extended to the temporal correlation of the source signals and clarified in [6], where it is proved that Toeplitzation and the augmentation techniques are very sensitive to this correlation. Abed Meraim *et al.* [1] presented an asymptotic performance analysis of subspace methods for blind identification of single-input multiple-output FIR systems where it is shown that the higher than second-order statistics of the input signals do not affect the asymptotic covariance of the estimated impulse response. Besides these works, most asymptotic performance analyses rest on the assumption that the sample covariance matrix of the data has a Wishart distribution (see, for example, [15]). But this assumption is valid only if the signals are Gaussian and i.i.d..

It is thus of importance to determine if the performance is affected by the joint distribution of the signals. The purpose of this contribution is to provide a unifying framework to investigate the asymptotic performance of second-order methods for parameter estimation in a general context under the stochastic model assumption in which both the waveforms and noise signals are possibly temporally correlated, possibly non-Gaussian, real or complex (possibly noncircular) random processes. In this context, the performance is *a priori* expected to depend on the joint distribution of the signals involved. We adopt a functional approach in which the Gaussian asymptotic distribution of the covariance-based parameter estimates is derived from the Gaussian asymptotic distribution of the sample covariance matrix which is proved for this general model. This allows us to give closed-form expressions for the asymptotic covariance matrices of parameter estimates. We then examine under which conditions the asymptotic covariance of parameter estimators are dependent (or not) on the probability distribution and on the temporal correlation of the signals involved. In particular, for the DOA estimation, it is established that under mild assumptions and under the condition that the noise is temporally uncorrelated, the asymptotic covariance matrix of parameter estimates is independent of the distribution and of the temporal correlation of the waveforms. On the other hand, this asymptotic covariance is sensitive to the temporal correlation of the signals involved when the noise is temporally correlated, which is the case when the observed signal are oversampled or when the noise includes jammers. This result belies an often accepted (see, for example, [8]) asymptotic robustness property.

This paper is organized as follows. After the general data model and some assumptions are introduced, some examples are given in Section 2. In Section 3, some regularity conditions assumed for the algorithms under study are specified and a general functional approach providing a unifying framework for asymptotic performance analysis is presented. In Section 4, the asymptotic normality of the sample covariance matrix is established for these general data models where a new central limit theorem is given. This allows us to question the classical asymptotic robustness property. This methodology is then applied to DOA, FIR, and sinusoidal frequency estimators for mixed spectra times series and for sinusoidal signals with very lowpass envelopes.

The following notations are used throughout the paper. Superscript $+$ stands for transpose or conjugate transpose according to the vectors and matrices are real- or complex-valued. Vec(.) is the vectorization operator that turns a matrix into a vector consisting of the columns of the matrix stacked one below another. It will be used in conjunction with the Kronecker product $\mathbf{A} \otimes \mathbf{B}$ as the block matrix, the (i,j) block element of which is $a_{i,j}\mathbf{B}$ (resp. $b^*_{i,j}\mathbf{A}$) for real (resp. complex) valued matrices and with the vec-permutation matrix $\mathbf{K}$ which transforms $\mathrm{Vec}(\mathbf{A})$ to $\mathrm{Vec}(\mathbf{A}^T)$ for any square matrix $\mathbf{A}$. (This unusual convention makes it easier to deal with complex matrices for which $\mathrm{Vec}(\mathbf{ABC}) = (\mathbf{A} \otimes \mathbf{C}^H)\mathrm{Vec}(\mathbf{B})$.) $\mathrm{Diag}(a_1, \ldots, a_n)$ is a diagonal matrix with diagonal elements a_i and $\mathrm{Diag}(\mathbf{A}_1, \ldots, \mathbf{A}_n)$ is a block-diagonal matrix with block-diagonal matrices $\mathbf{A}_i$.

2 Data model and algorithms under study

2.1 General hypotheses

In many applications, it is of interest to estimate the parameter $\Theta \in \Theta \subset R^q/C^q$ from the following p-variate real or complex (possibly noncircular) valued wide sense stationary time series

$$\mathbf{x}_t = \mathbf{E}_s(\Theta)\mathbf{s}_t + \mathbf{n}_t \qquad t = 1, \ldots, T. \tag{2.1}$$

$\mathbf{E}_s(\Theta)\mathbf{s}_t$ and $\mathbf{n}_t$ model the signals of interest and additive measurement noise, respectively. It is assumed that $\mathbf{E}_s(\Theta)$ is deterministic and known as a function of the unknown signal parameters Θ. Naturally, the probability distribution of $(\mathbf{x}_t)_{t=1,\ldots,T}$ depends on extra parameters which are also unknown, but we are only interested here in the estimation of parameters Θ. In this general model, $\mathbf{s}_t$ and $\mathbf{n}_t$ are multivariate independent, zero-mean, second-order stationary time series with spatial covariance matrices $\mathbf{R}_s \stackrel{\text{def}}{=} \mathrm{E}(\mathbf{s}_t\mathbf{s}_t^+)$ and $\mathbf{R}_n \stackrel{\text{def}}{=} \mathrm{E}(\mathbf{n}_t\mathbf{n}_t^+)$. Thus, the covariance matrix of $\mathbf{x}_t$ is

$$\mathbf{R}_x \stackrel{\text{def}}{=} \mathrm{E}(\mathbf{x}_t\mathbf{x}_t^+) = \mathbf{E}_s(\Theta)\mathbf{R}_s\mathbf{E}_s^+(\Theta) + \mathbf{R}_n. \tag{2.2}$$

We suppose that the parameterization used is identifiable to the second-order, in the following sense:

$$\mathbf{E}_s(\Theta_1)\mathbf{R}_s^{(1)}\mathbf{E}_s^+(\Theta_1) + \mathbf{R}_n^{(1)} = \mathbf{E}_s(\Theta_2)\mathbf{R}_s^{(2)}\mathbf{E}_s^+(\Theta_2) + \mathbf{R}_n^{(2)} \Rightarrow \Theta_1 = \Theta_2.$$

Some examples of this data model are briefly described in the following section.

2.2 Examples of application

(1) *Narrow-band DOA estimation*: $\mathbf{x}_t$ represents the p-vector of observed complex envelope at the sensor output. $\mathbf{E}_s(\Theta)$ is the $p \times K$ steering matrix, Θ the spatial parameters of the K sources referred to as the DOAs of the sources, $\mathbf{s}_t \stackrel{\text{def}}{=} (s_{t,1}, \ldots, s_{t,K})^T$ where $s_{t,k}$ is the complex envelope of the emitted signal by the source k at time t. $\mathbf{R}_s$ is the spatial covariance matrix

(which is assumed positive definite and diagonal for the algorithms that require the sources spatially uncorrelated). $\mathbf{n}_t$ and $\mathbf{R}_n = \sum_{l=1}^{L} a_l \mathbf{Q}_l$ are, respectively, the complex envelope and the spatial covariance matrix of the sensor output additive noise where a_l are unknown parameters and $\mathbf{Q}_l$ are known Hermitian weighting matrices (see, for example, [9]). By choosing $L = 1$, $a_1 = \sigma^2$ and $\mathbf{Q}_1 = \mathbf{I}_p$, the special case where the noise is spatially white is obtained.

(2) *Blind identification of FIR channels*: $\mathbf{x}_t$ represents the p-vector of observed complex envelope of the channel output. $\mathbf{E}_s(\Theta)$ is the $p \times K(L+1)$ Sylvester (resp., generalized Sylvester) matrix $T(\mathbf{h})$ (resp., $T(\mathbf{H})$) for the SIMO (resp., MIMO) channel. The components of Θ are the FIR coefficients of the channel. $\mathbf{s}_t \stackrel{\text{def}}{=} (s_{t,1}, \ldots, s_{t,K}, \ldots, s_{t-L,1}, \ldots, s_{t-L,K})^T$ where $s_{t,k}$ is the kth input signal at time t. $\mathbf{R}_s$ is the positive definite covariance matrix of the inputs (which is assumed diagonal for the linear prediction-based algorithms). $\mathbf{n}_t$ and $\mathbf{R}_n$ are, respectively, the complex envelope and the covariance matrix of the channel output additive noise. $\mathbf{R}_n = \sigma^2\mathbf{I}$ if the noise is spatially and temporally uncorrelated. This latter assumption does not include jammers and supposes that the temporal correlation of noise due to the oversampling is not taken into account.

(3) *Sinusoidal frequency estimation for mixed spectra times series*: $\mathbf{x}_t = (x_t, \ldots, x_{t-p+1})^T$ where, in the complex case, x_t is a sum of sinusoid signals and a linear stationary process $n_t = \sum_{l=0}^{\infty} b_l u_{t-l}$ with $\sum_{l=0}^{\infty} |b_l| < \infty$ and u_t is a sequence of zero-mean i.i.d. r.v. where $\mathrm{E}|u^4| < \infty$. $\mathbf{E}_s(\Theta) = (\mathbf{e}_1, \ldots, \mathbf{e}_K)$ with $\mathbf{e}_k = (1, e^{i2\pi f_k}, \ldots, e^{i2\pi(p-1)f_k})^H$. Θ represents the K distinct frequencies in $[-1/2, +1/2]$. $\mathbf{s}_t = (a_1 e^{i\phi_1} e^{i2\pi f_1 t}, \ldots, a_K e^{i\phi_K} e^{i2\pi f_K t})^T$ where $(a_k)_{k=1,\ldots,K}$ are fixed positive real numbers and $(\phi_k)_{k=1,\ldots,K}$ are r.v.s uniformly distributed on $[0, 2\pi]$. $\mathbf{n}_t = (n_t, \ldots, n_{t-p+1})^T$, $\mathbf{R}_n = \sigma^2 \mathbf{B}\mathbf{B}^H$ where $\mathbf{B}$ is the $p \times \infty$ Toeplitz filtering matrix with first row $\mathbf{b}^T = (b_0, b_1, \ldots)$ and σ^2 is the power of the noise innovation. $\mathbf{B}$ is generally unknown, except in the whitening approach.

(4) *Frequency estimation of sinusoidal signals with very lowpass envelopes*: The envelopes $a_{t,k}$ of the sinusoids are stationary time series and are slowly varying (w.r.t. the frequencies of the sinusoids f_k), so are considered constant during the window of size p. The signal model is identical to the previous one with $\mathbf{s}_t = (a_{t,1} e^{i2\pi f_1 t}, \ldots, a_{t,K} e^{i2\pi f_K t})^T$. Contrary to [3] which analyses the degradation of performance induced by the aforementioned mismodelling, our paper is only devoted to the asymptotic covariance of the estimates.

3 Algorithms under study

3.1 Functional approach

To consider the asymptotic performance of a second-order algorithm, we adopt a functional analysis which consists in recognizing that the whole process of constructing an estimate $\Theta(T)$ of Θ is equivalent to defining a functional relation linking this estimate $\Theta(T)$ to the statistics $\mathbf{R}_x(T) = \frac{1}{T}\sum_{t=1}^{T} \mathbf{x}_t\mathbf{x}_t^+$ from which it is inferred. This functional dependence is denoted $\Theta(T) = \mathrm{alg}(\mathbf{R}_x(T))$. Clearly, $\Theta = \mathrm{alg}(\mathbf{R}_x)$, therefore the different algorithms alg(.) constitute distinct extensions of the mapping $\mathbf{R}_x \to \Theta$ generated by (2.2) to any unstructured real symmetric or Hermitian matrix $\mathbf{R}_x(T)$. In the following, we consider regular algorithms. More specifically, we assume the following conditions.

3.2 Regular algorithms

(1) The function alg(.) is differentiable in a neighbourhood of $\mathbf{R}_x$: that is, if $\mathbf{D}^{\mathrm{alg}}_{\Theta,\mathbf{R}_x}$ (Expressions of $\mathbf{D}^{\mathrm{alg}}_{\Theta,\mathbf{R}_x}$ are ordinarily deduced from perturbation calculus.) denotes the $q \times p^2$ matrix of this differential evaluated at point $\mathbf{R}_x$

$$\mathrm{alg}(\mathbf{R}_x + \delta\mathbf{R}) = \Theta + \mathbf{D}^{\mathrm{alg}}_{\Theta,\mathbf{R}_x} \mathrm{Vec}(\delta\mathbf{R}) + o(\delta\mathbf{R}). \tag{3.1}$$

(2) For any $\Theta \in \Theta$ and any covariance matrices $\mathbf{R}_s$ and $\mathbf{R}_n$ (structured if the algorithm relies on this structure)

$$\mathrm{alg}\left(\mathbf{E}_s(\Theta)\mathbf{R}_s\mathbf{E}_s^+(\Theta) + \mathbf{R}_n\right) = \Theta. \tag{3.2}$$

These two requirements are met by most second-order algorithms including the covariance matching estimation techniques [9]. We note that to fulfil the requirement (1), the extension to $\mathbf{R}_x(T)$ of the mapping $\mathbf{R}_x \to \Theta$ sometimes needs regularization techniques (see, for example, [2]) for the linear prediction method in blind identification of FIR). The requirement (2) means that most second-order algorithms do not require the knowledge of covariance matrices $\mathbf{R}_s$ and $\mathbf{R}_n$. However, specified structures are sometimes needed. For example, $\mathbf{R}_s$ is assumed positive definite and diagonal for the DOA algorithms that suppose the sources are spatially uncorrelated (Toeplization and augmentation techniques) and for linear prediction-based methods used in blind identification algorithms of FIR. $\mathbf{R}_n$ is often assumed to be structured as a linear combination of known matrices $(\mathbf{Q}_l)_{l=1,\dots,L}$, proportional to the identity matrix or to a known matrix for noise whitening approaches.

3.3 Constraints upon the differential of the algorithm

To specify the conditions under which the asymptotic distribution of the estimated parameter Θ is invariant with respect to the distribution and the temporal correlation of $\mathbf{s}_t$ and $\mathbf{n}_t$, we need to prove the following lemma.

Lemma 3.1 *Under conditions 1 and 2, one has the constraints upon* $\mathbf{D}^{\mathrm{alg}}_{\Theta,\mathbf{R}_x}$ *respectively for covariance matrices* $\mathbf{R}_s$ *unstructured,* $\mathbf{R}_s$ *structured diagonal,* $\mathbf{R}_s$ *structured proportional to the identity matrix, and* $\mathbf{R}_n$ *structured as a linear combination of* $(\mathbf{Q}_l)_{l=1,\dots,L}$*;*

$$\mathbf{D}^{\mathrm{alg}}_{\Theta,\mathbf{R}_x}(\mathbf{E}_s(\Theta)\otimes\mathbf{E}_s(\Theta)) = \mathbf{0}, \tag{3.3}$$

$$\mathbf{D}^{\mathrm{alg}}_{\Theta,\mathbf{R}_x}(\mathbf{e}_{s,k}(\Theta)\otimes\mathbf{e}_{s,k}(\Theta)) = \mathbf{0}, \quad k=1,\dots,K, \tag{3.4}$$

$$\mathbf{D}^{\mathrm{alg}}_{\Theta,\mathbf{R}_x}\mathrm{Vec}(\mathbf{E}_s(\Theta)\mathbf{E}_s(\Theta)^+) = \mathbf{0}, \tag{3.5}$$

$$\mathbf{D}^{\mathrm{alg}}_{\Theta,\mathbf{R}_x}\mathrm{Vec}(\mathbf{Q}_l) = \mathbf{0}, \quad l=1,\dots,L, \tag{3.6}$$

with $\mathbf{E}_s(\Theta) = (\mathbf{e}_{s,1}(\Theta),\dots,\mathbf{e}_{s,K}(\Theta))$*.*

Proof

$$\begin{aligned}
\Theta &= \mathrm{alg}\Big(\,\mathbf{E}_s(\Theta)(\mathbf{R}_s+\delta\mathbf{R})\mathbf{E}_s^+(\Theta) + \sum_{l=1}^{L}(a_l+\delta a_l)\mathbf{Q}_l\,\Big)\\
&= \Theta + \mathbf{D}^{\mathrm{alg}}_{\Theta,\mathbf{R}_x}\Big(\,\mathrm{Vec}(\mathbf{E}_s(\Theta)\delta\mathbf{R}\mathbf{E}_s^+(\Theta)) + \mathrm{Vec}\left(\sum_{l=1}^{L}\delta a_l\mathbf{Q}_l\right) + o(\delta\mathbf{R}) + o(\delta\mathbf{a})\\
&= \Theta + \mathbf{D}^{\mathrm{alg}}_{\Theta,\mathbf{R}_x}\left((\mathbf{E}_s(\Theta)\otimes\mathbf{E}_s(\Theta))\mathrm{Vec}(\delta(\mathbf{R})) + \sum_{L=1}^{L}\delta a_l\mathrm{Vec}(\mathbf{Q}_l)\right)\\
&\qquad\qquad\qquad\qquad\qquad\qquad\qquad\qquad + o(\delta\mathbf{R}) + o(\delta\mathbf{a}),
\end{aligned}$$

with $\delta\mathbf{a} \stackrel{\mathrm{def}}{=} (\delta a_1,\dots,\delta a_L)^T$. When $\delta\mathbf{R}$ is diagonal, $\delta\mathbf{R} = \mathrm{Diag}(\delta\sigma_1^2,\dots,\delta\sigma_K^2)$, thus $\mathrm{Vec}\left(\mathbf{E}_s(\Theta)\delta\mathbf{R}\mathbf{E}_s^+(\Theta)\right) = \sum_{k=1}^{K}\delta\sigma_k^2\mathrm{Vec}(\mathbf{e}_{s,k}(\Theta)\mathbf{e}_{s,k}^+(\Theta)) = \sum_{k=1}^{K}\delta\sigma_k^2\left(\mathbf{e}_{s,k}(\Theta)\otimes\mathbf{e}_{s,k}(\Theta)\right)$. When $\delta\mathbf{R} = \delta\sigma_s^2\mathbf{I}$, $\mathrm{Vec}\left(\mathbf{E}_s(\Theta)\delta\mathbf{R}\mathbf{E}_s^+(\Theta)\right) = \delta\sigma_s^2\mathrm{Vec}(\mathbf{E}_s(\Theta)\mathbf{E}_s(\Theta)^+)$. □

Interpretation. We note that $\mathrm{Vec}(\mathbf{E}_s(\Theta)\mathbf{E}_s(\Theta)^+)$ is a linear combination of the vectors $\mathrm{Vec}(\mathbf{e}_{s,k}(\Theta)\otimes\mathbf{e}_{s,k}(\Theta))$, $k=1,\dots,K$, the latter vectors being in the column space of $\mathbf{E}_s(\Theta)\otimes\mathbf{E}_s(\Theta)$. The larger the a priori knowledge about $\mathbf{R}_s$ is needed, the less severe the constraints (3.3), (3.4), (3.5) upon $\mathbf{D}^{\mathrm{alg}}_{\Theta,\mathbf{R}_x}$ become.

To derive the asymptotic distribution of second-order estimators, we need to know the asymptotic distribution of the sample covariance matrix $\mathbf{R}_x(T)$.

4 Robustness of parameter estimates

For stationary processes $\mathbf{s}_t$ and $\mathbf{n}_t$ with finite fourth-order moments, the following theorem is proved.

Theorem 4.1 $\sqrt{T}\;(\mathrm{Vec}(\mathbf{R}_x(T)) - \mathrm{Vec}(\mathbf{R}_x))$ *converges in distribution to the zero-mean real (resp., complex) Gaussian distribution of covariance* $\mathbf{C}_{R_x}$ *(resp.,*

$\mathbf{C}_{R_x}$, $\mathbf{C}_{R_x}\mathbf{K}$) *in the real case (resp. in the complex case).*

$$\sqrt{T}\ (\mathrm{Vec}(\mathbf{R}_x(T)) - \mathrm{Vec}(\mathbf{R}_x)) \xrightarrow{L} N(\mathbf{0};\mathbf{C}_{R_x}) \quad (\textit{resp.},\ N(\mathbf{0};\mathbf{C}_{R_x},\mathbf{C}_{R_x}\mathbf{K})). \tag{4.1}$$

Furthermore (In the complex case, $\mathrm{Cov}\,(\mathrm{Vec}(\mathbf{R}_x(T)))$ *denotes* $\mathrm{E}(\mathrm{Vec}(\mathbf{R}_x(T) - \mathbf{R}_x)\mathrm{Vec}^H(\mathbf{R}_x(T) - \mathbf{R}_x))$. *We note that* $\mathrm{Vec}^T(\mathbf{R}_x(T) - \mathbf{R}_x) = \mathrm{Vec}^H(\mathbf{R}_x(T) - \mathbf{R}_x)\mathbf{K}$, *so* $\mathrm{E}(\mathrm{Vec}(\mathbf{R}_x(T) - \mathbf{R}_x)\mathrm{Vec}^T(\mathbf{R}_x(T) - \mathbf{R}_x)) = \mathrm{E}(\mathrm{Vec}(\mathbf{R}_x(T) - \mathbf{R}_x)\mathrm{Vec}^H(\mathbf{R}_x(T) - \mathbf{R}_x))\mathbf{K}$. *Therefore the noncircular complex Gaussian asymptotic distribution of* $\mathbf{R}_x(T)$ *is characterized by* $\mathbf{C}_{R_x}$ *only.),*

$$\lim_{T\to\infty} T\ \mathrm{Cov}\,(\mathrm{Vec}(\mathbf{R}_x(T))) = \mathbf{C}_{R_x} \tag{4.2}$$

where $\mathbf{C}_{R_x}$ *reads:*

$$\begin{aligned}\mathbf{C}_{R_x} &= (\mathbf{E}_s(\Theta)\otimes\mathbf{E}_s(\Theta))\,\mathbf{C}_{R_s}\left(\mathbf{E}_s^+(\Theta)\otimes\mathbf{E}_s^+(\Theta)\right) + \mathbf{C}_{R_n}\\ &+ (\mathbf{E}_s(\Theta)\otimes\mathbf{I}_p)\,\mathbf{C}_{R_{s,n}}\left(\mathbf{E}_s^+(\Theta)\otimes\mathbf{I}_p\right)\\ &+ (\mathbf{I}_p\otimes\mathbf{E}_s(\Theta))\,\mathbf{C}_{R_{n,s}}\left(\mathbf{I}_p\otimes\mathbf{E}_s^+(\Theta)\right)\end{aligned} \tag{4.3}$$

with

$$\mathbf{C}_{R_s} = \lim_{T\to\infty} T\ \mathrm{Cov}\,(\mathrm{Vec}(\mathbf{R}_s(T)))\,, \qquad \mathbf{C}_{R_n} = \lim_{T\to\infty} T\ \mathrm{Cov}\,(\mathrm{Vec}(\mathbf{R}_n(T)))\,,$$

$$\mathbf{C}_{R_{s,n}} = \begin{cases} \lim_{T\to\infty}\frac{1}{T}\sum_{t=1}^{T}\sum_{t=1}^{T}\mathbf{R}_s^{t-t}\otimes\mathbf{R}_n^{t-t} \\ \qquad \textit{in the circular complex case} \\ \lim_{T\to\infty}\frac{1}{T}\sum_{t=1}^{T}\sum_{t=1}^{T}\mathbf{R}_s^{t-t}\otimes\mathbf{R}_n^{t-t} \\ +\lim_{T\to\infty}\frac{1}{T}\sum_{t=1}^{T}\sum_{t=1}^{T}(\mathbf{R}_s'^{\,t-t}\otimes\mathbf{R}_n'^{\,t-t})\mathbf{K} \\ \qquad \textit{in the noncircular complex and real case}\end{cases}$$

and

$$\mathbf{C}_{R_{n,s}} = \begin{cases} \lim_{T\to\infty}\frac{1}{T}\sum_{t=1}^{T}\sum_{t=1}^{T}\mathbf{R}_n^{t-t}\otimes\mathbf{R}_s^{t-t} \\ \qquad \textit{in the circular complex case} \\ \lim_{T\to\infty}\frac{1}{T}\sum_{t=1}^{T}\sum_{t=1}^{T}\mathbf{R}_n^{t-t}\otimes\mathbf{R}_s^{t-t} \\ +\lim_{T\to\infty}\frac{1}{T}\sum_{t=1}^{T}\sum_{t=1}^{T}(\mathbf{R}_n'^{\,t-t}\otimes\mathbf{R}_s'^{\,t-t})\mathbf{K} \\ \qquad \textit{in the noncircular complex and real case}\end{cases}$$

where $\mathbf{R}_s(T) \stackrel{\mathrm{def}}{=} \frac{1}{T}\sum_{t=1}^{T}\mathbf{s}_t\mathbf{s}_t^+$, $\mathbf{R}_n(T) \stackrel{\mathrm{def}}{=} \frac{1}{T}\sum_{t=1}^{T}\mathbf{n}_t\mathbf{n}_t^+$, $\mathbf{R}_s^{t-t} \stackrel{\mathrm{def}}{=} \mathrm{E}(\mathbf{s}_t\mathbf{s}_t^+)$, $\mathbf{R}_n^{t-t} \stackrel{\mathrm{def}}{=} \mathrm{E}(\mathbf{n}_t\mathbf{n}_t^+)$, $\mathbf{R}_s'^{\,t-t} \stackrel{\mathrm{def}}{=} \mathrm{E}(\mathbf{s}_t\mathbf{s}_t^T)$ *and* $\mathbf{R}_n'^{\,t-t} \stackrel{\mathrm{def}}{=} \mathrm{E}(\mathbf{n}_t\mathbf{n}_t^T)$.

Proof In example of application (3), where x_t is a sum of sinusoid signals and an MA process, this theorem is proved in [7]. The generalization to the data model (2.1) follows the same lines. First, (4.2) is straightforwardly proved after tedious but simple manipulations. Then, to prove (4.1), we adapt the steps of [4, Section 7.3] to each model. □

Remark 1. This theorem extends theorems following the classic stochastic model assumption (see, for example, [8, 12], (Ottersten *et al* 1992)) to accommodate non-Gaussian and temporally correlated noise. For Gaussian temporally uncorrelated noise, $\mathbf{C}_{R_n}$ and the cross-terms $\mathbf{C}_{R_{s,n}}$ and $\mathbf{C}_{R_{n,s}}$ reduce to $\mathbf{R}_n \otimes \mathbf{R}_n$ for circular complex case (resp. $(\mathbf{R}_n \otimes \mathbf{R}_n)(\mathbf{I}_p + \mathbf{K})$ for real case), $\mathbf{R}_s \otimes \mathbf{R}_n$ and $\mathbf{R}_n \otimes \mathbf{R}_s$, respectively. The non-Gaussian assumption simply adds a fourth-order term in the expression of $\mathbf{C}_{R_n}$. But the temporal correlation assumption completely modifies the expression of $\mathbf{C}_{R_n}$, $\mathbf{C}_{R_{s,n}}$, and $\mathbf{C}_{R_{n,s}}$. When the noise is possibly non-Gaussian and temporally correlated, $\mathbf{C}_{R_n}$ becomes in the circular complex ARMA case (see, [6]):

$$\mathbf{C}_{R_n} = \int_{-1/2}^{+1/2} \mathbf{S}_n(f) \otimes \mathbf{S}_n(f) df + \mathbf{Q}_n, \tag{4.4}$$

where $\mathbf{S}_n(f)$ denotes the power cross-spectral density $p \times p$ matrix of $\mathbf{n}_t$. If the fourth-order polyspectrum of the components $(n_{t,k})_{k=1,\dots,p}$ of $\mathbf{n}_t$ for $k_1, k_2, k_3, k_4 = 1, \dots, p$ is defined as

$$\rho_{k_1,k_2,k_3,k_4}(f,f,f) \stackrel{\text{def}}{=} \sum_{\tau,\tau,\tau} \text{Cum}(n_{0,k_1}, n^*_{\tau,k_2}, n_{\tau,k_3}, {n^*}_{tau,k_4}) e^{i2\pi(f\tau+f\tau+f\tau)},$$

then

$$[\mathbf{Q}_n]_{p(j-1)+i,p(l-1)+k} = \int_{-1/2}^{+1/2} \int_{-1/2}^{+1/2} \rho_{i,j,l,k}(f,f,-f) df df$$

denotes the $p^2 \times p^2$ fourth-order cumulant matrix. Simplified formulae of the expressions of the cross-terms $\mathbf{C}_{R_{s,n}}$ and $\mathbf{C}_{R_{s,n}}$ given in Theorem 4.1 can be obtained if the sequence $(\mathbf{R}_s^t \otimes \mathbf{R}_n^t)_{t=\dots,-1,0,+1,\dots}$ is assumed absolutely summable. (This condition is satisfied if the sequences $\mathbf{R}_s^t$ and $\mathbf{R}_n^t$ are absolutely summable or if one of the two sequences is absolutely summable and the other bounded.) In this case, we get from [10, p.411]:

$$\mathbf{C}_{R_{s,n}} = \begin{cases} \sum_{t=-\infty}^{+\infty} (\mathbf{R}_s^t \otimes \mathbf{R}_n^t)(\mathbf{I}_{p^2} + \mathbf{K}) & = \int_{-1/2}^{+1/2} \mathbf{S}_n(f) \otimes \mathbf{S}_s(f) df)(\mathbf{I}_{p^2} + \mathbf{K}) \\ & \text{in the real case} \\ \sum_{t=-\infty}^{+\infty} \mathbf{R}_s^t \otimes \mathbf{R}_n^t & = \int_{-1/2}^{+1/2} \mathbf{S}_s(f) \otimes \mathbf{S}_n(f) df \\ & \text{in the complex case} \end{cases} \tag{4.5}$$

Remark 2. Detailed expressions of $\mathbf{C}_{R_s}$, $\mathbf{C}_{R_n}$, $\mathbf{C}_{R_{s,n}}$, and $\mathbf{C}_{R_{n,s}}$ depend on the application. For example,

$$\mathbf{C}_{R_n} = \int_{-1/2}^{+1/2} S_n^2(f)[\mathbf{e}_f\mathbf{e}_f^H \otimes \mathbf{e}_f\mathbf{e}_f^H]df + \kappa_u \mathrm{Vec}(\mathbf{B}\mathbf{B}^H)\mathrm{Vec}^H(\mathbf{B}\mathbf{B}^H), \qquad (4.6)$$

$$\mathbf{C}_{R_{n,s}} = \mathrm{Diag}(|a_1|^2\mathbf{e}_1\mathbf{e}_1^H S_n(f_1), \ldots, |a_K|^2\mathbf{e}_K\mathbf{e}_K^H S_n(f_K)),$$

and $\mathbf{C}_{R_{s,n}}$ is the block matrix whose (i,j) block element is

$$[\mathbf{C}_{R_{s,n}}]_{i,j} = \mathrm{Diag}(|a_1|^2[\mathbf{e}_1\mathbf{e}_1^H]_{i,j}S_n(f_1), \ldots, |a_K|^2[\mathbf{e}_K\mathbf{e}_K^H]_{i,j}S_n(f_K))$$

for sinusoidal frequency estimation with circular complex additive noise n_t of spectral density $S_n(f)$, $\kappa_u \stackrel{\text{def}}{=} \mathrm{Cum}(u_t, u_t^*, u_t, u_t^*)$ and $\mathbf{e}_f \stackrel{\text{def}}{=} (1, e^{i2\pi f}, \ldots, e^{i2(p-1))\pi f})^H$. A similar expression is obtained for $\mathbf{C}_{R_s}$ for blind identification of SIMO FIR channels when the input s_t is circular complex temporally uncorrelated of power σ_s^2 with $\kappa_s \stackrel{\text{def}}{=} \mathrm{Cum}(s_t, s_t^*, s_t, s_t^*)$ and $\mathbf{e}_f \stackrel{\text{def}}{=} (1, e^{i2\pi f}, \ldots, e^{i2L\pi f})^H$:

$$\mathbf{C}_{R_s} = \int_{-1/2}^{+1/2} \sigma_s^2[\mathbf{e}_f\mathbf{e}_f^H \otimes \mathbf{e}_f\mathbf{e}_f^H]df + \kappa_s \mathrm{Vec}(\mathbf{I})\mathrm{Vec}^H(\mathbf{I}). \qquad (4.7)$$

We note that, unlike the classic Bartlett formulation which is concerned with the sample correlation coefficients sequence, Theorem 4.1 is devoted to the sample covariance matrix. This formulation is better adapted to deriving the asymptotic distribution of estimated as is derived in the next section.

4.1 Asymptotic distribution of the estimated parameter

By the regularity condition (3.1), the asymptotic behaviours of $\Theta(T)$ and $\mathbf{R}_x(T)$ are directly related. The standard result on regular functions of asymptotically normal statistics (see, for example, [11, p. 122]) applies:

$$\sqrt{T}\ (\Theta(T) - \Theta) \xrightarrow{L} N(\mathbf{0}; \mathbf{C}_\Theta) \qquad \text{for } \Theta \in R^q$$
$$(\text{resp. } N(\mathbf{0}; \mathbf{C}_\Theta, \mathbf{C}_\Theta) \text{ for } \Theta \in C^q) \qquad (4.8)$$

with

$$\begin{aligned} \mathbf{C}_\Theta &= \lim_{T\to\infty} T\mathrm{E}\left((\Theta(T) - \Theta)(\Theta(T) - \Theta)^+\right) \\ &= \mathbf{D}_{\Theta,\mathbf{R}_x}^{\mathrm{alg}} \mathbf{C}_{R_x} \left(\mathbf{D}_{\Theta,\mathbf{R}_x}^{\mathrm{alg}}\right)^+ \qquad (4.9) \\ \mathbf{C}_\Theta &= \lim_{T\to\infty} T\mathrm{E}\left((\Theta(T) - \Theta)(\Theta(T) - \Theta)^T\right) \\ &= \mathbf{D}_{\Theta,\mathbf{R}_x}^{\mathrm{alg}} \mathbf{C}_{R_x}\mathbf{K} \left(\mathbf{D}_{\Theta,\mathbf{R}_x}^{\mathrm{alg}}\right)^T. \qquad (4.10) \end{aligned}$$

We can now state our main result: For second-order algorithms satisfying the regularity conditions of Section 3.2, the asymptotic covariance of estimators is invariant to the distribution and to the temporal correlation of $\mathbf{s}_t$ (and $\mathbf{n}_t$) depending on the application and $\mathbf{C}_\Theta = \mathbf{D}^{\mathrm{alg}}_{\Theta,\mathbf{R}_x}\mathbf{C}^*_{R_x}\left(\mathbf{D}^{\mathrm{alg}}_{\Theta,\mathbf{R}_x}\right)^+$, $\mathbf{C}_\Theta = \mathbf{D}^{\mathrm{alg}}_{\Theta,\mathbf{R}_x}\mathbf{C}^*_{R_x}\mathbf{K}\left(\mathbf{D}^{\mathrm{alg}}_{\Theta,\mathbf{R}_x}\right)^T$, where $\mathbf{C}^*_{R_x}$ is deduced from the expression of $\mathbf{C}_{R_x}$ (4.3) by suppression of the one or two first terms.

Proof and interpretation. This results immediately from the constraints upon $\mathbf{D}^{\mathrm{alg}}_{\Theta,\mathbf{R}_x}$ of Lemma 3.1. The larger the *a priori* knowledge about $\mathbf{R}_s$ and $\mathbf{R}_n$ is needed, the less severe the constraints upon $\mathbf{D}^{\mathrm{alg}}_{\Theta,\mathbf{R}_x}$ are and the less robust to the distribution of the signals the second-order estimators become.

Applied to the examples described in Section 2.2, the following robustness results hold.

4.2 Examples of applications

(1) *Narrow-band DOA estimation.* If $\mathbf{n}_t$ is assumed temporally uncorrelated (an assumption admitted in all papers devoted to performance analysis), *the algorithms that do not suppose the sources to be spatially uncorrelated are robust to the distribution and to the temporal correlation of the sources* $\mathbf{s}_t$ because $\mathbf{C}^*_{R_x}$ is deduced from the expression of $\mathbf{C}_{R_x}$ by suppression of its first term thanks to constraint (3.3) and because the terms $\mathbf{C}_{R_{s,n}}$ and $\mathbf{C}_{R_{n,s}}$ reduce respectively to the spatial terms $\mathbf{R}_s \otimes \mathbf{R}_n$ and $\mathbf{R}_n \otimes \mathbf{R}_s$. This extends the results by Cardoso and Moulines [5] that have shown that the asymptotic performance of most high-resolution covariance-based DOA estimators is independent of the distribution of the source signals for independent snapshots.
It is shown [6] that the Toeplization and augmentation techniques, which are based on the source spatial uncorrelation assumption, are very sensitive to the distribution and to the temporal correlation of the sources in the case of several sources because the constraint (3.3) is not satisfied. For only one source, the robustness is preserved thanks to constraint (3.4).
On the contrary, if $\mathbf{n}_t$ *is assumed temporally correlated, all the second-order algorithms are sensitive to the temporal correlation of the sources* due of the contribution of terms $\mathbf{C}_{R_{s,n}}$ and $\mathbf{C}_{R_{n,s}}$ in $\mathbf{C}^*_{R_x}$.

(2) *Blind identification of FIR channels.* From the general methodological viewpoint, the second-order algorithms may be classified as methods that do not suppose the inputs $s_{t,k}$ temporally uncorrelated (for example, the subspace methods which exploit low-rank space-time properties) and methods that explicitly suppose that the inputs $s_{t,k}$ are temporally uncorrelated (for example, the linear prediction methods, using specific invertibility properties of FIR models).
The first methods are robust to the distribution of the inputs because the contribution of $\mathbf{C}_{R_s}$ is cancelled in (4.9) and (4.10) thanks to constraint

(3.3). However, these methods are sensitive to the temporal correlation of the inputs because the terms $\mathbf{C}_{R_{s,n}}$ and $\mathbf{C}_{R_{n,s}}$ depend on the temporal correlation of the inputs. We note that this sensitivity remains even if the noise is temporally uncorrelated because $\mathbf{R}_s$ includes temporal correlation in this space-time application.
For the second methods, the robustness to the distribution of the sources is preserved because the constraint (3.5) applied to (4.7) cancels the contribution of the cumulant κ_s of the input signal thanks to the equality

$$(\mathbf{E}_s(\Theta) \otimes \mathbf{E}_s^+(\Theta))\text{Vec}(\mathbf{I}) = \text{Vec}(\mathbf{E}_s(\Theta)\mathbf{I}(\mathbf{E}_s^+(\Theta)) = \text{Vec}(\mathbf{E}_s(\Theta)(\mathbf{E}_s^+(\Theta)).$$

This robustness property extends to any second-order algorithm the robustness result of [1] proved for some subspace methods after calculus of $\mathbf{D}_{\Theta,\mathbf{R}_x}^{\text{MUSIC}}$. We note that proving this robustness property directly from the expression of $\mathbf{D}_{\Theta,\mathbf{R}_x}^{\text{alg}}$ for each specific algorithm would be tedious and cumbersome; see, for example, the intricate expression of $\mathbf{D}_{\Theta,\mathbf{R}_x}^{\text{LP}}$ in [2]).
Furthermore, as a byproduct of our results we note that the asymptotic covariance $\mathbf{C}_\Theta$ has the same expression under the assumptions that $\mathbf{s}_t$ is Gaussian i.i.d. and s_t are temporally uncorrelated with any distribution .
Therefore our result validates the asymptotic performance and limitations results of Zeng and Tong [15] (which were based on the i.i.d. Gaussian assumption of $\mathbf{s}_t$) for an input s_t temporally uncorrelated with any distribution.

(3) *Sinusoidal frequency estimation for mixed spectra times series. The second-order algorithms are robust to the distribution of the noise* $\mathbf{n}_t$ because the contribution of the cumulant κ_u of the noise innovation in (4.6) is cancelled thanks to constraint (3.6) where $\mathbf{Q}_1 = \mathbf{B}\mathbf{B}^H$ in (4.9) and (4.10).
This results apparently contradicts a Monte Carlo simulation recently presented in [14] in which the frequency estimators degrade with an heavy-tailed probability distribution of the noise. In fact, this simulation is presented with the number of observations $T = 300$ and the number of lags $p = 30$ and furthermore with a complex circular α-stable distribution of the noise with $\alpha = 1$ for which $\text{E}|n_t|^2 = \infty$. In contrast, our result is asymptotic in the number of data samples T and it is well known that T has to be very large for the asymptotic expressions to describe the estimated MSEs adequately for large values of p. For example, for $p = 8$ and $SNR = 20$ dB a good agreement requires $T = 10,000$ in [13]. Therefore our asymptotic results do not contradict the Monte Carlo simulation of [14] and the need for robust algorithms for large values of p.

(4) *Frequency estimation of sinusoidal signals with very lowpass envelopes.* Using the proof of examples (1) and (3), *the second-order algorithms are robust to the distribution of the envelopes $a_{t,k}$ of the sinusoidal signals and of* $\mathbf{n}_t$ but sensitive to their temporal correlation.

Bibliography

[1] Abed Meraim, K., Cardoso, J.F., Gorokhov, A.Y. and Loubaton, P. On subspace methods for blind identification of single-input multiple output FIR systems, *IEEE Trans. on Signal Processing*, **45**, 42–55, 1997.

[2] Abed Meraim, K., Moulines, E. and Loubaton, P. Prediction error method for second-order blind identification, *IEEE Trans. on Signal Processing*, **45**, 694–705, 1997.

[3] Besson, O. and Stoica, P. Analysis of MUSIC and ESPRIT frequency estimates for sinusoidal signals with lowpass envelopes, *IEEE Trans. on Signal Processing*, **44**, 2359–2364, 1996.

[4] Brockwell, P.J. and Davis, R.A. *Time Series, Theory and Methods*, 2nd Ed., Springer, 1991.

[5] Cardoso, J.F. and Moulines, E. A robustness property of DOA estimators based on covariance, *IEEE Trans. on Signal Processing*, **42**, 3285–3287, 1994.

[6] Delmas, J.P. and Meurisse, Y. Asymptotic performance analysis of DOA algorithms with temporally correlated narrow-band signals, *IEEE Trans. on Signal Processing*, **48**, 2669–2674, 2000.

[7] Delmas, J.P. Asymptotic normality of sample covariance matrix for mixed spectra time series: application to sinusoidal frequencies estimation, *IEEE Trans. on Information Theory*, **47**, 1681–1687, 2001.

[8] Ottersten, B., Viberg, M. and Kailath, T. Analysis of subspace fitting and ML techniques for parameter estimation from sensor array data, *IEEE Trans. on Signal Processing*, **40**, 590–599, 1992.

[9] Ottersten, B., Stoica, P. and Roy, R. Covariance matching estimation techniques for array signal processing, *Digital Signal Processing*, **8**, 185–210, 1998.

[10] Porat, B. *Digital processing of random signals, Theory and Methods*, Prentice-Hall, Englewood Cliff, NJ, 1993.

[11] Serfling, R.J. *Approximation Theorems of Mathematical Statistics*, Wiley, New York, 1980.

[12] Stoica, P. and Nehorai, A. Performance study of conditional and unconditional direction of arrival estimation, *IEEE Trans. on Acoustics Speech and Signal Processing*, **38**, 1783–1795, 1990.

[13] Stoica, P. and Söderström, T. Statistical analysis of MUSIC and subspace rotation estimates of sinusoidal frequencies, *IEEE Trans. Signal Processing*, **39**, 1836–1847, 1991.

[14] Visuri, S., Oja, H. and Koivunen, V. Nonparametric statistics for subspace based frequency estimation. In *Proc. Conf. EUSIPCO*, (Tampere), 2000.

[15] Zeng, H.H. and Tong, L. Blind channel estimation using the second-order statistics: asymptotic performance and limitations, *IEEE Trans. on Signal Processing*, **45**, 2060–2071, 1997.

Bias/Variance Trade-Offs in Direction of Arrival Estimation Using Sensor Arrays

Stephen D. Hayward
Defence Evaluation and Research Agency, St. Andrews Road, Malvern, WR14 3PS, UK

Abstract

In this work we develop bounds on the accuracy with which the directions of arrival of spatially distributed sources can be estimated. We consider complex targets, composed of multiple scattering centres which, when illuminated by a radar or sonar, give rise to scattered wave-fields which are distorted due to constructive and destructive interference. The Cramer–Rao bounds are often employed to put a limit on the performance of angle-estimation algorithms, but it is usually only the bounds on the variance of an unbiased estimator that are considered. Here we show that

- unbiased estimators may not exist for complex targets when the Fisher information matrix becomes singular;
- a more general form of the Cramer–Rao bounds can be used to explore the trade-off between bias and variance.

1 Introduction

It is well known that arrays of sensors can be used to extract information about multiple targets from measurements of their superimposed transmitted or scattered waveforms. Of particular importance in many radar, sonar and communications applications is the extraction of information relating to the direction of arrival (DOA) of each target. In general a single sensor cannot be used to obtain such DOA information. For a linear array of equispaced sensors the problem is closely related to the spectral analysis of discretely sampled time series, and it was from the latter field that so-called high-resolution or super-resolution techniques were adopted. A great deal of material has been published in the literature on the subject of DOA estimation using sensor arrays, many algorithms have been described, and bounds on the achievable accuracy of DOA estimates have been published. However, attention has been mostly given to single-point targets, and the accuracy achievable by unbiased estimators. Recently there has been some interest in the problem of estimating the DOA of spatially distributed targets, and a few algorithms have been described in the literature [1], [2]. This work has been stimulated by recent requirements to accurately locate mobile phone users, as an aid to the emergency services. In an urban environment scattering from buildings in the vicinity of a mobile phone user transforms the problem into one of locating a distributed source. The work described in this paper has been motivated by the observation that conventional methods of DOA estimation (by which we mean beamscanning or monopulse techniques) do not perform well against complex targets, due to interference effects, or glint. In this paper we investigate the use of Cramer–Rao (CR) bounds in determining the achievable accuracy with which a complex target DOA can be estimated. The paper is structured as follows. In Section 2 we define a simple model of a complex target, and introduce the concepts of likelihood, Fisher information, and the CR bounds on estimation accuracy. In Section 3 we investigate properties of the CR bounds for single point target and two-point targets, and in Section 4 we show how the uniform CR bound can be used to give a lower limit on the mean-square error (MSE) of both biased and unbiased estimators. Our conclusions are made in Section 5. We will be concerned with DOA estimation using a single snapshot of data, but most of what follows can be easily extended to the case where we have multiple snapshots available. Throughout the paper x^H denotes the conjugate transpose of vector x, x^T its transpose, and I an identity matrix of suitable dimension. Statistical expectation is denoted by $E\langle . \rangle$, and $[X]_{jk}$ refers to the element of matrix X in the jth row and kth column. The Moore–Penrose pseudo-inverse of X is denoted X^+.

2 Maximum likelihood estimation and the Cramer–Rao bound

We will assume that we have a planar array of m sensors, and that the target is composed of n sources, with steering vectors $a(u_j, v_j)$ for $j = 1 \dots n$, and complex amplitudes α_j. Here $\{u_j\}$ and $\{v_j\}$ are the unknown direction-cosines which we wish to estimate, while α is a vector of unknown nuisance parameters, in which we have no interest. Let a single vector observation at the array be denoted by

$$z = A\alpha + \epsilon, \tag{2.1}$$

where $A \equiv \begin{bmatrix} a(u_1, v_1) & \cdots & a(u_n, v_n) \end{bmatrix}$, and ϵ is a vector of noise samples drawn from a zero-mean multivariate normal distribution, with covariance $\sigma^2 I$. The jth steering vector is defined to be $[a(u_j, v_j)]_k = \exp(2\pi i(x_k u_j + y_k v_j))$, where the kth sensor has coordinates (x_k, y_k), measured in wavelengths. We define a parameter vector $\theta \equiv \begin{bmatrix} \phi^T & \eta^T \end{bmatrix}^T$, with ϕ the vector of desired unknowns and η a vector of nuisance parameters. In our problem $\phi \equiv \begin{bmatrix} u^T & v^T \end{bmatrix}^T$ and $\eta \equiv \alpha$. An estimate of θ is denoted $\hat{\theta} = g(z; \theta)$. The estimate is said to be unbiased if the mean, $m(\theta) \equiv E\left\langle \hat{\theta} \right\rangle = \theta$. The error covariance is defined to be $E\left\langle (\hat{\theta} - m(\theta))(\hat{\theta} - m(\theta))^H \right\rangle$. From the definition of z in (2.1), the probability density function for z given θ is

$$F_z(z; \theta) = \frac{1}{\pi^m \sigma^{2m}} \exp\left\{ -\frac{1}{\sigma^2}(z - A\alpha)^H (z - A\alpha) \right\}. \tag{2.2}$$

Viewed as a function of θ for given z, we refer to $F_z(z;\theta)$ as the likelihood function, and $\ln F_z(z;\theta)$ as the log-likelihood. For a given likelihood function the Fisher information matrix is defined to be

$$[J(\theta)]_{jk} = -E\left\langle \frac{\partial^2}{\partial\theta_j\partial\theta_k}(\ln F_z(z;\theta))\right\rangle. \tag{2.3}$$

$J(\theta)$ contains information about the average curvature of the log-likelihood function, and so describes the sensitivity of the likelihood function to small changes in the value of θ. A large average curvature implies high sensitivity, and consequently good estimation accuracy. For an unbiased estimator the Fisher information can be used to compute the CR lower bound on the error covariance matrix [3]:

$$E\left\langle(\hat{\theta}-\theta)((\hat{\theta}-\theta))^H\right\rangle \geq J^{-1}(\theta). \tag{2.4}$$

Other bounds exist, and the CR bound is not necessarily the highest lower bound, but is the easiest to compute. The CR bound is a local bound and may fail if the likelihood is multi-modal. It is often desirable to eliminate the nuisance parameters from the computation of $J(\theta)$. This can be achieved by using the profile likelihood [4], which is defined for the desired parameter vector to be

$$F_z^P(z;\phi) \equiv \sup_\eta F_z(z;\theta). \tag{2.5}$$

$F_z^P(z;\phi)$ has many of the properties of $F_z(z;\theta)$, and in particular leads to a bound on the variance of estimates of ϕ. Following Hayward [5] we obtain the profile Fisher information by first proving the following lemmas.

Lemma 2.1 *The profile likelihood for* $\phi \equiv \begin{bmatrix} u_1 & \cdots & u_n & v_1 & \cdots & v_n \end{bmatrix}^T$ *is*

$$F_z^P(z;\phi) = \text{const} - \frac{1}{\sigma^2}z^H\left(I - AA^+\right)z.$$

Proof The observation z is parametrized by the vector $\theta \equiv \begin{bmatrix} \phi^T & \eta^T \end{bmatrix}^T$, with log-likelihood [7]

$$F_z(z;\theta) = \text{const} - \frac{1}{\sigma^2}(z - A\alpha)^H(z - A\alpha). \tag{2.6}$$

By definition $\alpha^* = A^+z$ minimizes the 2-norm of $z - A\alpha$. It follows directly that (2.6) is maximized for all A by α^*, and substitution into (2.6) yields

$$F_z^P(z;\phi) = \text{const} - \frac{1}{\sigma^2}z^H\left(I - AA^+\right)z.$$

When A has full column rank, $A^+ = (A^HA)^{-1}A^H$ and the profile likelihood takes the more familiar form

$$F_z^P(z;\phi) = \text{const} - \frac{1}{\sigma^2}z^H\left(I - A(A^HA)^{-1}A^H\right)z.$$

□

Lemma 2.2

$$A^H \frac{\partial (AA^+)}{\partial \phi_j} = \frac{\partial A^H}{\partial \phi_j}(I - AA^+),$$

wherever AA^+ is differentible.

Proof From the Moore–Penrose conditions [6] we have

$$A^H AA^+ = A^H.$$

Differentiating both sides with respect to ϕ_j gives

$$\frac{\partial A^H}{\partial \phi_j} AA^+ + A^H \frac{\partial (AA^+)}{\partial \phi_j} = \frac{\partial A^H}{\partial \phi_j},$$

and the result follows directly. AA^+ is differentiable at a point ϕ^* providing that AA^+, and consequently A, has locally constant rank [8]. □

Lemma 2.3

$$\frac{\partial^2 (AA^+)}{\partial \phi_j \partial \phi_k} = -\frac{\partial (AA^+)}{\partial \phi_j}\frac{\partial A}{\partial \phi_k} - \frac{\partial (AA^+)}{\partial \phi_k}\frac{\partial A}{\partial \phi_j} + (I - AA^+)\frac{\partial^2 A}{\partial \phi_j \partial \phi_k},$$

wherever AA^+ is differentible.

Proof Starting with

$$AA^+A = A,$$

differentiate both sides with respect to ϕ_j and ϕ_k, and the result follows directly. □

Lemma 2.4 *If ε is drawn from a m-variate, zero-mean, normal distribution with covariance $\sigma^2 I$, then $\varepsilon^H AA^+ \varepsilon$ is chi-squared distributed with $2r$ degrees of freedom, where $r = rank(AA^+) = rank(A)$.*

Proof The result follows directly from the fact that AA^+ is idempotent (that is, $AA^+AA^+ = AA^+$) via application of Theorem 1.4.2 in [9]. □

Theorem 2.5 *The Fisher information matrix, $J^p(\phi)$, corresponding to the profile log-likelihood $F_z^P(z;\phi)$ can be written as*

$$J^p(\phi) = \frac{2}{\sigma^2}\mathrm{Re}\begin{bmatrix} X^H {D_u}^H (I - AA^+) D_u X & X^H {D_u}^H (I - AA^+) D_v X \\ X^H {D_v}^H (I - AA^+) D_u X & X^H {D_v}^H (I - AA^+) D_v X \end{bmatrix},$$

where $X \equiv \mathrm{diag}(\alpha)$, $D_u \equiv \frac{\partial A}{\partial u}$, $D_v \equiv \frac{\partial A}{\partial v}$.

Proof

$$[J^p(\phi)]_{jk} \equiv E\left\langle \frac{\partial^2 F_z^P(z;\phi)}{\partial\phi_j \partial\phi_k} \right\rangle.$$

According to Lemma 2.1

$$[J^p(\phi)]_{jk} \equiv \frac{1}{\sigma^2}\left[\alpha^H A^H \frac{\partial^2 AA^+}{\partial\phi_j\partial\phi_k} A\alpha + E\left\langle \varepsilon^H \right\rangle \frac{\partial^2 AA^+}{\partial\phi_j\partial\phi_k} A\alpha \right.$$
$$\left. + \alpha^H A^H \frac{\partial^2 AA^+}{\partial\phi_j\partial\phi_k} E\left\langle \varepsilon \right\rangle + E\left\langle \varepsilon^H \frac{\partial^2 AA^+}{\partial\phi_j\partial\phi_k} \varepsilon \right\rangle \right]. \tag{2.7}$$

According to Lemmas 2.2 and 2.3 the first term in (2.7) can be written as

$$\alpha^H \frac{\partial A^H}{\partial\phi_j}(I - AA^+)\frac{\partial A^H}{\partial\phi_k}\alpha + \alpha^H \frac{\partial A^H}{\partial\phi_k}(I - AA^+)\frac{\partial A^H}{\partial\phi_i}\alpha$$
$$+ \alpha^H A^H (I - AA^+) \frac{\partial^2 A^H}{\partial\phi_j\partial\phi_k}\alpha. \tag{2.8}$$

Since $A^H AA^+ = A^H$, the last term in (2.8) is zero, and the first two terms have the more compact form

$$2\mathrm{Re}\begin{bmatrix} X^H D_u{}^H (I - AA^+) D_u X & X^H D_u{}^H (I - AA^+) D_v X \\ X^H D_v{}^H (I - AA^+) D_u X & X^H D_v{}^H (I - AA^+) D_v X \end{bmatrix}_{jk}.$$

The second and third terms in (2.7) are zero because ε has zero mean. The last term in (2.7) can be written as

$$E\left\langle \varepsilon^H \frac{\partial^2 AA^+}{\partial\phi_j\partial\phi_k}\varepsilon \right\rangle = \frac{\partial^2}{\partial\phi_j\partial\phi_k} E\left\langle \varepsilon^H AA^+ \varepsilon \right\rangle.$$

According to Lemma 2.4, $\varepsilon^H AA^+ \varepsilon$ is distributed with mean and variance which are independent of A and consequently of ϕ. The last term in (2.7) is therefore equal to zero, and the theorem is proved. □

Theorem 2.5 is a more general result than that obtained in [7] since it allows the columns of A to be linearly dependent, and thus accommodates complex targets composed of very closely spaced scatterers. However, the result only applies where $A^+(\theta)$ is continuous, which implies that $A(\theta)$ must have locally constant rank [8]. The result therefore does not hold for two coincident point scatterers. Given a model for the observed data, the parameters can be estimated using the maximum likelihood method. This involves finding the set of parameters which maximizes the likelihood or profile likelihood. Maximum likelihood estimates are known to be asymptotically (with increasing numbers of observed snapshots) unbiased and efficient [3].

3 Examples

3.1 Single-point target

For a linear array and a single-point scatterer the bound on estimator variance from (2.4) and Theorem 2.5 reduces to

$$E\left\langle (\hat{u}-u)(\hat{u}-u)^H \right\rangle \geq \frac{\sigma^2}{2\alpha^2 \parallel a_u(u) \parallel_2^2}.$$

It can be easily shown that the maximum likelihood estimate for the target DOA is obtained by finding the peak of the beamscan function

$$b(u) = \frac{\mid a^H(u)z \mid^2}{a^H(u)a(u)}.$$

In turn it can be shown [10] that given an initial estimate of the target DOA, u_0, the first step in an iterative solution for the beamscan peak gives

$$\hat{u} = u_0 + \mathrm{Re}\left[\frac{a_u{}^H(u_0)za^H(u_0)a(u_0)}{a^H(u_0)za_u{}^H(u_0)a_u(u_0)}\right], \tag{3.1}$$

which is equivalent to the one-dimensional phase-comparison monopulse method. The monopulse estimator is unbiased, and the variance is plotted as a function of signal-to-noise-ratio (SNR) in Fig. 1, for comparison with the CR bound. It is clear that the monopulse estimator meets the bound. Such an estimator is said to be efficient.

3.2 Two-point target

For a linear array and a two-point target we can obtain an approximate form for the Fisher information, and consequently the CR bound. Consider a line array

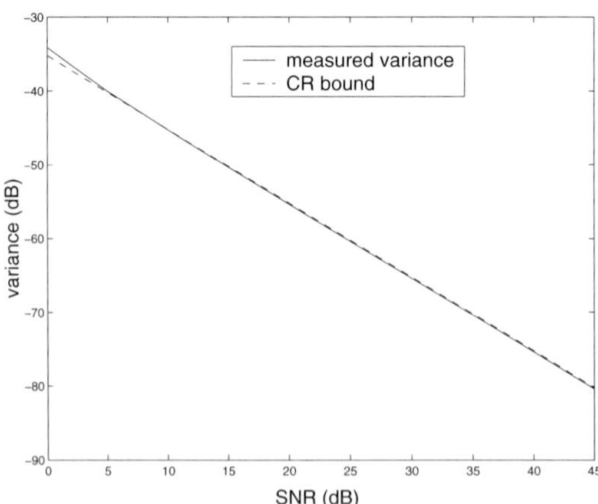

FIG. 1. Variance of monopulse estimator as a function of the single target SNR.

of m sensors with coordinates $\{x_k\}$ such that $\sum_{k=1}^{m} x_k = 0$. Let there be two sources with angular coordinates $u_1 = r$ and $u_2 = -r$, where we have assumed without loss of generality that the coordinate of the mean source position is zero. (Translation of the mean source position is achieved through pre-multiplication of A and D_u by a unitary matrix Q. It can easily be shown that $J^p(\phi)$ in (3.2) is invariant to such a transformation of coordinates.) From Theorem 2.5 the Fisher information matrix can be expressed as

$$J^p(\phi) = \frac{2}{\sigma^2}\mathrm{Re}\left(X^H {D_u}^H (I - A(A^H A)^{-1} A^H) D_u X\right) \equiv \frac{2}{\sigma^2}\mathrm{Re}(X^H P X). \quad (3.2)$$

Theorem 3.1 *For the array and source geometry defined above, the inverse of the matrix P defined in (3.2) has the following structure:*

$$P^{-1} = \frac{1}{\sqrt{2}}\begin{bmatrix} 1 & 1 \\ -1 & 1 \end{bmatrix}\begin{bmatrix} \frac{r^{-2}}{2k_1} + O(r^0) & 0 \\ 0 & \frac{r^{-4}}{2k_2} + O(r^{-2}) \end{bmatrix}\frac{1}{\sqrt{2}}\begin{bmatrix} 1 & -1 \\ 1 & 1 \end{bmatrix},$$

where k_1 and k_2 are functions of the array geometry but not the source separation.

For a proof of Theorem 3.1 see the Appendix. When the sources are in phase with equal amplitude α, $X = \alpha \begin{bmatrix} 1 & 0 \\ 0 & 1 \end{bmatrix}$ and

$$J^p(\phi)^{-1} = \frac{\sigma^2}{2\alpha^2}P^{-1}.$$

If we define two scalar functions of the parameter vector $\phi = \begin{bmatrix} u_1 & u_2 \end{bmatrix}^T$, to be $t_s(\phi) = \frac{u_1+u_2}{2}$ and $t_s(\phi) = \frac{u_1-u_2}{2}$, then it follows from Theorem 3.1 that the variances of unbiased estimators of t_s, the mean source position, and t_d, half the difference, satisfy

$$E\left\langle |\hat{t}_d - t_d|^2 \right\rangle \geq \frac{\sigma^2}{8\alpha^2 k_1} r^{-2} + O(r^0); \quad (3.3)$$

$$E\left\langle |\hat{t}_s - t_s|^2 \right\rangle \geq \frac{\sigma^2}{8\alpha^2 k_2} r^{-4} + O(r^{-2}). \quad (3.4)$$

The CR bounds computed directly from (3.2) for targets with a SNR of 0 dB are plotted in Fig. 2 as functions of the target separation, for comparison with the approximations of (3.3) and (3.4). When r is small the bound on the accuracy of $\hat{t}_d$ is very much less than that on $\hat{t}_s$, although both bounds approach ∞ as $r \to 0$, indicating that unbiased estimators do not exist for small values of r. It is known that unbiased estimators may not exist for singular problems [11]. When the targets are combined in anti-phase, so that $X = \alpha \begin{bmatrix} 1 & 0 \\ 0 & -1 \end{bmatrix}$, the behaviour of the two estimators as $r \to 0$ is reversed.

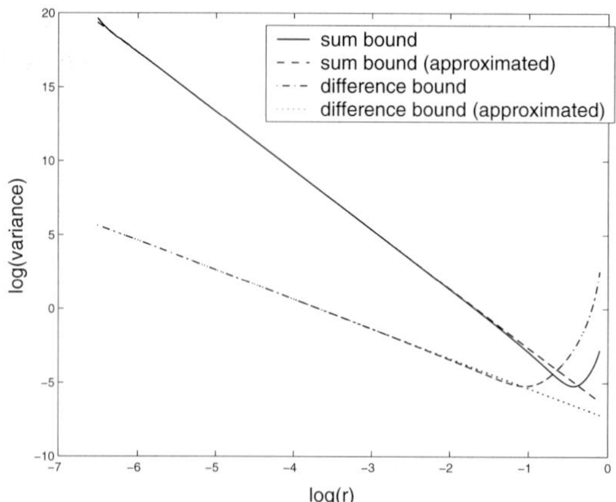

FIG. 2. Computed CR bounds on sum and difference estimators as functions of the separation, compared to approximations.

4 Bias/variance trade-off and the uniform CR bound

In many instances a biased estimator may be preferable to an unbiased estimator, even if the latter is available, as we discuss below. The effects of bias on estimator performance may be accounted for by measuring the MSE,

$$MSE = (bias)^2 + variance.$$

The variance of a biased estimator is lower bounded by the biased CR bound in [12], which takes the form

$$E\left\langle(\hat{t}-t)^2\right\rangle \geq (\nabla t(\phi) + \nabla b(\phi))^H J^p(\phi)^{-1}(\nabla t(\phi) + \nabla b(\phi)), \tag{4.1}$$

where ∇ denotes the gradient operator, and $t(\phi)$ and $b(\phi)$ are the scalar function to be estimated and the bias function respectively. For the two-point target example $\nabla t_s(\phi) = \frac{1}{2}\begin{bmatrix} 1 & 1 \end{bmatrix}^T$, and $\nabla t_d(\phi) = \frac{1}{2}\begin{bmatrix} 1 & -1 \end{bmatrix}^T$. It is clear from (4.1) that the bound on the variance of a biased estimator can be trivially reduced to zero by choosing $b(\phi) = -t(\phi)$, but this solution does not in general minimize the MSE. It is also true that a constant bias (that is, one that is independent of the parameters being estimated) is not significant, since it can be simply subtracted off. It is noted in [12] that the 'unbiased bound' is not of immediate use in bounding the performance of estimators with arbitrary bias functions, since (4.1) requires the gradient of the bias function to be known. Instead, a bound (referred to as the uniform CR bound) on the variance of a biased estimator has been developed as a function of the bias-gradient norm. The bias-gradient norm is defined to be $\|\nabla b(\theta)\|_C$, where $\|x\|_C{}^2 \equiv x^H Cx$, and C is an arbitrary symmetric positive definite matrix. It can be shown that if $\mathcal{C}$ is the set of all points in the elliptic region of the parameter space surrounding the point θ, defined by $\mathcal{C} = \{\psi : (\psi-\theta)^T C^{-1}(\psi-\theta)) \leq 1\}$, then $\|\nabla b(\theta)\|_C{}^2 \approx \sup_{\psi\in\mathcal{C}} |b(\psi) - b(\theta)|^2$; that is, $\|\nabla b(\theta)\|_C$ approximates the maximal bias variation in $\mathcal{C}$.

Theorem 4.1 *Let $\|\nabla b(\theta)\|_C \leq \delta$, for a fixed $\delta \geq 0$. Define P to be the matrix that projects onto the column space of $C^{-\frac{1}{2}} J(\theta)^{-1} C^{-\frac{1}{2}}$. Then*

$$E\left\langle (\hat{t}-t)^2 \right\rangle \geq B(\theta, \delta), \tag{4.2}$$

where $B(\theta,\delta) = 0$ when $\delta^2 \geq \nabla^H t(\theta) C^{\frac{1}{2}} P C^{\frac{1}{2}} \nabla t(\theta)$; and

$$B(\theta,\delta) = \lambda^2 \nabla^H t(\theta) C \left[\lambda C + J(\theta)^{-1}\right]^{-1} J(\theta)^{-1} \left[\lambda C + J(\theta)^{-1}\right]^{-1} C \nabla t(\theta), \tag{4.3}$$

otherwise. In (4.3) $\lambda > 0$ is the positive solution of

$$\delta^2 = \nabla^H t(\theta) J(\theta)^{-1} \left[\lambda C + J(\theta)^{-1}\right]^{-1} C \left[\lambda C + J(\theta)^{-1}\right]^{-1} J(\theta)^{-1} \nabla t(\theta). \tag{4.4}$$

When $J(\theta)^{-1}$ does not exist the above results hold for $J(\theta)^{+}$.

For a proof of Theorem 4.1 see, [12]. By treating δ as a function of λ we can plot the minimum achievable $MSE = B(\theta,\lambda) + \delta^2(\lambda)$ against $bias^2 = \delta^2$ by varying λ over the range $(0, \infty)$, using (4.3) and (4.4). The region below the resulting curve is the unachievable region.

To use Theorem 4.1 we need to define the region of parameter space in which we are interested. In the two-point target problem we were able to reduce the dimension of the parameter space to two, by eliminating the nuisance parameter, η, through the use of the profile likelihood. But as we have shown in Theorem 3.1 that both the singular values of the profile Fisher information matrix $J^p(\phi)$ approach zero as the source separation becomes small, there is clearly no bias function $b(\phi)$ that will prevent the right-hand side of (4.1) becoming large. Instead we use the uniform bound to explore bias/variance tradeoffs in a six-dimensional space, comprising; $\alpha_s \equiv \alpha_1 + \alpha_2$; $\alpha_d \equiv \frac{\alpha_1 - \alpha_2}{\alpha_s}$; $t_{ws} \equiv (\alpha_1 u_1 + \alpha_2 u_2)/\alpha_s$; $t_{wd} \equiv \frac{u_1 - u_2}{2}$; ψ_1; ψ_2; where u_1 is the direction-cosine of the first source , α_1 is its (real) amplitude, and ψ_1 its phase.

The parameter we are interested in estimating is now t_{ws}, the amplitude-weighted mean source direction-cosine. The results presented below were obtained for an elliptic region of interest having a width equal to 0.5 (one beamwidth) in the t_{ws} direction, 0.01 in the t_{wd} direction, and 1000, 2, 2π, and 2π in the α_s, α_d, ψ_1, and ψ_2 directions respectively. These values reflect our uncertainty about the true values of the parameters. In particular we are restricting attention to scenarios for which the uncertainty in the source separation is small, which is appropriate for targets comprising closely spaced sources.

In Fig. 3(a) we plot the uniform bound as a function of δ^2 for the case of two in-phase 20 dB scatterers with a separation of 0.01, which for the 4 element array represents around 0.02 beamwidths. The uniform bound agrees with the unbiased bound in the limit as $\delta^2 \to 0$, but falls to around -40 dB as δ^2 increases, showing that a very significant advantage can be obtained through the use of a biased estimator. The MSE follows the uniform bound to this point. In Fig. 3(b)

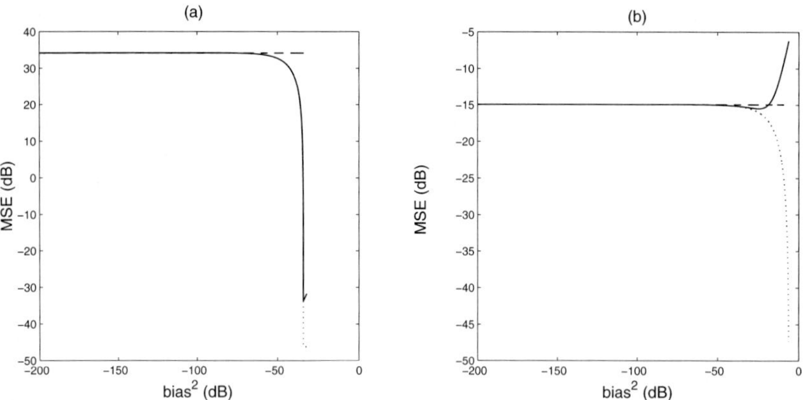

FIG. 3. Uniform CR bound (short dashes) plotted as a function of the squared bias-gradient norm, for comparison with the unbiased bound (long dashes) and the MSE (solid curve); (a) two 20 dB in-phase scatterers; (b) two 20 dB anti-phase scatterers.

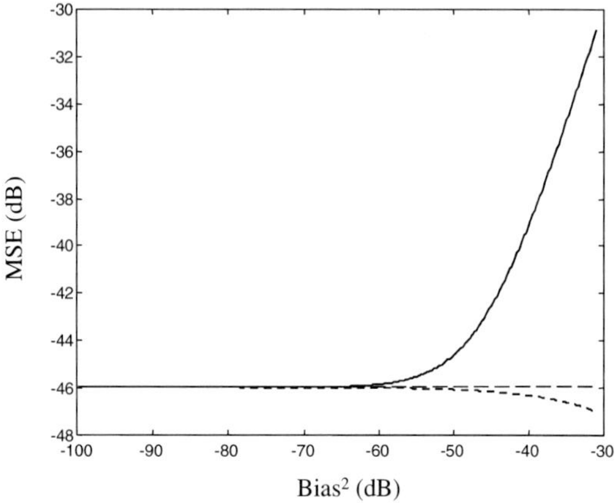

FIG. 4. Uniform CR bound (short dashes) plotted as a function of the squared bias-gradient norm, for comparison with the unbiased bound (long dashes) and the MSE (solid curve) for a single 20 dB scatterer.

we show the same information for anti-phase scatterers, where the minimum achievable MSE is around −16 dB, indicating that there is only a minor benefit in using a biased estimator in this case. Figure 4 shows that when a single scatterer is present the uniform and unbiased CR bounds agree.

5 Conclusions

Unbiased estimators need not exist, and will not in general exist for the DOAs of multiple closely spaced targets, since, as we have shown, the Fisher information matrix becomes approximately singular under these conditions. Analysis of the CR bound for unbiased estimators of the mean position and the difference in position of two closely spaced in-phase sources shows that the bound on the error in estimating the difference varies with the second power of the source separation, while the bound on the error in estimating the mean varies with the fourth power. This result is reversed for sources in anti-phase.

MSE is a better measure of estimator performance under these conditions, since it allows us to compare biased and unbiased estimators. The uniform CR bound provides a means to evaluate the bias/variance trade-off, and we have shown that for targets composed of two equal amplitude scatterers, significant reductions in the MSE can be achieved by using a biased estimator of the mean DOA. The uniform bound indicates that much better estimation accuracy should be achievable for in-phase scatterers than anti-phase scatterers, in contrast to the results from the unbiased bound. This agrees with practice, since monopulse is known to work very well for closely spaced in-phase sources, but poorly for out of phase (glinting) targets.

Acknowledgements

This work was carried out as part of Technology Group 3 of the MoD Corporate Research Programme.

Bibliography

[1] Meng, Y., Stoica, P. and Wong, K.M. Estimation of the directions of arrival of spatially dispersed signals in array processing, *IEE Proc.-Radar, Sonar Navig.*, **143**, 1–9, 1996.

[2] Goldberg, J. and Messer, H. Inherent limitations in the localization of a coherently scattered source, *IEEE Trans. Sig. Proc.*, **46**, 3441–3444, 1998.

[3] Mendel, J.M. *Lessons in Estimation Theory for Signal Processing, Communications and Control*, Prentice-Hall, Englewood Cliffs, NJ, 1995.

[4] Barndorff-Nielson, O.E. and Cox, D.R. *Inference and Asymptotics*, Chapman and Hall, London, 1994.

[5] Hayward, S.D. and Green, A.R. *Direction finding for tracking radar using a simplified subspace-based approach*, 8th Adaptive Sensor Array Processing Workshop, MIT Lincoln Labs, Lexington MA, 2000.

[6] Golub, G.H. and Van Loan, C.F. *Matrix Computations, 2nd ed.*, Johns Hopkins University Press, Baltimore, MD, 1989.

[7] Stoica, P. and Nehorai, A. MUSIC, maximum likelihood and the Cramer–Rao bound, *IEEE Trans.* **37**, 720–741, 1989.

[8] Magnus, J.R. and Neudecker, H. *Matrix Differential Calculus with Applica-*

tions in Statistics and Econometrics, Wiley, New York, 1991.

[9] Muirhead, R.J. *Aspects of Multivariate Statistical Theory*, Wiley, New York, 1982.

[10] Nickel, U. Monopulse estimation with adaptive arrays, *IEE Proc. F*, **140**, 303–308, 1993.

[11] Liu, R.C. and Brown, L.D. Nonexistance of informative unbiased estimators in singular problems, *Ann. Stat.*, **21**, 1–13, 1993.

[12] Hero, A.O., Fessler, J.A. and Usman, M. Exploring estimator bias-variance tradeoffs ssing the uniform CR bound, *IEEE Trans. Sig. Proc.*, **44**, 2026–2041, 1996.

Appendix: Proof of Theorem 3.1

Writing $A = \begin{bmatrix} a_1 & a_2 \end{bmatrix}$, and $D_u = \begin{bmatrix} d_1 & d_2 \end{bmatrix}$, we get

$$P = \begin{bmatrix} \delta_0 & \delta_1 \\ \delta_1^* & \delta_0 \end{bmatrix} - \frac{1}{m^2 - \beta\beta^*} \begin{bmatrix} \gamma\gamma^* & \gamma\gamma\beta^* \\ \gamma^*\gamma^*\beta & \gamma\gamma^* \end{bmatrix}, \tag{A.1}$$

where $\beta \equiv {a_1}^H a2$, $\gamma \equiv {d_1}^H a2$, $\delta_0 \equiv {d_1}^H d1$, and $\delta_1 \equiv {d_1}^H d2$. Expanding the inner products we obtain the following identities;

$$\beta = \sum_k^m \exp -4\pi i r x_k, \quad \beta\beta^* = \sum_k^m \sum_j^m \exp -4\pi i r(x_j - x_k),$$

$$\gamma\gamma^* = 4\pi^2 \sum_k^m \sum_j^m x_k x_j \exp 4\pi i r(x_j - x_k),$$

$$\gamma\gamma = -4\pi^2 \sum_k^m \sum_j^m x_k x_j \exp -4\pi i r(x_j + x_k),$$

$$\delta_0 = 4\pi^2 \sum_k^m {x_k}^2, \quad \delta_1 = 4\pi^2 \sum_k^m {x_k}^2 \exp -4\pi i r x_k.$$

Writing

$$(m^2 - \beta\beta^*)\,[P]_{11} = m^2(2\pi)^2 \sum_k^m {x_k}^2$$

$$-(2\pi)^2 \sum_k^m \sum_j^m \sum_l^m {x_l}^2 \exp\{4\pi i r(x_j - x_k)\}$$

$$-m(2\pi)^2 \sum_k^m \sum_j^m x_k x_j \exp\{4\pi i r(x_j - x_k)\},$$

and expanding the right-hand side as a Taylor series in powers of r, all terms in odd powers of r are zero because of the symmetry of the array. After some manipulation, the terms in r^0 and r^2 can be shown to be zero also. Consequently we can write

$$(m^2 - \beta\beta^*)[P]_{11} = \frac{(4\pi r)^4 (2\pi)^2 6\overline{x^2}(m\overline{x^4} - \overline{x^2}^2)}{4!} + O(r^6), \tag{A.2}$$

where $\overline{x^n} \equiv \sum_k^m x^n$. Expanding $m^2 - \beta\beta^*$ in a similar manner we get $m^2 - \beta\beta^* = (4\pi r)^2 m\overline{x^2} + O(r)$. Substituting in (A.2) we get

$$[P]_{11} = \frac{r^2 2\pi^4 (m\overline{x^4} - \overline{x^2}^2)}{m} + O(r^4).$$

Writing

$$\begin{aligned}(m^2 - \beta\beta^*)[P]_{12} &= -m^2(2\pi)^2 \sum_k^m {x_k}^2 \exp\{-4\pi i r x_k\} \\ &-(2\pi)^2 \sum_k^m \sum_j^m \sum_l^m {x_l}^2 \exp\{4\pi i r(x_j - x_k - x_l)\} \\ &+(2\pi)^2 \sum_k^m \sum_j^m \sum_l^m x_k x_j \exp\{4\pi i r(x_j + x_k - x_l)\},\end{aligned}$$

and again expanding the right-hand side as a power series in r, we get

$$[P]_{12} = -\frac{r^2(2\pi)^4(m\overline{x^4} - \overline{x^2}^2)}{m} + O(r^4).$$

Therefore

$$P = \frac{r^2(2\pi)^4(m\overline{x^4} - \overline{x^2}^2)}{m} \begin{bmatrix} 1 & -1 \\ -1 & 1 \end{bmatrix} + O(r^4) \equiv k_1 r^2 \begin{bmatrix} 1 & -1 \\ -1 & 1 \end{bmatrix} + O(r^4). \tag{A.3}$$

The eigendecomposition of P follows immediately from (A.3):

$$P = \frac{1}{\sqrt{2}} \begin{bmatrix} 1 & 1 \\ -1 & 1 \end{bmatrix} \begin{bmatrix} 2k_1 r^2 + O(r^4) & 0 \\ 0 & 2k_2 r^4 + O(r^6) \end{bmatrix} \frac{1}{\sqrt{2}} \begin{bmatrix} 1 & -1 \\ 1 & 1 \end{bmatrix},$$

where k_2 does not depend on r. On inverting P we obtain

$$P^{-1} = \frac{1}{\sqrt{2}} \begin{bmatrix} 1 & 1 \\ -1 & 1 \end{bmatrix} \begin{bmatrix} \frac{r^{-2}}{2k_1} + O(r^0) & 0 \\ 0 & \frac{r^{-4}}{2k_2} + O(r^{-2}) \end{bmatrix} \frac{1}{\sqrt{2}} \begin{bmatrix} 1 & -1 \\ 1 & 1 \end{bmatrix},$$

and the theorem is proved.

It can be shown that $k_2 = \frac{(2\pi)^6(\overline{x^2 x^6} - \overline{x^4}^2)}{9\overline{x^2}}$.

Robustness of Narrowband DOA Algorithms with Respect to Signal Bandwidth

Yann Meurisse and Jean-Pierre Delmas
Institut National des Télécommunications. 9 rue Charles Fourier, 91011 Evry Cedex, France.

Abstract
The purpose of this paper is to determine the domain of validity of spatial covariance-based narrowband DOA algorithms processing non-narrowband data. By focusing on the case of one source, order detection and asymptotic (w.r.t. the bandwidth and the number of snapshots) bias and variance given by any narrowband algorithm are considered. An order detector based on numerical analysis arguments introduced in channel order detection is proposed. Closed-form expressions are given for asymptotic bias and variance of the DOA estimated by the MUSIC algorithm where we show the key role of the symmetry of the bandwidth w.r.t. the frequency of interest. Furthermore, thanks to a separability property and a reparametrization, a closed-form expression of the Cramer–Rao bound is given for the DOA of a source of symmetric spectrum in the case of an arbitrary linear array. This allows us to prove that the MUSIC algorithm remains efficient in a large domain of bandwidth in these conditions.

1 Introduction

The problem of estimating the directions of arrival (DOAs) of multiple plane waves impinging on an array of sensors may be classified into narrowband and broadband data processing according to whether the complex envelope of the received signals can be considered as time constant or not along the array. As the broadband approaches generally require an increased computational complexity in comparison with the narrowband ones, it is of interest to examine if the narrowband methods can be used for a sufficiently wide bandwidth without sacrificing performance. This question, interestingly has not been adequately treated in the literature.

The purpose of this contribution is to determine the domain of validity of spatial covariance-based narrowband DOA algorithms processing non-narrowband data. We prove that the vague definition of the narrowband assumption often given in the literature, namely that the array aperture is much less than the inverse relative bandwidth (that is, $\frac{Mb}{f_0} \ll 1$ where M denotes the number of sensors) is much too severe. To study the robustness of the narrowband DOA algorithms with respect to signal bandwidth, we concentrate to the case of one source.

This paper is organized as follows. After the data model and some notations are introduced in Section 2, the performance of the eigendecomposition-based order detectors are considered where a criterion based on numerical analysis arguments [5] is proposed in Section 3. Then the asymptotic (w.r.t. the number of snapshots and the signal bandwidth) bias and variance of the estimated DOA are studied and illustrated in the case of the standard MUSIC algorithm in Section 4. Thanks to a separability property, a closed-form expression of the Cramer–Rao bound (CRB) is given for the DOA parameter of a single source of any symmetric power spectral density in the case of an arbitrary linear array in Section 5. Finally, Section 6 is devoted to a comparison between narrowband and broadband algorithms.

2 Data model

Consider K radiating sources observed by an arbitrary array of M sensors. The received signals are bandpass filtered (with bandwidth B) around the frequency f_0 of interest (with $B < 2f_0$). (We suppose that their power spectral density is zero in the neighbourhood of the frequency 0.) After frequency down-shifting the sensor signals to baseband, the complex envelope is generated. The M-vector $\mathbf{n}_t$ of complex envelope of the noise is assumed throughout the paper to be temporally and spatially uncorrelated with $\mathrm{E}(\mathbf{n}_t\mathbf{n}_t^H) = \sigma_n^2\mathbf{I}_M$ and independent of the sources. If s_t^k denotes the complex envelope of the kth source w.r.t. the frequency f_0 of interest and $\mu_k(f)$ its spectral measure, the complex envelope of this source observed at the mth sensor is $s_{t-\tau_{k,m}}^k e^{-i2\pi f_0\tau_{k,m}}$, where $\tau_{k,m}$ are delays associated with the signal propagation time from the kth source to the mth sensor. These parameters contain information about the source location Θ_k relative to the array. Then the M-vector of observed complex envelope at the array output is $\mathbf{y}_t = \sum_{k=1}^K \int_{-B/2}^{+B/2} e^{i2\pi ft}\mathbf{a}(\Theta_k, f_0 + f)d\mu_k(f) + \mathbf{n}_t$, where $\mathbf{a}(\Theta_k, \nu) \stackrel{\text{def}}{=} [e^{-i2\pi\nu\tau_{k,1}}, \ldots, e^{-i2\pi\nu\tau_{k,M}}]^T$. If the sources are spatially uncorrelated, the spatial covariance matrix may be written as

$$\mathbf{R}_b = \mathrm{E}(\mathbf{y}_t\mathbf{y}_t^H) = \sum_{k=1}^K \int_{-B/2}^{+B/2} S_k(f)\mathbf{a}(\Theta_k, f_0 + f)\mathbf{a}^H(\Theta_k, f_0 + f)df + \sigma_n^2\mathbf{I}_M, \tag{2.1}$$

where $S_k(f)$ denotes the power spectral density of the kth source. For the very narrowband case, this matrix becomes

$$\mathbf{R}_0 = \sum_{k=1}^K \sigma_k^2\mathbf{a}(\Theta_k, f_0)\mathbf{a}^H(\Theta_k, f_0) + \sigma_n^2\mathbf{I}_M, \tag{2.2}$$

where σ_k^2 denotes the power of the kth source. We consider that the power spectral density $S_k(f)$ of the complex envelope of the kth broadband source is parametrized by its centred frequency f_m, its standard deviation f_σ, and by its power σ_k^2. If $S(f)$ denotes a normalized function, that is, a function satisfying

$$\int S(f)df = 1 \quad \int fS(f)df = 0 \quad \int f^2 S(f)df = 1, \tag{2.3}$$

the kth source spectrum is expressed by $S_k(f) = \frac{\sigma_k^2}{f_\sigma} S\left(\frac{f-f_m}{f_\sigma}\right)$, and if $R(t)$ denotes the correlation function associated with $S(f)$, the kth source correlation is $R_k(t) = \sigma_k^2 R(f_\sigma t)e^{i2\pi f_m t}$. The covariance matrix $\mathbf{R}_b$ may be written as

$$\mathbf{R}_b = \sum_{k=1}^{K} \mathbf{a}(\Theta_k, f_0)\mathbf{a}^H(\Theta_k, f_0) \odot \mathbf{R}_{s_k} + \sigma_n^2 \mathbf{I}_M$$

where $\mathbf{R}_{s_k}$ is the $M \times M$ matrix whose (m,n)th term is the source correlation

$$\begin{aligned}[\mathbf{R}_{s_k}]_{m,n} = \mathrm{E}(s^k_{t-\tau_{k,m}} {s^k_{t-\tau_{k,n}}}^*) &= \sigma_k^2 \int S(\nu) e^{i2\pi(\nu f_\sigma + f_m)(\tau_{k,n}-\tau_{k,m})} d\nu \\ &= \sigma_k^2 R\left(f_\sigma(\tau_{k,n} - \tau_{k,m})\right) e^{i2\pi f_m(\tau_{k,n}-\tau_{k,m})}.\end{aligned}$$

If $\mathbf{z}_m$ and $\mathbf{p}_k$ denote, respectively, the coordinate vector of the mth sensor referenced to a specific sensor and the unit wavevector associated with the kth source, $\tau_{k,m} = \frac{1}{f_0}\frac{\mathbf{z}_m^T \mathbf{p}_k}{\lambda_0}$. Because (see (2.3)) $R(0) = 1$ and for a symmetric spectrum $R'(0) = 0$, we have $\mathbf{R}_{s_k} = \sigma_k^2 \mathbf{1}\mathbf{1}^T + \delta\mathbf{R}_{s_k}$ where $\mathbf{1}$ denotes the M-element vector of ones with

$$\begin{aligned}[\delta\mathbf{R}_{s_k}]_{m,n} &= \sigma_k^2 \alpha_{m,n,k}\left[i2\pi\left(\frac{f_m}{f_0}\right) + R'(0)\left(\frac{f_\sigma}{f_0}\right)\right] \\ &+ O\left(\frac{f_\sigma^2}{f_0^2}\right) + O\left(\frac{f_m^2}{f_0^2}\right) + O\left(\frac{f_m f_\sigma}{f_0^2}\right) \text{ for an arbitrary spectrum} \\ &= \sigma_k^2 \alpha_{m,n,k}\left[i2\pi\left(\frac{f_m}{f_0}\right) + \frac{1}{2}\alpha_{m,n,k} R'(0)\left(\frac{f_\sigma^2}{f_0^2}\right)\right] + O\left(\frac{f_m^2}{f_0^2}\right) \\ &+ O\left(\frac{f_\sigma^4}{f_0^4}\right) + O\left(\frac{f_m f_\sigma^2}{f_0^3}\right) \text{ for a spectrum symmetric w.r.t. } f_0 + f_m \\ &= \frac{1}{2}\sigma_k^2 \alpha_{m,n,k}^2 R'(0)\left(\frac{f_\sigma^2}{f_0^2}\right) + O\left(\frac{f_\sigma^4}{f_0^4}\right) \\ &\stackrel{\text{def}}{=} \sigma_k^2 \left(\frac{f_\sigma^2}{f_0^2}\right)[\tilde{\mathbf{R}}_{s_k}]_{m,n} + O\left(\frac{f_\sigma^4}{f_0^4}\right) \text{ for a spectrum symmetric w.r.t. } f_0,\end{aligned}$$

with $\alpha_{m,n,k} \stackrel{\text{def}}{=} \frac{(\mathbf{z}_n - \mathbf{z}_m)^T \mathbf{p}_k}{\lambda_0}$ and where $\tilde{\mathbf{R}}_{s_k}$ depends only on the normalized spectrum $S(f)$ of the kth source and on the position of the sensors. Therefore $\mathbf{R}_b$ may be considered as a perturbation of $\mathbf{R}_0$:

$$\mathbf{R}_b = \mathbf{R}_0 + \delta\mathbf{R}_b \quad \text{with} \quad \delta\mathbf{R}_b = \sum_{k=1}^{K} \mathbf{a}(\Theta_k, f_0)\mathbf{a}^H(\Theta_k, f_0) \odot \delta\mathbf{R}_{s_k} \tag{2.4}$$

3 Performance of the detector

The most commonly used detectors are formulated in terms of the eigenvalues of the sample spatial covariance matrix $\mathbf{R}_b(T)$ derived from T independent snapshots $\mathbf{y}_t$. Typical examples are the Akaike information criterion (AIC) and the minimum description length (MDL) detectors. Unfortunately, if the bandwidth of the signal increases, the AIC and MDL criteria rapidly tend to overestimate the number of sources and in practice the estimated number of sources by these detectors may be far from the true number [9].

The detection of the number of non-narrowband signals from the sample spatial covariance matrix, however, is analogous to the detection of the 'effective' order of the impulse response of a channel from the sample spatio-temporal covariance matrix. In fact, the sample spatial covariance matrix $\mathbf{R}_b(T)$ observed in a non-narrowband scenario can be considered as the sum of an 'ideal' rank-K matrix $\mathbf{R}$ and a 'perturbation' matrix $\delta\mathbf{R}_{b,T}$,

$$\mathbf{R}_b(T) = \mathbf{R} + \delta\mathbf{R}_{b,T}$$

where K is the number of sources and $\mathbf{R}$ is the signal's noise-free spatial covariance matrix associated with K narrowband signals. The 'perturbation' matrix $\delta\mathbf{R}_{b,T}$ incorporates the influence of the non-narrowband assumption, the influence of the additive noise, and the influence of the estimated (inexact) statistics. Using the concept of canonical angles between subspaces and invariant subspace perturbation results, a 'maximally stable' decomposition of the range space of the sample covariance matrix into signal and noise subspace has been proposed in the channel order determination context by Liavas *et al.* [5]. This approach has led to the following criterion. The detected order $\widehat{K}$ is the value of k which minimizes

$$r(k) \stackrel{\text{def}}{=} \begin{cases} \frac{\lambda_{k+1}}{\lambda_k - 2\lambda_{k+1}}, & \text{if } \lambda_{k+1} \leq \frac{\lambda_k}{3} \\ 1, & \text{otherwise} \end{cases}$$

where $\lambda_1 \geq \lambda_2 \geq \cdots \geq \lambda_M$ denote the eigenvalues of the sample covariance matrix $\mathbf{R}_b(T)$. We propose to apply this criterion in the context of non-narrowband array processing. Contrary to the AIC and MDL criterion which base their detection on the similarity of the smallest eigenvalues, the proposed criterion is based on the existence of an eigenvalue gap. Because simulations (see, for example, Fig. 1 and [10] show that as signal's bandwidth is increased, eigenvalues pop up from the noise floor one at a time, the proposed criterion is potentially promising.

Fig. 1 shows that contrary to the AIC and MDL criteria, the Liavas criterion is not very sensitive to increasing the bandwidth for a finite number of snapshots. What can be said about the quality of the estimates obtained by these criteria is given by their behaviour as the snapshot size increases to infinity. Fig. 2 shows for one source, the domain of fractional bandwidth $\frac{Mb}{f_0}$ for which the MDL and Liavas criteria correctly detect the number of sources in the large-sample

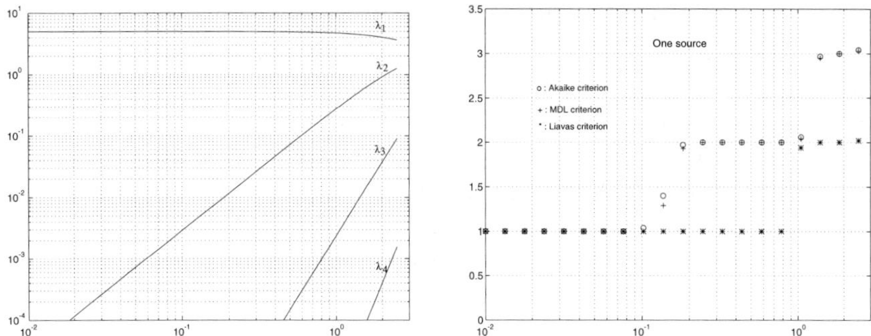

FIG. 1. Eigenvalues of the signal's noise-free spatial covariance matrix and detected mean number of sources (by 100 runs) by the Akaike, MDL and Liavas criteria for one source of centred flat spectrum and a uniform linear array of $M = 5$ sensors as a function of the fractional bandwidth $\frac{Mb}{f_0}$.

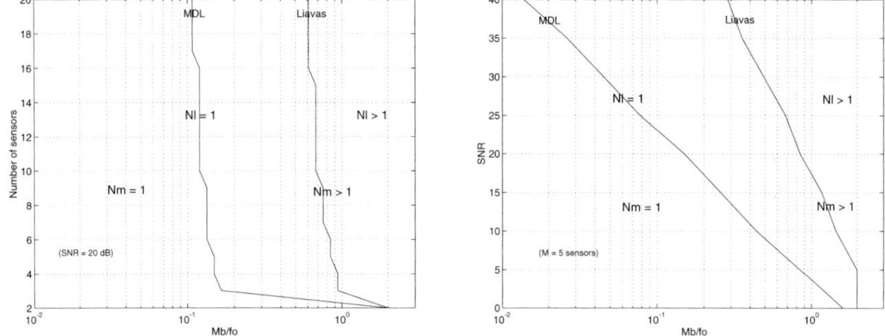

FIG. 2. Asymptotic (w.r.t. the number of snapshots) detected number of sources by the MDL (Nm) and Liavas (Nl) criteria as a function of the fractional bandwidth $\frac{Mb}{f_0}$, the number of sensors and the SNR for a uniform linear array and one source of centred flat spectrum.

limit (that is, with exact statistics). In this figure the Liavas criterion always outperforms the MDL criterion. We see that the performance of the MDL and Liavas criteria degrades when the SNR increases (but the Liavas criterion is more stable) and remains relatively invariant when the number of sensors increases.

4 Asymptotic bias and variance

To consider the asymptotic (w.r.t. to the number of snapshots and signal bandwidth) bias and variance of the DOA estimated by a narrowband second-order statistics (SOS)-based algorithm, we adopt a functional approach that consists of recognizing that the whole process of constructing an estimate $\Theta(T)$ of Θ is equivalent to defining a functional relation linking this estimate $\Theta(T)$ to the sample statistics $\mathbf{R}_b(T) = \frac{1}{T}\sum_{t=1}^{T} \mathbf{y}_t\mathbf{y}_t^H$ from which it is inferred. This functional dependence is denoted $\Theta(T) = \text{alg}(\mathbf{R}_b(T))$. Clearly, $\Theta = \text{alg}(\mathbf{R}_0)$, and therefore the different narrowband SOS-based algorithms $\text{alg}(\cdot)$ constitute distinct extensions of the mapping $\mathbf{R}_0 \to \Theta$ generated by (2.2) to any unstructured Hermitian matrix $\mathbf{R}_b(T)$. $\mathbf{R}_b(T)$ may be considered as a perturbation of $\mathbf{R}_b$:

$$\mathbf{R}_b(T) = \mathbf{R}_b + \delta\mathbf{R}_T, \tag{4.1}$$

where $\delta\mathbf{R}_T$ is the finite sample size error, verifying $\text{E}(\delta\mathbf{R}_T) = \mathbf{O}$ and $\text{Cov}(\delta\mathbf{R}_T) = O\left(\frac{1}{T}\right)$. Because the mapping $\text{alg}(\cdot)$ is sufficiently regular in a neighborhood of $\mathbf{R}_b$ for most SOS-based algorithms, we have from (4.1),

$$\Theta(T) = \text{alg}(\mathbf{R}_b) + (D^{\text{alg}}_{\mathbf{R}_b}, \delta\mathbf{R}_T) + O(\|\delta\mathbf{R}_T\|^2), \tag{4.2}$$

where $(D^{\text{alg}}_{\mathbf{R}_b}, \delta\mathbf{R}_T)$ denotes the differential of the mapping $\text{alg}(\cdot)$ evaluated at point $\mathbf{R}_b$ applied to $\delta\mathbf{R}_T$. Taking expectations, we obtain

$$\text{E}(\Theta(T)) = \text{alg}(\mathbf{R}_b) + O\left(\frac{1}{T}\right). \tag{4.3}$$

By considering $\mathbf{R}_b$ as a perturbation of $\mathbf{R}_0$ (see (2.4)), a first-order perturbation analysis of a narrowband SOS-based algorithm acting on $\mathbf{R}_b$ evaluated at point $\mathbf{R}_0$ gives

$$\begin{aligned}\text{alg}(\mathbf{R}_b) &= \text{alg}(\mathbf{R}_0) + (D^{\text{alg}}_{\mathbf{R}_0}, \delta\mathbf{R}_b) + O(\|\delta\mathbf{R}_b\|^2) \\ &= \Theta + \mathbf{D}^{\text{alg}}_{\mathbf{R}_0}\text{Vec}(\delta\mathbf{R}_b) + O(\|\delta\mathbf{R}_b\|^2), \end{aligned} \tag{4.4}$$

where $\mathbf{D}^{\text{alg}}_{\mathbf{R}_0}$ denotes the matrix associated with the differential of the narrowband SOS-based algorithm $\text{alg}(\cdot)$ at point $\mathbf{R}_0$. (Expressions of $\mathbf{D}^{\text{alg}}_{\mathbf{R}_0}$ and $\mathbf{D}^{\text{alg}}_{\mathbf{R}_b}$ are ordinarily deduced from perturbation calculus.) So from (4.3) and (4.4), the following result holds.

Theorem 4.1 *The asymptotic bias (w.r.t. to the number of snapshots and signal bandwidth) of the estimate $\Theta(T)$ given by a narrowband SOS-based algorithm is given by:*

$$\text{E}(\Theta(T)) - \Theta = \mathbf{D}^{\text{alg}}_{\mathbf{R}_0}\text{Vec}(\delta\mathbf{R}_b) + O\left(\frac{1}{T}\right) + O(\|\delta\mathbf{R}_b\|^2). \tag{4.5}$$

Then, from (4.2) and (4.3), the mapping $\text{alg}(\cdot)$ gives the deviation from the asymptotic mean $\text{E}(\Theta(T))$: $\Theta(T) - \text{E}(\Theta(T)) = \mathbf{D}^{\text{alg}}_{\mathbf{R}_b}\text{Vec}(\delta\mathbf{R}_T) + O\left(\frac{1}{T}\right) + O(\|\delta\mathbf{R}_b\|^2)$. We therefore have the following theorem.

Theorem 4.2 *The asymptotic variance (w.r.t. to the number of snapshots and signal bandwidth) of the estimate $\Theta(T)$ given by a narrowband SOS-based algorithm may be written as*

$$\text{Var}(\Theta(T)) = \frac{1}{T}\mathbf{D}^{\text{alg}}_{\mathbf{R}_b}\mathbf{C}_{R_b}\left(\mathbf{D}^{\text{alg}}_{\mathbf{R}_b}\right)^H + O\left(\frac{1}{T^2}\right) + O\left(\frac{1}{T}\right)O(\|\delta\mathbf{R}_b\|^2) + O(\|\delta\mathbf{R}_b\|^4)$$

with

$$\mathbf{C}_{R_b} = \lim_{T\to\infty} T\mathrm{Cov}(\mathrm{Vec}(\mathbf{R}_b(T)) = \lim_{T\to\infty} T\mathrm{E}(\mathrm{Vec}(\delta\mathbf{R}_T)\mathrm{Vec}^H(\delta\mathbf{R}_T))$$
$$= \mathbf{R}_b \otimes_c \mathbf{R}_b$$

for independent circular complex snapshots $\mathbf{y}_t$*, [2, p. 336]. (The Kronecker product* $\mathbf{A} \otimes_c \mathbf{B}$ *is the block matrix, the* (i, j) *block element of which is* $b^*_{i,j}\mathbf{A}$*, (this unusual convention makes it easier to deal with complex matrices for which* $\mathrm{Vec}(\mathbf{ABC}) = (\mathbf{A} \otimes_c \mathbf{C}^H)\mathrm{Vec}(\mathbf{B})$*.)*

These general results are illustrated in the case of the classic MUSIC algorithm in the following section.

4.1 MUSIC algorithm

First of all, we note that the differential matrix $\mathbf{D}^{\mathrm{alg}}_{\mathbf{R}_0}$ of the standard MUSIC algorithm deduced from perturbation calculus (see, for example, [3]) is

$$\mathbf{D}^{\mathrm{music}}_{\mathbf{R}_0} = \frac{1}{\alpha_{\theta_1}}\left(\mathbf{a}^H_{\theta_1}\Gamma_s \otimes_c \mathbf{a}^{'H}_{\theta_1}\Pi_n + \mathbf{a}^{'H}_{\theta_1}\Pi_n \otimes_c \mathbf{a}^H_{\theta_1}\Gamma_s\right), \tag{4.6}$$

where Π_n and Γ_s denote the orthogonal projection onto the noise space associated with $\mathbf{R}_0$ and the Moore–Penrose pseudoinverse $(\mathbf{R}_0 - \sigma^2_n\mathbf{I}_M)^{\#}$ respectively. In this section $\mathbf{a}_{\theta_1} \stackrel{\mathrm{def}}{=} \mathbf{a}(\Theta_1, f_0)$ and α_{θ_1} is the geometrical factor $2\mathbf{a}^{'H}_{\theta_1}\Pi_n\mathbf{a}^{'}_{\theta_1}$ with $\mathbf{a}^{'}_{\theta_1} \stackrel{\mathrm{def}}{=} \frac{d\mathbf{a}_{\theta_1}}{d\theta_1}$. Therefore the asymptotic bias becomes

$$\begin{aligned}[\mathbf{D}^{\mathrm{music}}_{\mathbf{R}_0}\mathrm{Vec}(\delta\mathbf{R}_b)]_1 &= \frac{2}{\alpha_{\theta_1}}\mathrm{Re}\left(\mathbf{a}^H_{\theta_1}\Gamma_s\delta\mathbf{R}_b\Pi_n\mathbf{a}^{'}_{\theta_1}\right)\\ &= \frac{2}{\alpha_{\theta_1}}\int_{-B/2}^{+B/2} S_1(f)\mathrm{Re}\left(\mathbf{a}^H_{\theta_1}\Gamma_s\Delta_{\theta_k,f}\mathbf{a}_{\theta_1}\mathbf{a}^H_{\theta_1}\Delta^H_{\theta_1,f}\Pi_n\mathbf{a}^{'}_{\theta_1}\right)df\end{aligned}$$

where $\Delta_{\theta_1,f} \stackrel{\mathrm{def}}{=} \mathrm{Diag}(e^{-i2\pi f\tau_{1,1}}, \ldots, e^{-i2\pi f\tau_{1,M}})$. We prove the following results.

Theorem 4.3 *The asymptotic bias (w.r.t. to the number of snapshots and* *signal bandwidth) is given by*

$$\begin{aligned}
\mathrm{E}(\Theta_1(T)) - \Theta_1 &= \frac{4\pi}{M\alpha_{\theta_1}} \sum_{n=1}^{M} \frac{{\mathbf{z}_n}^T \mathbf{p}_1^{'}}{\lambda_0} \\
&\quad \mathrm{Im}\left[\sum_{m=1}^{M} R\left(\frac{f_\sigma}{f_0}\frac{(\mathbf{z}_n-\mathbf{z}_m)^T\mathbf{p}_1}{\lambda_0}\right) e^{i2\pi\frac{f_m}{f_0}\frac{(\mathbf{z}_n-\mathbf{z}_m)^T\mathbf{p}_1}{\lambda_0}}\right] \\
&- \frac{4\pi}{M^2\alpha_{\theta_1}} \sum_{n=1}^{M} \frac{{\mathbf{z}_n}^T \mathbf{p}_1^{'}}{\lambda_0} \mathrm{Im}\left[\sum_{l,m=1}^{M} R\left(\frac{f_\sigma}{f_0}\frac{(\mathbf{z}_l-\mathbf{z}_m)^T\mathbf{p}_1}{\lambda_0}\right) e^{i2\pi\frac{f_m}{f_0}\frac{(\mathbf{z}_l-\mathbf{z}_m)^T\mathbf{p}_1}{\lambda_0}}\right] \\
&+ O\left(\frac{1}{T}\right) + O\left(\frac{f_\sigma^2}{f_0^2}\right) + O\left(\frac{f_m^2}{f_0^2}\right) + O\left(\frac{f_\sigma f_m}{f_0^2}\right) \quad \textit{for an arbitrary spectrum} \quad (4.7) \\
&= \frac{8\pi^2}{M\alpha_{\theta_1}}\left(\frac{f_m}{f_0}\right) \sum_{n,m=1}^{M} \frac{{\mathbf{z}_n}^T \mathbf{p}_1^{'}}{\lambda_0} \frac{(\mathbf{z}_n-\mathbf{z}_m)^T\mathbf{p}_1}{\lambda_0} + O\left(\frac{1}{T}\right) \\
&+ O\left(\frac{f_\sigma^2}{f_0^2}\right) + O\left(\frac{f_m^2}{f_0^2}\right) + O\left(\frac{f_\sigma f_m}{f_0^2}\right) \quad \textit{for an symmetric spectrum w.r.t. } f_0 + f_m \quad (4.8) \\
&= O\left(\frac{1}{T}\right) + O\left(\frac{f_\sigma^4}{f_0^4}\right) \quad \textit{for a symmetric spectrum w.r.t. } f_0 \quad (4.9)
\end{aligned}$$

with $\mathbf{p}_1^{'} \stackrel{\text{def}}{=} \frac{d\mathbf{p}_1}{d\Theta_1}$ *In the specific case of an arbitrary linear array and for a symmetric spectrum w.r.t.* $f_0 + f_m$

$$\mathrm{E}(\phi_1(T)) - \phi_1 = \phi_1 \frac{f_m}{f_0} + O\left(\frac{1}{T}\right) + O\left(\frac{f_\sigma^4}{f_0^4}\right) \quad (4.10)$$

where ϕ_1 *denotes here the spatial parameter* $\phi_1 = \pi \sin\theta_1$, *with* θ_1 *the DOA relative to the array broadside.*

The result (4.3) shows that the behaviour of narrowband DOA estimators stongly depends on the symmetry of the source spectrum w.r.t. its centred value and from the offset of this centred value w.r.t. the frequency f_0. Concerning the asymptotic (w.r.t. the number of snapshots and signal bandwidth) variance, we prove the following result for a symmetric spectrum w.r.t. f_0.

Theorem 4.4

$$\mathrm{Var}(\theta_1^b(T)) = \mathrm{Var}(\theta_1^0(T))\left(1 + c\left(\frac{f_\sigma^2}{f_0^2}\right) + O\left(\frac{f_\sigma^4}{f_0^4}\right)\right) + O\left(\frac{1}{T^2}\right) \quad (4.11)$$

where $\mathrm{Var}(\theta_1^0(T))$ *is the classic asymptotic variance of MUSIC algorithm given in narrowband scenario (see, for example, [7, (3.12)]).*

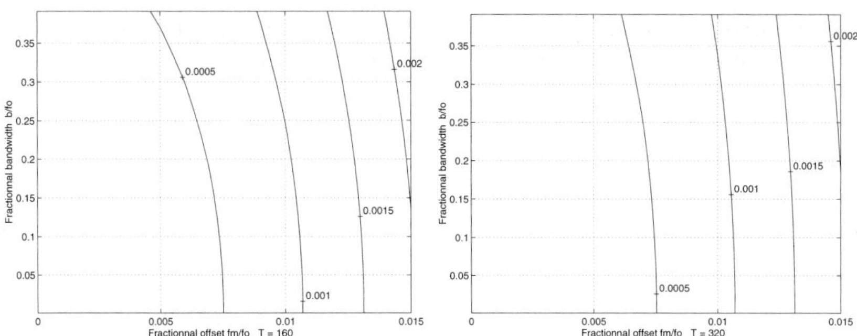

FIG. 3. Constant DOA mean square error contours as a function of the fractional bandwidth $\frac{b}{f_0}$ and the fractional offset $\frac{f_m}{f_0}$ for one source of centred flat spectrum impinging on a uniform linear array of five sensors with a SNR of 20 dB and $T = 160$ or 320 snapshots.

The crucial role played by the frequency offset of the spectrum on performance is illustrated by Fig. 3. This figure shows the MSE of the spatial DOA ϕ_1 estimated by the standard MUSIC algorithm: $\text{MSE}(\phi_1(T)) = \text{bias}^2(\phi_1(T)) + \text{Var}(\phi_1(T))$ in which bias and variance are respectively given by $\phi_1 \frac{f_m}{f_0}$ (see (4.10)) and by $\frac{1}{T}\mathbf{D}^{\text{music}}_{\mathbf{R}_b}(\mathbf{R}_0 \otimes_c \mathbf{R}_0 + \mathbf{R}_0 \otimes_c \delta\mathbf{R}_b + \delta\mathbf{R}_b \otimes_c \mathbf{R}_0)\left(\mathbf{D}^{\text{music}}_{\mathbf{R}_b}\right)^H$. The principal term of this MSE may be written as $\text{MSE}(\phi_1(T)) = \frac{c_0}{T} + \phi_1^2\left(\frac{f_m^2}{f_0^2}\right) + \frac{c_1}{T}\left(\frac{f_m}{f_0}\right) + \frac{c_2}{T}\left(\frac{f_\sigma^2}{f_0^2}\right)$, in which c_0/T denotes the asymptotic (w.r.t. the number of snapshots) variance of the MUSIC algorithm acting on very narrowband data. We see the key role of the frequency offset f_m: the narrowband SOS-based algorithms are much more sensitive to signal bandwidth than in the case of centred spectrum w.r.t. f_0. Furthermore, because the bias is constant in T, this sensitivity to f_m increases with the number of snapshots.

5 Cramer–Rao lower bound

In the case of T independent circular complex Gaussian zero-mean snapshots $\mathbf{y}_t$, the CRB for the unknown parameters $\Psi = (\theta_1, f_\sigma, \sigma_1^2, \sigma_n^2)^T$ of the spatial covariance matrix $\mathbf{R}_b$ is simply derived from inverting the Fisher information matrix (FIM) $\mathbf{I}(\Psi)$ (see, for example, [4, (15.52)]):

$$[\mathbf{I}(\Psi)]_{k,l} = T\ \text{Tr}\left(\mathbf{R}_b^{-1}\frac{\partial \mathbf{R}_b}{\partial \psi_k}\mathbf{R}_b^{-1}\frac{\partial \mathbf{R}_b}{\partial \psi_l}\right). \tag{5.1}$$

Extracting a closed-form expression of the CRB for the DOA's parameter θ_1 alone from the CRB matrix directly is far from trivial due to the need to invert the Fisher information matrix. To circumvent this difficulty, many authors (see, for example, [6] have exploited the asymptotic efficiency of the maximum likelihood (ML) estimator. So, the CRB of the entire parameter vector Ψ is given by the asymptotic covariance of the ML estimator. Deducing directly the asymptotic covariance matrix of the ML estimator of the DOA parameter Θ supposes that the likelihood function may be separable. Unfortunately, the separability of the likelihood function used in the case of narrowband sources (see, for example, [6] cannot be extended outside the narrowband scenario.

Thanks to a reparametrization, we show in the following that in the special case of one broadband source with a symmetric spectrum w.r.t. its centred frequency f_0, far enough from an arbitrary observing linear array, a closed-form expression of the DOA's parameter θ_1 alone can be derived. We consider a broadband source whose symmetric power spectral density $S_1(f)$ is parametrized by its standard deviation f_σ and the source power σ_1^2. The spatial covariance matrix (2.1) is

$$\mathbf{R}_b = \Delta_{\theta_1} \mathbf{R}_\Phi \Delta_{\theta_1}^H, \tag{5.2}$$

with $\Delta_{\theta_1} \stackrel{\text{def}}{=} \text{Diag}(e^{-i2\pi f_0 \tau_{1,1}}, \ldots, e^{-i2\pi f_0 \tau_{1,M}})$ and $\mathbf{R}_\Phi = \mathbf{R}_{s_1} + \sigma_n^2 \mathbf{I}_M$ where $\mathbf{R}_{s_1}$ is parametrized by $(f_\sigma \tau_{1,m})_{m=1,\ldots,M}$ and σ_1^2. In the specific case of a linear array, because $\tau_{1,m} = \frac{d_m}{c} \sin\theta_1$, $\mathbf{R}_b$ is uniquely parametrized by the parameter $\Psi' \stackrel{\text{def}}{=} (\theta_1, \Phi)$ where $\Phi = (\phi_2, \phi_3, \phi_4)$ with $\phi_2 \stackrel{\text{def}}{=} f_\sigma \sin\theta_1$, $\phi_3 \stackrel{\text{def}}{=} \sigma_1^2$ and $\phi_4 \stackrel{\text{def}}{=} \sigma_n^2$ parametrize $\mathbf{R}_\Phi$.

Thanks to the one-to-one mapping $\Psi \leftrightarrow \Psi'$, the CRB of the parameter Ψ is related to the FIM of the parameter Ψ' by (see, for example, [4, (3.30)])

$$CRB_\Psi = \frac{\partial \Psi}{\partial \Psi'^T} \mathbf{I}^{-1}(\Psi') \frac{\partial \Psi^T}{\partial \Psi'}. \tag{5.3}$$

Following the same lines as [1, Appendix], we prove that

$$[\mathbf{I}(\Psi')]_{1,k} = 0, \quad k = 2,3,4 \tag{5.4}$$

$$[\mathbf{I}(\Psi')]_{k,l} = T \; \text{Tr}\left(\mathbf{R}_\Phi^{-1} \frac{\partial \mathbf{R}_\Phi}{\partial \phi_k} \mathbf{R}_\Phi^{-1} \frac{\partial \mathbf{R}_\Phi}{\partial \phi_l}\right), \quad k,l = 2,3,4 \tag{5.5}$$

$$[\mathbf{I}(\Psi')]_{1,1} = T \; (2\text{Tr}\left(\Delta'_{\theta_1} \mathbf{R}_\Phi^{-1} \Delta'_{\theta_1} \mathbf{R}_\Phi) - 2\text{Tr}\left(\Delta'^{2}_{\theta_1}\right)\right), \tag{5.6}$$

where $\Delta'_{\theta_1} \stackrel{\text{def}}{=} \text{Diag}(-2\pi f_0 \frac{d_1}{c} \cos\theta_1), \ldots, -2\pi f_0 \frac{d_M}{c} \cos\theta_1)$. Consequently the FIM of the parameter Ψ' is block-diagonal structured:

$$\begin{bmatrix} I(\theta_1) & \mathbf{0}^T \\ \mathbf{0} & \mathbf{I}(\Phi) \end{bmatrix}$$

where $I(\theta_1)$ denotes the Fisher information of the parameter θ_1 considered as the only unknown parameter and $\mathbf{I}(\Phi)$ the FIM of the parameter Φ derived from T independent circular complex Gaussian zero-mean observations of covariance matrix $\mathbf{R}_\Phi$. Since $\frac{\partial \Psi}{\partial \Psi'^T}$ is structured as

$$\begin{bmatrix} 1 & 0 & 0 & 0 \\ -\frac{f_\sigma}{\tan\theta_1} & \frac{1}{\sin\theta_1} & 0 & 0 \\ 0 & 0 & 1 & 0 \\ 0 & 0 & 0 & 1 \end{bmatrix}$$

the CRB of the parameter Ψ becomes

$$\begin{bmatrix} I^{-1}(\theta_1) & aI^{-1}(\theta_1) & 0 & 0 \\ aI^{-1}(\theta_1) & a^2I^{-1}(\theta_1)+b^2[\mathbf{I}^{-1}(\Phi)]_{1,1} & b[\mathbf{I}^{-1}(\Phi)]_{1,2} & b[\mathbf{I}^{-1}(\Phi)]_{1,3} \\ 0 & b[\mathbf{I}^{-1}(\Phi)]_{2,1} & [\mathbf{I}^{-1}(\Phi)]_{2,2} & [\mathbf{I}^{-1}(\Phi)]_{2,3} \\ 0 & b[\mathbf{I}^{-1}(\Phi)]_{3,1} & [\mathbf{I}^{-1}(\Phi)]_{3,2} & [\mathbf{I}^{-1}(\Phi)]_{3,3} \end{bmatrix}$$

with $a \stackrel{\text{def}}{=} -\frac{f_\sigma}{\tan\theta_1}$ and $b \stackrel{\text{def}}{=} \frac{1}{\sin\theta_1}$. Therefore the CRB of the parameter θ_1 is

$$CRB_{\theta_1} = I^{-1}(\theta_1) = \frac{1}{2T}\left(\text{Tr}(\Delta'_{\theta_1}\mathbf{R}_\Phi^{-1}\Delta'_{\theta_1}\mathbf{R}_\Phi) - \text{Tr}\left(\Delta'^{\,2}_{\theta_1}\right)\right)^{-1}, \tag{5.7}$$

and because $[\mathbf{I}^{-1}(\Psi)]_{1,1} = [\mathbf{I}(\Psi)]^{-1}_{1,1}$, CRB_{θ_1} is not reduced with the knowledge of parameters $(f_\sigma, \sigma_1^2, \sigma_n^2)$. Furthermore, we note that when the spectral band of the source is known, that is, f_σ known, $\mathbf{I}(\Psi)$ is block diagonal, so the parameters θ_1 and (σ_1^2, σ_n^2) are decoupled. Furthermore, thanks to $\mathbf{R}_\Phi = \sigma_1^2\mathbf{1}\mathbf{1}^T + \left(\sigma_n^2\mathbf{I}_M + \delta\mathbf{R}_{s_1}\right)$, we prove the following result.

Theorem 5.1 *The CRB of the parameter θ_1 issued from a broadband source is given by*

$$CRB_{\theta_1^b} = CRB_{\theta_1^0}\left(1 + c'\left(\frac{f_\sigma^2}{f_0^2}\right) + O\left(\frac{f_\sigma^4}{f_0^4}\right)\right) \tag{5.8}$$

where $CRB_{\theta_1^0}$ is the classic CRB given in narrowband scenario (see, for example, [6, (17)]).

6 Comparison between narrowband and broadband algorithm

Naturally, a thorough comparison between narrowband and broadband algorithms would need a large quantity of scenarios (various arrays, DOAs, source spectra and SNR), and is beyond the scope of this paper. In the following, we simply exhibit a situation in which a narrowband algorithm outperforms a broadband one in a very large domain of bandwidth in the previous conditions of comparison. Figure 4 explores the case of one source of centred flat spectrum impinging on a linear array. It shows the mean square errors of the DOA estimated by the standard MUSIC and by a classic focusing algorithm [8] with respect to the bandwidth compared with the CRB. We notice a good agreement between the exact CRB and the asymptotic CRB $CRB_{\theta_1^0}\left(1 + c'\left(\frac{f_\sigma^2}{f_0^2}\right)\right)$ up to $\frac{Mb}{f_0} = 0.6$ and between the theoretical MSE and the estimated MSE given by the standard MUSIC algorithm up to $\frac{Mb}{f_0} = 0.5$. In addition, Fig. 4 shows that the asymptotic CRB coincides with the theoretical MSE given by the standard MUSIC algorithm. $\text{Var}(\theta_1^0(T)) = CRB_{\theta_1^0}$ and $\text{Var}(\theta_1^b(T)) \approx CRB_{\theta_1^b}$ for the standard MUSIC algorithm when $\frac{Mb}{f_0} \leq 0.5$. Consequently, the standard MUSIC algorithm remains efficient with increasing bandwidth up to $\frac{Mb}{f_0} = 0.5$. Considering

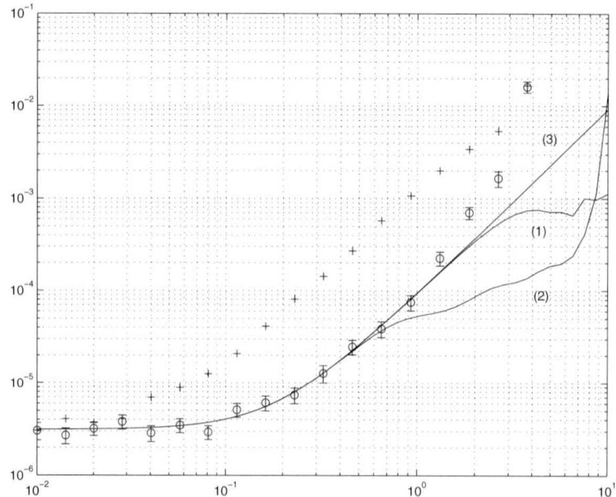

FIG. 4. Analytical asymptotically (w.r.t. b and T) (1) and estimated (100 runs) with 99% confidence interval (circles with error bars) by the MUSIC algorithm and by a focusing algorithm (+) MSE, CRB_{θ_1} (2) and asymptotically (w.r.t. b) CRB_{θ_1} (3) of the spatial DOA $\phi_1 = \pi \sin\theta_1$ of one source of centred flat spectrum impinging on a linear array of five sensors, SNR of 20 dB, $T = 160$ (10 sections of 16 frequencies for the focusing algorithm), versus $\frac{Mb}{f_0}$.

the comparison with the focusing algorithm by Wang and Kaveh, Fig. 4 shows that the standard MUSIC algorithm outperforms this algorithm for all bandwidths. Consequently, the definition of a narrowband situation, based on the second eigenvalue of the signal's noise-free spatial covariance being smaller than the noise power as given in [10] (for example, $Mb/f_0 < 0.2$ in this simulation) is much too severe.

7 Conclusion

In this paper we considered the domain of validity of spatial covariance-based narrowband DOA algorithms processing non-narrowband data, with particular attention paid to the standard MUSIC algorithm. By focusing on the cases of one source, order detection and asymptotic (w.r.t. the bandwidth and the number of snapshots) bias and variance were studied. We found that the behaviour of MUSIC estimators strongly depends on the symmetry of the source spectrum w.r.t. its centred value and on the offset of this centred value w.r.t. the frequency f_0. A closed-form expression of the Cramer–Rao bound was given for the DOA of a single source of symmetric spectrum in the case of an arbitrary linear array and we showed that the standard MUSIC algorithm remains efficient with increasing source bandwidth and thus outperforms all broadband algorithms.

Consequently, the narrowband DOA algorithms indeed are robust with respect to signal bandwidth, which certainly explains their popularity in practical conditions. However, questions relating to spatially correlated sources and fair comparisons with broadband algorithms require further investigation.

Bibliography

[1] Besson, O. and Stoica, P. Decoupled estimation of DOA and angular spread for spatially distributed source, *IEEE Trans. on Signal Processing*, **48**, 1872–1882, 2000.

[2] Brillinger, D.R. *Times Series, Data Analysis and Theory.* San Francsico, Holden-Day, 1980.

[3] Cardoso, J.F. and Moulines, E. Asymptotic performance analysis of direction-finding algorithms based on fourth-order cumulants, *IEEE Trans. on Signal Processing*, **43**, 214–224, 1995.

[4] Kay, S.M. *Fundamentals of Statistical Signal Processing*, New York, Prentice-Hall, 1993.

[5] Liavas, A.P., Regalia, P.A. and Delmas, J.P. Blind channel approximation: effective channel order determination, *IEEE Trans. on Signal Processing*, **47**, 3336–3344, 1999.

[6] Ottersten, B., Viberg, M. and Kailath, T. Analysis of subspace fitting and ML techniques for parameter estimation from sensor array data, *IEEE Trans. on Signal Processing*, **40**, 590–599, 1992.

[7] Stoica, P. and Nehorai, A. MUSIC, maximum likelihood, and Cramer–Rao bound, *IEEE Trans. on ASSP*, **37**, 720–741, 1989.

[8] Wang, H. and Kaveh, M. Coherent signal subspace processing for the detection and estimation of angles of arrival of multiple wide-band sources, *IEEE Trans. on ASSP*, **33**, 823–831, 1985.

[9] Xu, W. and Kaveh, M. Analysis of the performance and sensitivity of eigendecomposition-based detectors, *IEEE Trans. on Signal Processing*, **43**, 1413–1426, 1995.

[10] Zatman, M. How narrow is narrowband, *IEE Proc. Radar, Sonar Navig.*, **145**, 85–91, 1998.

A Mathematical Representation and Comparison of Detectors for Wireless Communication using Multiple Antennas

Catherine Z. W. Hassell Sweatman, John S. Thompson, Bernard Mulgrew and Peter M. Grant

Department of Electronics and Electrical Engineering, University of Edinburgh, The King's Buildings, Edinburgh EH9 3JL, UK

Abstract

Detection algorithms for single-user wireless communications in a Rayleigh flat-fading environment, using multiple antennas at both the transmitter and receiver are described and compared, assuming repetition coding. The linear decorrelating detector and minimum mean squared error detector are compared with nonlinear decision feedback detectors based on BLAST. Due to the effects of error propagation in the BLAST-type schemes, the MMSE detector performs best.

1 Introduction

It has been shown that, for the case of single-user wireless communication in a Rayleigh flat-fading environment, extraordinary capacity is available if multiple antennas are used at both the transmitter and receiver. These results [1] assume that the channel is not known at the transmitter but is tracked at the receiver. In this paper, detection algorithms for such systems are described and compared.

Sending the data stream for a single user from multiple transmit antennas ensures multipath transmission. Using multiple antennas at the receiver and appropriate digital signal processing multistream detection techniques, the scattering can be exploited to improve system capacity [1, 2, 3]. Standard multiuser detection techniques [4] may be adapted to multistream detection since these techniques are designed to enhance performance by reducing multiple access interference.

The linear decorrelating (zero-forcing) detector and minimum mean squared error detector (MMSED), borrowed from multiuser detection, are compared with their D-BLAST versions. Comparisons are made using bit error ratio (BER) versus signal-to-noise ratio (SNR) simulation. The effect of error propagation in detectors based on BLAST (Bell Laboratories Layered Space-Time) schemes is investigated. A novel singular value decomposition (SVD) representing the MMSED is given.

Detector comparisons are made assuming repetition coding at the transmitter. Spatial and temporal diversity at the receiver are obtained by transmitting

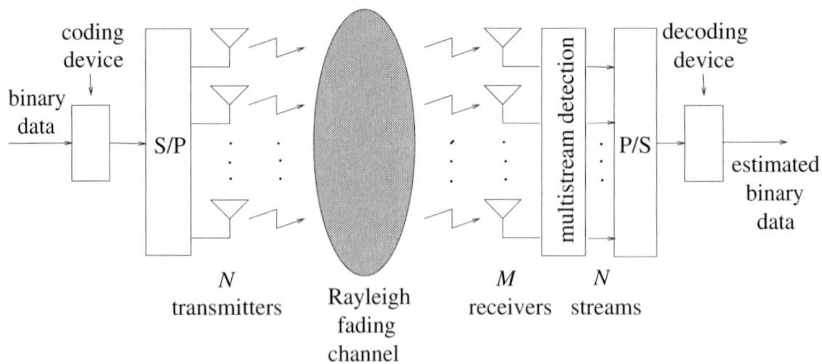

FIG. 1. The (N, M) single-user wireless communication system.

delayed versions of a single stream through multiple antennas (as in a simple D-BLAST system [2]).

2 System model and assumptions

Let N denote the number of transmitting antennas and let M denote the number of receiving antennas. The (N, M) single-user system under consideration is depicted in Fig. 1.

It is assumed that there are a large number of scatterers in the environment and that all the antennas are spaced for uncorrelated Rayleigh fading. The channel model assumes flat fading between each pair of transmitting and receiving antennas. The transmitter does not have knowledge of the channel, but the receiver can track the channel characteristics perfectly. Various modulation schemes could be considered.

Assume that transmissions from the N antennas are synchronous and let T s be the symbol period. Let $\mathcal{C}_{L,K}$ be the set of $L \times K$ matrices over $\mathbb{C}$. Let $\boldsymbol{x}(t_1)$, $\boldsymbol{x}(t_2)$, ... , $\boldsymbol{x}(t_J) \in \mathcal{C}_{N,1}$ correspond to J consecutive transmissions, $J \geq N$. Here $t_j = t_1 + (j-1)T$, $1 \leq j \leq J$. It is assumed that the channel fading is quasi-static. For the time interval under consideration, the combined operation of coding, serial-to-parallel (S/P) conversion and modulation at the transmitter (see Fig. 1) is represented by $\boldsymbol{X} = [\boldsymbol{x}(t_1)\ \boldsymbol{x}(t_2) \ldots \boldsymbol{x}(t_J)] \in \mathcal{C}_{N,J}$ and the discrete-time system equation is

$$\boldsymbol{Y} = \sqrt{\frac{P}{N}}\boldsymbol{H}\boldsymbol{X} + \sigma\boldsymbol{N} \tag{2.1}$$

where

- $\boldsymbol{H} \in \mathcal{C}_{M,N}$ is the channel matrix, components are assumed independent, each one is assumed distributed $\mathbb{C}N(0,1)$;
- the columns of $\boldsymbol{Y} \in \mathcal{C}_{M,J}$ represent the received signal vectors corresponding to transmissions $\boldsymbol{x}(t_1)$, $\boldsymbol{x}(t_2)$, ... , $\boldsymbol{x}(t_J)$;
- $\sigma\boldsymbol{N} \in \mathcal{C}_{M,J}$ represents additive white Gaussian noise (AWGN) at the receiver, each component of $\boldsymbol{N}$ is assumed distributed $\mathbb{C}N(0,1)$, σ^2 is the

noise variance and

- the total transmitted signal power at time t_j is $\frac{P}{N} \parallel \boldsymbol{x}(t_j) \parallel^2$, $1 \leq j \leq J$.

Denote by x_k the kth component of $\boldsymbol{x}$, $1 \leq k \leq N$. The average transmitted power is $P\rho$ where

$$\rho = \mathcal{E}[x_k(t)x_k^*(t)] \tag{2.2}$$

depends only upon the chosen modulation scheme. The quantity ρ represents the number of bits per symbol (so $\rho = 1$ for BPSK and $\rho = 3$ for 8-PSK).

In this paper it is assumed that the binary input data are independent and identically distributed and no encoding is implemented before the S/P conversion. The S/P conversion comprises repetition coding on a space-time diagonal, a form of delay diversity [5]. Explicitly, let the serial symbol stream to be transmitted by the first antenna be represented by $s_1, s_2, \ldots, s_Q$, $Q \geq 2N - 1$; where s_q is a constellation point from the signal space diagram corresponding to the modulation scheme used, $1 \leq q \leq Q$. Assume that the symbol stream is sent by the kth transmit antenna with delay T_k s, $1 \leq k \leq N$, and that delay $T_k = (k-1)T$, $1 \leq k \leq N$, so that, for example if $J = Q + (N-1)$ and $N = 3$, then

$$\boldsymbol{X} = \begin{pmatrix} s_1 & s_2 & s_3 & s_4 & s_5 & \ldots & s_Q & 0 & 0 \\ 0 & s_1 & s_2 & s_3 & s_4 & \ldots & s_{Q-1} & s_Q & 0 \\ 0 & 0 & s_1 & s_2 & s_3 & \ldots & s_{Q-2} & s_{Q-1} & s_Q \end{pmatrix}.$$

This is a simple repetition coding scheme of rate $1/N$, it ensures redundancy over both space and time. In the simulations performed for this paper, only the symbol s_N (s_3 in the example above) is estimated and demodulated and the others are treated as interference. This process is repeated 10^5 times to estimate BER performance.

3 Detection algorithms for (N, M) systems

Let the detection at the receiver of multiple transmitted data streams (from any number of users) be denoted multistream detection. Standard multiuser detectors [4] such as the linear decorrelating detector (LDD) [6, 7] and the MMSED [8, 9] may be applied to the problem of multistream detection. More recently proposed for single-user detection are the D-BLAST [2] versions of these detectors. The detection algorithms for single-user (N, M) systems using repetition coding studied in this paper are described below. Firstly, a multistream detector (see Fig. 1) is applied to each received signal vector to produce an estimate of the corresponding transmitted signal vector. This process is implemented multiple times to produce a soft estimate of the N transmitted symbol streams. Next, a parallel-to-serial (P/S) conversion is performed to produce a soft estimate of the serial symbol stream transmitted from the first antenna. Note that each symbol is sent once by each transmitting antenna, on a space-time diagonal. The N soft estimates for one transmitted symbol from the multistream detector are combined using maximal ratio combining (MRC) [10]. Then a hard estimate for the transmitted symbol is made.

The rest of this section is devoted to a description of the LDD, the MMSED and detectors based on the D-BLAST algorithm. Since each component of $\boldsymbol{H}$ is assumed distributed $\mathbb{C}N(0,1)$, $\boldsymbol{H}$ has full rank. For the LDD (Section 3.1) and the D-BLAST algorithm (Section 3.3), the restriction $N \leq M$ is adopted, since in this case $N-1$ linearly independent received signal vectors may be successfully nulled. In practice, since the D-BLAST algorithm includes the subtraction of estimated signals, it is necessary to consider channel matrices of rank less than or equal to N. In the following descriptions, dependence on time is often omitted for clarity in presentation.

3.1 Linear decorrelating detector

The LDD is represented by a matrix $\boldsymbol{C} \in \mathcal{C}_{N,M}$, namely

$$\boldsymbol{C}_{\mathrm{LD}} = \left(\sqrt{\frac{P}{N}}\boldsymbol{H}\right)^{+}, \tag{3.1}$$

where $^+$ denotes the Moore–Penrose inverse [11]. If $N \leq M$ and $\boldsymbol{H}$ has rank N, then $\boldsymbol{C}_{\mathrm{LD}}\left(\sqrt{\frac{P}{N}}\boldsymbol{H}\right) = \boldsymbol{I}_N$, the identity matrix of size N. The kth row of $\boldsymbol{C}_{\mathrm{LD}}$ nulls out (is orthogonal to) the received signals from all but the kth transmitting antenna, $1 \leq k \leq N$. Hence it may null out some of the received signal from the kth transmitting antenna too. This detector takes no account of the noise variance.

3.2 Minimum mean squared error detector

The MMSED is represented by a matrix $\boldsymbol{C} \in \mathcal{C}_{N,M}$ which minimizes

$$\Sigma_{k=1}^{N}\mathcal{E}[\| (\boldsymbol{C}\boldsymbol{y} - \boldsymbol{x})_k \|^2]. \tag{3.2}$$

The MMSED is expected to perform better than the LDD since it takes the noise variance into account; it is obtained by adjusting the LDD weights to reduce the effects of noise at the expense of inter-antenna interference, finding the optimal compromise.

3.2.1 *Matrix representations of the MMSED*

Consider signals and systems as described in Section 2 for which the statistics of the binary input data and the combined operation of coding, serial-to-parallel conversion, and modulation (see Fig. 1) are such that, at any chosen time t,

$$\mathcal{E}[x_i(t)x_k^*(t)] = 0 \tag{3.3}$$

if $i \neq k$, $1 \leq i,\ k \leq N$. For example, it might be that the binary input data are independent and identically distributed and that delayed streams are sent to each transmitting antenna in turn.

Lemma 3.1 *If $\sigma^2 > 0$ then the MMSED may be represented by*

$$\boldsymbol{C}_{MMSE} = \sqrt{\frac{P}{N}}\boldsymbol{H}^H\Big[\frac{P}{N}\boldsymbol{H}\boldsymbol{H}^H + \frac{\sigma^2}{\rho}\boldsymbol{I}_M\Big]^{-1}. \tag{3.4}$$

Corollary 3.2 *If $\sigma^2 > 0$ and if $\sqrt{\frac{P}{N}}\boldsymbol{H} = \boldsymbol{U}\boldsymbol{D}\boldsymbol{V}^H$ is a SVD, where unitary $\boldsymbol{U} \in \mathcal{C}_{M,M}$, unitary $\boldsymbol{V} \in \mathcal{C}_{N,N}$ and matrix $\boldsymbol{D} \in \mathcal{C}_{M,N}$ has the form*

$$\boldsymbol{D} = \begin{pmatrix} \boldsymbol{S} & \boldsymbol{0} \\ \boldsymbol{0} & \boldsymbol{0} \end{pmatrix}$$

where

$$\boldsymbol{S} = \begin{pmatrix} \lambda_1 & 0 & \dots & 0 \\ 0 & \lambda_2 & \dots & 0 \\ \vdots & \vdots & \ddots & \vdots \\ 0 & 0 & \dots & \lambda_r \end{pmatrix},$$

$r \geq 1$ is the rank of $\boldsymbol{H}$ and $\{\lambda_i\}_{i=1}^r$ are the non-zero singular values of $\sqrt{\frac{P}{N}}\boldsymbol{H}$ then

$$\boldsymbol{C}_{MMSE} = \boldsymbol{V}\begin{pmatrix} \frac{\alpha_1}{\sqrt{P}({\alpha_1}^2+\frac{\sigma^2}{P\rho})} & 0 & \dots & 0 & \\ 0 & \frac{\alpha_2}{\sqrt{P}({\alpha_2}^2+\frac{\sigma^2}{P\rho})} & \dots & 0 & \\ & & & & \boldsymbol{0} \\ \vdots & \vdots & \ddots & \vdots & \\ 0 & 0 & \dots & \frac{\alpha_r}{\sqrt{P}({\alpha_r}^2+\frac{\sigma^2}{P\rho})} & \\ & & & & \\ & \boldsymbol{0} & & & \boldsymbol{0} \end{pmatrix}\boldsymbol{U}^H, \tag{3.5}$$

where $\alpha_i = \frac{\lambda_i}{\sqrt{P}}$, $1 \leq i \leq r$, are the non-zero singular values of $\frac{\boldsymbol{H}}{\sqrt{N}}$.

In a sense, by considering $\frac{\boldsymbol{H}}{\sqrt{N}}$, one has normalized the channel matrix since, if the components of $\boldsymbol{H}$ are independent and identically distributed ($\mathbb{C}N(0,1)$), then $\mathcal{E}[\frac{\boldsymbol{H}\boldsymbol{H}^H}{N}] = \boldsymbol{I}_M$. The quantity $\frac{P\rho}{\sigma^2}$ is the received symbol SNR for a $(1,1)$ system. Expression (3.5) is a SVD for $\boldsymbol{C}_{\text{MMSE}}$ and hence provides an alternative means of computing an estimate for the matrix representing the MMSED (if ρ is known and P, σ^2 and $\boldsymbol{H}$ can be estimated). Expression (3.5) shows the effect of P and $\frac{P\rho}{\sigma^2}$ on the singular values of $\boldsymbol{C}_{\text{MMSE}}$. As the noise level ($\sigma^2$) increases, the singular values of $\boldsymbol{C}_{\text{MMSE}}$ decrease. From (3.4) or (3.5), it follows that $\lim_{\sigma^2\to\infty}\boldsymbol{C}_{\text{MMSE}}$ is the zero matrix.

Lemma 3.1 and Corollary 3.2 are proved in the Appendix. It follows from Corollary 3.2 that $\lim_{\sigma^2\to 0}\boldsymbol{C}_{\text{MMSE}} = \boldsymbol{C}_{\text{LD}} = \left(\sqrt{\frac{P}{N}}\boldsymbol{H}\right)^+$.

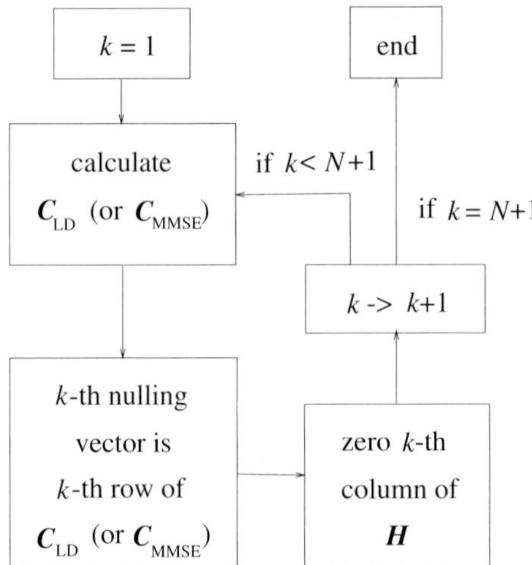

FIG. 2. Calculation of DFD nulling vectors.

3.3 D-BLAST

The D-BLAST detector for (N, M) systems, $N \leq M$, was proposed by Foschini [2]. A simple version of this decision feedback detector (DFD) is described here. At each time step, a sum of N M-dimensional signals is received, one from each transmitting antenna. The N transmitted symbols are estimated one by one. Since the receiver has knowledge of the channel, estimates of transmitted symbols provide estimates of received signals. Received M-dimensional signals not yet estimated are nulled after those previously estimated are subtracted from the sum.

The nulling vectors may be chosen according to various criteria (see Fig. 2). In this paper, both zero-forcing (denoted as DZF-DFD) and minimum mean squared error criteria (denoted as DMMSE-DFD) are considered (see (3.1) and (3.4)). The restriction $N \leq M$ is adopted since, in this case, $\boldsymbol{H}$ has rank N with probability one and $N - 1$ linearly independent received signal vectors may be successfully nulled.

Since the detector incorporates the subtraction of estimated signals, it is affected by error propagation.

4 Derivation of SNR formulae

In this section, expressions for the received symbol SNR at the point of making the decision are derived for the LDD and the DZF-DFD (assuming perfect cancellation by subtraction) in terms of P, σ^2, ρ, N and M. Using the notation of Section 2, setting $Q = 2N - 1$, for each channel ($\boldsymbol{H}$) simulated, one symbol stream $s_1, s_2, \ldots, s_{2N-1}$ was simulated and one symbol (s_N) was estimated. In practice, all nulling vectors were normalized to unit length (to preserve the noise power). For any matrix $\boldsymbol{A}$, with $i_{\boldsymbol{A}}$ columns, let $\boldsymbol{A}_i$ denote the ith column of $\boldsymbol{A}$,

$1 \le i \le i_{\boldsymbol{A}}$. For a given $\boldsymbol{H}$, for each detector, if the normalized kth nulling vector is $\boldsymbol{w}_k^H$, then the kth MRC coefficient is $\alpha_k{}^*$ where $\alpha_k = \boldsymbol{w}_k^H \boldsymbol{H}_k$, $1 \le k \le N$, and hence a soft estimate for the symbol s_N (after MRC and up to scaling by the real positive term $\sqrt{\frac{P}{N}}(\Sigma_{k=1}^N \parallel \alpha_k \parallel^2)$) is

$$\begin{aligned} \hat{s_N} &= \Sigma_{k=1}^N \alpha_k{}^* \boldsymbol{w}_k^H \boldsymbol{Y}_{(k+N-1)} \\ &= \Sigma_{k=1}^N \alpha_k{}^* \boldsymbol{w}_k^H \left[\sqrt{\frac{P}{N}} \boldsymbol{H}_k \boldsymbol{X}_{k,(k+N-1)} + \sigma \boldsymbol{N}_{(k+N-1)} \right]. \end{aligned}$$

Hence, after MRC, the received signal power (averaged over the constellation set) is $\frac{P}{N}\rho(\Sigma_{k=1}^N \parallel \alpha_k \parallel^2)^2$ and the received noise power is $\sigma^2(\Sigma_{k=1}^N \parallel \alpha_k \parallel^2)$ and so, for the channel sampled, the received SNR at the point of making the decision is $\frac{P\rho}{\sigma^2 N}(\Sigma_{k=1}^N \parallel \alpha_k \parallel^2)$.

Recall that $\boldsymbol{H}$ is an $M \times N$ matrix, and each component is assumed distributed $\mathbb{C}N(0,1)$, so that $\mathcal{E}[\parallel \boldsymbol{H}_k \parallel^2] = M$, $1 \le k \le N$. When using the LDD, one expects to retain fraction $\frac{M-(N-1)}{M}$ of signal power at each of the N nulling steps, since at each step $N-1$ interfering signals (in $\mathbb{C}^M$) must be nulled. Hence, $\mathcal{E}[\parallel \alpha_k \parallel^2] = M[\frac{M-(N-1)}{M}] = [M-(N-1)]$. When using the DZF-DFD, one expects to retain fraction $\frac{M-(N-k)}{M}$ of signal power at the kth nulling step, since at this step $N-k$ interfering signals must be nulled, $1 \le k \le N$. Assuming perfect cancellation by subtraction, $\mathcal{E}[\parallel \alpha_k \parallel^2] = [M-(N-k)]$. Hence, averaging over channels, the received symbol SNR at the point of making the decision is equal to

$$\frac{P\rho}{\sigma^2}[M-(N-1)] \tag{4.1}$$

for the LDD and

$$\frac{P\rho}{\sigma^2}\left[M - \frac{(N-1)}{2}\right] \tag{4.2}$$

for the DZF-DFD. More detailed statistics may be found in [2].

Hence, for (N, M) systems, $M \ge N \ge 2$, incorporating zero-forcing into a DFD scheme should improve the error rate. Since the MMSED generally performs better than zero-forcing, incorporating it into a DFD scheme should further improve the error rate. However, in practice, the error rate suggested by (4.2) must be regarded as a lower bound since the DFD schemes are affected by error propagation.

5 Simulation parameters and results

Bit error ratio versus SNR simulations were carried out in order to compare detector performances and to investigate the effect of error propagation (EP) in the DFD schemes (see Figs 3–5). The curves labelled NEP (no error propagation) were obtained by implementing perfect cancellation by subtraction. The

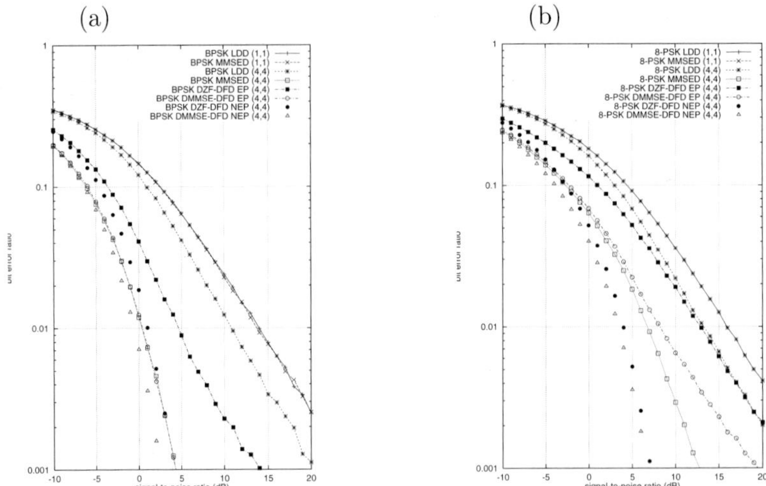

FIG. 3. Detector comparison, $N = 4$, $M = 4$; (a) BPSK and (b) 8-PSK modulation.

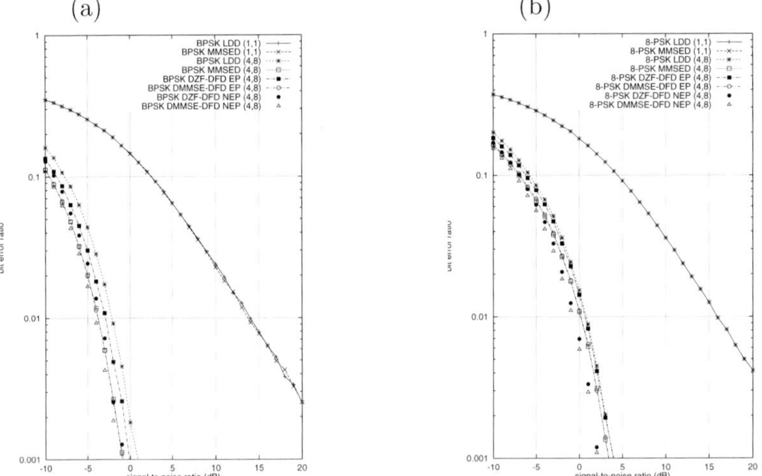

FIG. 4. Detector comparison, $N = 4$, $M = 8$; (a) BPSK and (b) 8-PSK modulation.

modulation schemes considered were BPSK and 8-PSK (with Gray encoding). Signal-to-noise ratios between -10 dB and 20 dB were considered, where a SNR of ν dB corresponds to $P/\sigma^2 = 10^{\nu/10}$. For each point plotted, 10^5 channels were simulated. Included for reference are the $(1, 1)$ curves.

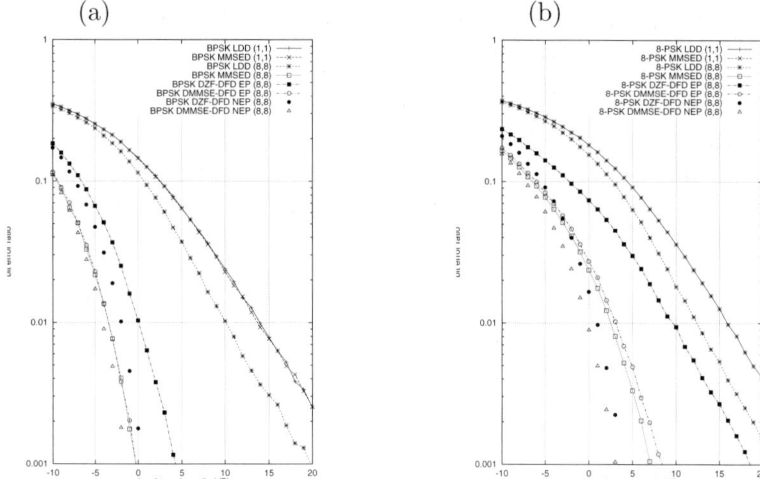

FIG. 5. Detector comparison, $N = 8$, $M = 8$; (a) BPSK and (b) 8-PSK modulation.

6 Observations and discussion

6.1 Detector comparison for BPSK modulation

For BPSK modulation, the MMSED and the DMMSE-DFD EP (with error propagation) version perform best and yield almost indistinguishable BER curves over the range of SNRs considered (Figs 3–5). The DZF-DFD EP performs better than the LDD. For example, for a $(4, 4)$ system, to achieve a BER equal to 0.01, the MMSED and the DMMSE-DFD EP version require a SNR of approximately 0.4 dB, the DZF-DFD EP requires a SNR of approximately 4.7 dB, and the LDD requires a SNR of approximately 10.8 dB.

6.2 Detector comparison for 8-PSK modulation

For 8-PSK modulation, over the range of SNRs considered, the MMSED and the DMMSE-DFD EP version perform best. For the $(4, 4)$ and $(8, 8)$ systems, the BER versus SNR curves for these detectors diverge with increasing SNR due to the effect of EP in the DFD technique. The LDD performs worst at the lower SNRs. However, the BER versus SNR curves for LDD and the DZF-DFD EP version converge with increasing SNR due to the effect of EP in the DFD algorithm.

6.3 General observations

In all the cases examined, adding antennas at the receiver lowers BER and reduces the differences in detector performance (Figs 3 and 4).

For (N, N) systems, detector performance improves with increasing N: marginally so for the LDD, markedly so for the MMSED. Hence, for (N, N) systems, detector performance diverges with increasing N (Figs 3 and 5).

For any (N, M) system, at sufficiently high SNR, the MMSED will converge to the LDD and so the detector ranking may change.

6.4 The effect of error propagation

The effect of EP in all the DFD schemes considered is non-negligible, especially at higher SNR. The DZF-DFD suffers more than the DMMSE-DFD. For a fixed modulation scheme, at a fixed BER, for example 0.01, the effect becomes less significant for (N, M) systems as M increases (Figs 3 and 4) and less significant for (N, N) systems as N increases (Figs 3 and 5), especially for the LDD and DZF-DFD. For fixed N and M, the effect becomes more significant as the signal space increases; that is, for higher-order modulation schemes.

7 Conclusions

Detection algorithms for single-user wireless communication using multiple antennas at both the transmitter and receiver in a Rayleigh (flat) fading environment, in the presence of AWGN, are compared. It is assumed that the binary input data are independent and identically distributed and no encoding is implemented before a S/P conversion. The S/P conversion comprises repetition coding on a space-time diagonal, a form of delay diversity. The modulation schemes considered are BPSK and 8-PSK (with Gray encoding).

The LDD and the MMSED are compared with their DFD versions using BER versus SNR simulation. The effect of EP in the DFD schemes is studied.

For (N, M) systems, $M \geq N \geq 2$, incorporating zero-forcing into a DFD scheme improves performance in the absence of EP. Since the MMSED generally performs better than zero-forcing, incorporating it into a DFD scheme further improves performance. However, the effect of EP in all the DFD schemes considered is non-negligible, especially at higher SNRs. Taking EP into account, when considering the LDD, the MMSED and their DFD versions, together with repetition coding, the MMSED appears to be the best practical choice since, in all the cases examined, it performed best or nearly best, it does not suffer from EP, and it is the simplest of all the detectors considered to implement (if training data are available).

Acknowledgements

Catherine Z. W. Hassell Sweatman and John S. Thompson gratefully acknowledge support from the EPSRC and Nortel Networks.

Appendix: Proofs for Section 3.2

Lemma A.1 *For signals and systems as described in Section 2 for which (3.3) holds, the expectation (3.2) can be written as a trace, namely*

$$\rho \, Tr \left[\left(\boldsymbol{I}_N - \boldsymbol{C}\sqrt{\frac{P}{N}}\boldsymbol{H} \right) \left(\boldsymbol{I}_N - \boldsymbol{C}\sqrt{\frac{P}{N}}\boldsymbol{H} \right)^H + \frac{\sigma^2}{\rho}\boldsymbol{C}\boldsymbol{C}^H \right]. \tag{A.1}$$

The proof of Lemma A.1 is obtained by straightforward algebra. Lemmas A.1 and A.2 yield a closed form expression for $\boldsymbol{C}_{\text{MMSE}}$ following a method suggested in [8] for a similar real-valued problem.

Lemma A.2 *For any $\boldsymbol{W} \in \mathcal{C}_{N,M}$, and any positive definite $\boldsymbol{Q} \in \mathcal{C}_{M,M}$, let $f : \mathcal{C}_{N,M} \to \mathbb{R}$ be defined by*

$$f(\boldsymbol{C}) = \operatorname{Tr}[\boldsymbol{C}\boldsymbol{Q}\boldsymbol{C}^H - \boldsymbol{W}\boldsymbol{C}^H - \boldsymbol{C}\boldsymbol{W}^H]. \tag{A.2}$$

Then, $\boldsymbol{C} = \boldsymbol{W}\boldsymbol{Q}^{-1}$ minimizes (A.2).

Proof of Lemma A.2 For any $\boldsymbol{Z} \in \mathcal{C}_{N,M}$, consider $f(\boldsymbol{W}\boldsymbol{Q}^{-1} + \boldsymbol{Z})$. Now

$$f(\boldsymbol{W}\boldsymbol{Q}^{-1} + \boldsymbol{Z}) = f(\boldsymbol{W}\boldsymbol{Q}^{-1}) + \operatorname{Tr}[\boldsymbol{Z}\boldsymbol{Q}(\boldsymbol{Q}^{-1})^H\boldsymbol{W}^H - \boldsymbol{Z}\boldsymbol{W}^H + \boldsymbol{Z}\boldsymbol{Q}\boldsymbol{Z}^H].$$

Since $\boldsymbol{Q}$ is positive definite, $\boldsymbol{Q}$ is Hermitian and so $(\boldsymbol{Q}^{-1})^H = (\boldsymbol{Q})^{-1}$. Hence

$$f(\boldsymbol{W}\boldsymbol{Q}^{-1} + \boldsymbol{Z}) = f(\boldsymbol{W}\boldsymbol{Q}^{-1}) + \operatorname{Tr}[\boldsymbol{Z}\boldsymbol{Q}\boldsymbol{Z}^H].$$

Since $\boldsymbol{Q}$ is positive definite, $\boldsymbol{Z}\boldsymbol{Q}\boldsymbol{Z}^H$ is non-negative definite, and so

$$f(\boldsymbol{W}\boldsymbol{Q}^{-1} + \boldsymbol{Z}) \geq f(\boldsymbol{W}\boldsymbol{Q}^{-1}).$$

Hence, $\boldsymbol{C} = \boldsymbol{W}\boldsymbol{Q}^{-1}$ minimizes $\operatorname{Tr}[\boldsymbol{C}\boldsymbol{Q}\boldsymbol{C}^H - \boldsymbol{W}\boldsymbol{C}^H - \boldsymbol{C}\boldsymbol{W}^H]$.

Proof of Lemma 3.1 and Corollary 3.2 By Lemma A.1, given N, M, P, $\boldsymbol{H} \in \mathcal{C}_{M,N}$, $\sigma^2 > 0$ and a fixed modulation scheme (and hence ρ), the MMSED is represented by a matrix $\boldsymbol{C} \in \mathcal{C}_{N,M}$ which minimizes (A.1). Hence, the MMSED is represented by a matrix $\boldsymbol{C}$ which minimizes the function $f : \mathcal{C}_{N,M} \to \mathbb{R}$ defined by

$$\begin{aligned} f(\boldsymbol{C}) &= \operatorname{Tr}\left[\left(\boldsymbol{I}_N - \boldsymbol{C}\sqrt{\frac{P}{N}}\boldsymbol{H}\right)\left(\boldsymbol{I}_N - \boldsymbol{C}\sqrt{\frac{P}{N}}\boldsymbol{H}\right)^H + \frac{\sigma^2}{\rho}\boldsymbol{C}\boldsymbol{C}^H - \boldsymbol{I}_N\right] \\ &= \operatorname{Tr}\left[-\boldsymbol{C}\sqrt{\frac{P}{N}}\boldsymbol{H} - \sqrt{\frac{P}{N}}\boldsymbol{H}^H\boldsymbol{C}^H + \frac{P}{N}\boldsymbol{C}\boldsymbol{H}\boldsymbol{H}^H\boldsymbol{C}^H + \frac{\sigma^2}{\rho}\boldsymbol{C}\boldsymbol{C}^H\right]. \end{aligned}$$

Set $\boldsymbol{W} = \sqrt{\frac{P}{N}}\boldsymbol{H}^H$ and $\boldsymbol{Q} = \frac{P}{N}\boldsymbol{H}\boldsymbol{H}^H + \frac{\sigma^2}{\rho}\boldsymbol{I}_M$. Since $\sigma^2 > 0$, $\boldsymbol{Q}$ is positive definite and the proof of Lemma 3.1 is completed by applying Lemma A.2.

Substituting the SVD $\sqrt{\frac{P}{N}}\boldsymbol{H} = \boldsymbol{U}\boldsymbol{D}\boldsymbol{V}^H$ into (3.4) one obtains

$$\boldsymbol{C}_{\text{MMSE}} = \boldsymbol{V}\boldsymbol{D}\boldsymbol{U}^H\left[\boldsymbol{U}\begin{pmatrix} \boldsymbol{S}^2 + \frac{\sigma^2}{\rho}\boldsymbol{I}_r & \boldsymbol{0} \\ \boldsymbol{0} & \frac{\sigma^2}{\rho}\boldsymbol{I}_{M-r} \end{pmatrix}\boldsymbol{U}^H\right]^{-1}$$

$$= \boldsymbol{V} \left(\begin{array}{ccccc} \frac{\lambda_1}{{\lambda_1}^2+\frac{\sigma^2}{\rho}} & 0 & \cdots & 0 & \\ 0 & \frac{\lambda_2}{{\lambda_2}^2+\frac{\sigma^2}{\rho}} & \cdots & 0 & \\ & & & & \mathbf{0} \\ \vdots & \vdots & \ddots & \vdots & \\ 0 & 0 & \cdots & \frac{\lambda_r}{{\lambda_r}^2+\frac{\sigma^2}{\rho}} & \\ & & & & \\ & \mathbf{0} & & & \mathbf{0} \end{array} \right) \boldsymbol{U}^H.$$

The proof of Corollary 3.2 is completed by substituting $\alpha_i = \frac{\lambda_i}{\sqrt{P}}$, $1 \leq i \leq r$.

Bibliography

[1] Foschini, G.J. and Gans, M.J. (1998). On limits of wireless communication in a fading environment when using multiple antennas. *Wireless Personal Comm.*, **6**, 311–335.

[2] Foschini, G.J. (1996). Layered space-time architecture for wireless communication in a fading environment when using multi-element antennas. *Bell Labs Technical J., Autumn 1996*, 41–59.

[3] Golden, G.D., Foschini, G.J., Valenzuela, R.A. and Wolniansky, P.W. (1999). Detection algorithm and initial laboratory results using V-BLAST space-time communication architecture. *Electron. Lett.*, **35**, 14–16.

[4] Moshavi, S. (1996). Multi-user detection for DS-CDMA communications. *IEEE Personal Commun.*, **34**, 124–135.

[5] Tarokh, V., Seshadri, N. and Calderbank, A.R. (1998). Space-time codes for high data rate wireless communication: performance criterion and code construction. *IEEE Trans. on Inform. Theory*, **44**, 744–765.

[6] Schneider, K.S. (1979). Optimum detection of code division multiplexed signals. *IEEE Trans. Aerospace Electron. Sysems*, **15**, 181–185.

[7] Kohno, R., Imai, H. and Hatori, M. (1983). Cancellation techniques of co-channel interference in asynchronous spread sprectrum multiple access systems. *Trans. Electron. and Commun. in Japan*, **66**, 416–423.

[8] Honig, M., Madhow, U. and Verdu S. (1995). Blind adaptive multiuser detection. *IEEE Trans. Inform. Theory*, **41**, 994–960.

[9] Xie, Z., Short, R.T. and Rushforth, C.K. (1990). A family of suboptimum detectors for coherent multi-user communications. *IEEE J. Select. Areas Commun.*, **8**, 683–690.

[10] Proakis, J.G. (1995). *Digital Communications, Third ed.*, McGraw-Hill, New York.

[11] Lancaster, P. and Tismenetsky, M. (1985). *The Theory of Matrices, Second Edition with Applications*, Academic, New York.

Likelihood Ratio Methods for Underwater Acoustics Signal Detection

C. R. Baker
Department of Statistics, University of North Carolina, Chapel Hill, NC 27599-3260, USA

A. Climescu-Haulica
Communications Research Center Canada, 3701 Carling Ave.,Ottawa, Ontario K2H 8S2

A. F. Gualtierotti
IDHEAP, 21 Route de la Maldière, CH-1022 Chavannes-près-Lausanne

1 Introduction

Underwater acoustics is about sonars, that is, techniques that use waves of mechanical vibration to transmit and receive information, under water, for such waves are the ones that propagate most easily in that environment. One would thus expect that detection would transit through the wave equation whose prototype is given by $Dp = dN$, where p is pressure, N is noise, and D is an operator of the form

$$D = \Delta - \frac{1}{c^2}\frac{\partial^2}{\partial t^2} + \frac{b}{c^2}\frac{\partial}{\partial t}\Delta,$$

where $b, c > 0$, and $\Delta = \frac{\partial^2}{\partial x^2} + \frac{\partial^2}{\partial y^2} + \frac{\partial^2}{\partial z^2}$. b is viscosity, and c, velocity.

One would solve the equation and compute the law of the process. There would be a process for noise only, and a process for the signal obscured by noise. Computation of the likelihood[1] would follow. The crucial difficulty is that b and c are themselves stochastic fields and that "determination of the probability distributions of the wave fields in a stochastic medium (also in a weakly inhomogeneous stochastic medium) is impossible [18]page 152."

A second approach that has been attempted could be defined as phenomenological in that an acoustic field is defined by a linear superposition of individual fields randomly produced by similar sources [15]. The randomness is described by various analytically tractable processes. Unfortunately such methods produce explicit laws only for the first order, implying reliance on independent observa-

[1] To shorten expressions, the term likelihood will be used, rather than "likelihood ratio," which is the correct expression, as one uses a ratio of likelihhods.

tions and consequently loss of information. They furthermore require estimation of parameters for which there seldom will be adequate information.

A third approach is basically statistical and it is that which is pursued here. It consists in positing a very general signal-and-noise model, then deriving a generic likelihood function, since the likelihood is the best detector in most practical instances [10], and then discretizing the likelihood to fit it to data. It should be noted that discretizing must follow derivation of the likelihood, as the law of the signal-in-noise process is typically unknown.

Indeed (see [12]Section 1.1),

> " ... in the operation of a sonar system the operator is repeatedly faced with the problem of detecting a signal which is obscured by noise. This signal may be an echo resulting from a transmitted signal over which the operator has some control, or it may have its origin in some external source. These two modes of operation are commonly distinguished as active and passive sonar. Similar situations arise in radar surveillance and in seismic exploration ... Signals come in all shapes and forms. In active sonar systems one may use simple sinusoidal signals of fixed duration and modulations thereof. There are impulsive signals such as those made with explosions or thumpers. At the other extreme one may make use of pseudorandom signals. In passive systems, the signals whose detection are sought may be noise in the conventional meaning of the word; noise produced by propellers or underwater swimmers, for example. ...
>
> ... When echoes are produced by extended targets such as submarines, there are two distinct ways in which the echo structure is affected. First, there is the interference between reflections from the different structural features on the hull of the submarine. This interference leads to a target strength that fluctuates rapidly with changes in the aspect. Secondly, there is the elongation of the composite echo due to the distribution of reflecting features along the submarine. ... A final source of pulse distortion is the Doppler shifts produced by the relative motions between the source, the medium, and the targets. Since the source, the medium and the target (or detector in passive listening) may each have a different vector velocity relative to the bottom, the variety of effects may be quite large."

The latter description shows that the signal can only be given, *a priori,* a broadly qualitative description. So it will be assumed that the noise N is a mean zero, continuous in quadratic mean process (it has thus paths of finite energy) whose law will be successively Gaussian, spherically invariant, and the convolution of a Gaussian process with a causally filtered Poisson process. The signal S will have paths in the reproducing kernel Hilbert space (RKHS) of the noise, which makes it quite smoother, and the observed waveform when a signal is received will be $S + N$. It can be shown that, in the Gaussian case at least, such an additive representation is not a restriction [2].

Previous work on this topic has mostly been a (difficult) mathematical travail with models whose bearing on really applied problems has been somewhat tenuous. A recent survey on what is available can be found in [14]. One will find below a general method to obtain likelihood ratios for fairly general and realistic noise models.

2 The detection problem

There are four operations that one must perform successfully to be able to claim that a likelihood detection problem is solved. To enumerate them, let P_N and P_{S+N} be, respectively, the probabilities induced on $L_2\,[0,T]$ by N and $S+N$. $[0,T]$ is the time span of observation for detection. One must then firstly ascertain that the likelihood exists for the problem at hand, which is not always the case as Slepian [17] noticed already in 1958! Technically, one must check that P_{S+N} is absolutely continuous with respect to P_N.

Secondly, when the likelihood exists, one must produce a functional Λ that is defined, and thus computable without knowing which of the P_N or P_{S+N} regimes obtains, for every observable noise or signal in noise path, and for which one has that

$$P_{S+N}\,(A) = \int_A \Lambda\,(f)\,P_N\,(df)\,.$$

The functional Λ being available, one must thirdly be able to solve for Λ_0, for every predifined probability of false alarm $\alpha \in \,]0,1[\,$, the equation ($f \in \mathcal{L}_2\,[0,T]$ means that $\int f^2 < \infty$)

$$\alpha = P_N\,(f \in \mathcal{L}_2\,[0,T] :\ \Lambda\,(f) > \Lambda_0)\,,$$

and then compute the probability of detection $1-\beta$ where

$$\beta = P_{S+N}\,(f \in \mathcal{L}_2\,[0,T] :\ \Lambda\,(f) \leq \Lambda_0)\,.$$

Fourthly, given observations of f at times $t_1 < \cdots < t_n$, one must be able to obtain approximations Λ_n to Λ and $\Lambda_0^{(n)}$ to Λ_0 such that, simultaneously,

$$\begin{aligned} P_N\left(f \in \mathcal{L}_2\,[0,T] :\ \Lambda_n\,(f\,(t_1)\,,\ldots,f\,(t_n)) > \Lambda_0^{(n)}\right) &\approx \alpha \\ P_{S+N}\left(f \in \mathcal{L}_2\,[0,T] :\ \Lambda_n\,(f\,(t_1)\,,\ldots,f\,(t_n)) > \Lambda_0^{(n)}\right) &\approx 1-\beta \end{aligned}$$

3 Likelihoods

One will find below a fairly general method to derive likelihoods as well as the results of its application to three different noise cases. The idea is to call upon the Cramer–Hida representation of second order processes [9, 11] and to interpret it as a path transformation.

3.1 The Cramer–Hida representation and its use in detection

The Cramer–Hida representation says that, in $L_2[P]$, with P the underlying overall probability,

$$N(\cdot,t)=\sum_{i=1}^{M}\int_0^t F_i(t,x)\,B_i(\cdot,dx).$$

M is the multiplicity of the representation: it can be finite or infinite. The processes $B_i,\ 1\le i\le M$, are mutually orthogonal processes, with orthogonal increments, mean zero and variance $\beta_i,\ 1\le i\le M$. The functions $F_i,\ 1\le i\le M$, are causal in that $F_i(t,x)=0,\ x>t$, and, for $t\in[0,T]$, fixed, but arbitrary, $F_i(t,\cdot)$ is in $L_2[\beta_i],\ 1\le i\le M$. The integrals are mean-square limits [19]. The representation can be so chosen that the L_2 spaces generated by the noise and its representation are identical: the representation is then said to be proper canonical. Finally, when N is Gaussian, the processes $B_i,\ 1\le i\le M$, are also Gaussian, and thus have independent increments and continuous paths.

The main insight [2] comes from seeing that the Cramer–Hida representation can be interpreted as a functional relating the paths of N to those of $\underline{B}$, the vector with components $B_i,\ 1\le i\le M$:

$$N=\Phi(\underline{B}).$$

Indeed, as S has paths in the RKHS of N, it follows [2] that

$$S(\cdot,t)=\sum_{i=1}^{M}\int_0^t F_i(t,x)\,s_i(\cdot,x)\,\beta_i(dx),$$

and consequently that

$$(S+N)(\cdot,t)=\sum_{i=1}^{M}\int_0^t F_i(t,x)\left[s_i(\cdot,x)\,\beta_i(dx)+B_i(\cdot,dx)\right].$$

If one writes

$$Z_i(\cdot,t)=\int_0^t s_i(\cdot,x)\,\beta_i(dx)+B_i(\cdot,dx),$$

one can in a number of cases prove [6] that $P_{\underline{Z}}$ is absolutely continuous with respect to $P_{\underline{B}}$. $\underline{Z}$ is the vector with components $Z_i,\ 1\le i\le M$, and $P_{\underline{Z}}$ and $P_{\underline{B}}$ are the probabilities induced on the appropriate sample space by respectively $\underline{Z}$ and $\underline{B}$. From there it follows [2] that $S+N=\Phi(\underline{Z})$ and consequently that the computation of the likelihood can be decoupled into two distinct operations, that of computing the likelihood of $P_{\underline{Z}}$ with respect to $P_{\underline{B}}$ and that of computing the conditional law of $\underline{B}$ with respect to N (a point mass):

$$\frac{dP_{S+N}}{dP_N}(f)=\frac{dP_{\underline{Z}}}{dP_{\underline{B}}}(m(f,\cdot)),$$

where m is a process on $\mathcal{L}_2[0,T] \times [0,T]$ whose probability law with respect to P_N is that of $\underline{B}$ with respect to P.

Computation of the likelihood of $P_{\underline{Z}}$ with respect to $P_{\underline{B}}$ is basically a stochastic calculus problem, for which plenty of technical knowhow is available, but its precise expression will depend on the particular law of the noise N that has been chosen for the model. The conditional law will mostly depend on the properties of the covariance operator R_N of N :

$$R_N = \sum_{i=1}^{\infty} \lambda_i e_i \otimes e_i,$$

where the e_i, $1 \leq i < \infty$, are a complete orthonormal set of eigenvetors of R_N, and the λ_i, $1 \leq i < \infty$, are positive, decreasing reals with a finite sum (such operators are said to belong to the trace class [1]).

The different known cases will now be stated.

3.2 The case of Gaussian noise

Let $L_2\left[\boxed{\beta}\right]$ be the direct sum of the Hilbert spaces $L_2[\beta_i]$:

$$L_2\left[\boxed{\beta}\right] = \oplus_{i=1}^{M} L_2[\beta_i].$$

An element of $L_2\left[\boxed{\beta}\right]$ can be represented as $\underline{f}$, a vector of components $f_i \in L_2[\beta_i]$ such that

$$\sum_{i=1}^{M} \int_0^T f_i^2(x)\, \beta_i(dx) < \infty.$$

Let K be the closure of the range of the square root of R_N and U_i be the operator from $L_2[0,T]$ into $L_2[\beta_i]$ that maps $f \in L_2[0,T]$ to the equivalence class in $L_2[\beta_i]$ of the function

$$t \mapsto \int_t^T F_i(x,t)\, f(x)\, dx.$$

U is then the operator from $L_2[0,T]$ to $L_2\left[\boxed{\beta}\right]$ such that the i-th component of Uf, $f \in L_2[0,T]$, is $U_i f$. One has that, $U^\star$ being the transpose of U,

$$R_N = U^\star U.$$

The polar decomposition [16]pages 4-6 says then that $U = JR_N^{\frac{1}{2}}$, where the operator $J : L_2[0,T] \longrightarrow L_2\left[\boxed{\beta}\right]$ is a partial isometry with initial space K and final space the closure L of the range of U. The fact that the decomposition is proper canonical yields that J is onto, so that, when N has full support, J is unitary. The ith component of the process m is then defined as follows:

$$m_i(f,t) = \sum_{k=1}^{M} \frac{1}{\lambda_k} \langle U_i^\star I_{[0,t]}, e_k \rangle_{L_2[0,T]} \langle f, e_k \rangle_{L_2[0,T]}.$$

Finally, with respect to $P_{\underline{B}}$ and $P_{\underline{Z}}$, almost surely,

$$\ln\left[\frac{dP_{\underline{Z}}}{dP_{\underline{B}}}(g)\right] = \int_0^T \langle \underline{s}(g,t), \underline{ev}(g,dt)\rangle - \frac{1}{2}\|\underline{s}(g,\cdot)\|^2_{L_2[\underline{\beta}]}.$$

$\underline{ev}$ is the evaluation map defined by $\underline{ev}(g,t) = g(t)$. It is a semi-martingale with respect to both $P_{\underline{B}}$ and $P_{\underline{Z}}$, and the integral must be understood as the value at T of a martingale as defined in [6].

The assumption that S has paths in the RKHS of N (denoted $H(N)$ below) is enough to insure that the likelihood exists, but to obtain its explicit expression as above, one must restrict the set of admissible signals: assuming for example that

$$E\left[e^{\frac{1}{2}\|S\|^2_{H(N)}}\right] < \infty.$$

3.3 The case of spherically invariant noise

If N is spherically invariant, it has a representation of the form AG [13], where G is a Gaussian process and A is a positive random variable, independent of G. N is thus a variance mixture. If, for instance, A has values 1 and $a \gg 1$, with respective probabilities $1-p$ and p, $0 < p < 1$, one has a contaminated Gaussian noise of impulsive form, as, from time to time, a very large value of the process will be observed.

The Cramer–Hida representation applied to G yields the equation

$$N(\cdot,t) = \sum_{i=1}^{M}\int_0^t F_i(t,x)\,\tilde{B}_i(\cdot,dx),$$

where

$$\tilde{B}_i(\omega,t) = A(\omega)\,B_i(\omega,t),$$

and B_i comes from the representation of G. The process $\tilde{B}_i$ is still a continuous martingale.

The likelihood exists and the process m is that of the purely Gaussain case, but the expression for $dP_{\underline{Z}}/dP_{\underline{\tilde{B}}}$ is given by the formula [4] (the expression for $\underline{Z}$ given in 3.1 must be adapted: B_i should be replaced with $\tilde{B}_i$)

$$\ln\left[\frac{dP_{\underline{Z}}}{dP_{\underline{\tilde{B}}}}(g)\right] = \int_0^T \langle\frac{\underline{s}(g,t)}{\sigma^2(g,t)}, \underline{ev}(g,dt)\rangle - \frac{1}{2}\left\|\frac{\underline{s}(g,\cdot)}{\sigma(g,\cdot)}\right\|^2_{L_2[\underline{\beta}]},$$

where

$$\sigma(g,t) = \frac{\operatorname{trace}\left[\underline{ev}(g,t)\,\underline{ev}(g,t)^\star - 2\int_0^t \underline{ev}(g,x)\,\underline{ev}(g,dx)^\star\right]}{\sum_{i=1}^M \beta_i(t)}.$$

The numerator is the path definition of the quadratic variation of $\underline{ev}$ and the denominator is finite because of the properties of the Cramer–Hida representation for Gaussian processes.

3.4 The case of a noise with a filtered Poisson component

One assumes a noise of the form

$$N(\cdot, t) = \int_0^t F(t, x) B(\cdot, dt),$$

that is a Cramer–Hida multiplicity of ($M =$) 1, and a component B of the form

$$B(\omega, t) = \frac{W(\omega, t) + [\Pi(\omega, t) - \beta(t)]}{\sqrt{2}},$$

with Π a Poisson process with expectation β and W a Gaussian process with independent increments and variance β. Consequently, the quadratic variation of B is β as well.

Again, the likelihood exits and m is as in the Gaussian noise case, but now the expression for dP_Z/dP_B, with the appropriate assumption, is given by the formula [7]:

$$\begin{aligned}\ln\left[\frac{dP_Z}{dP_B}\right](g) &= \int_0^T s(g, x)\, ev(g, dx) - \frac{1}{4}\int_0^T s^2(g, x)\, \beta(dx) \\ &- \frac{1}{\sqrt{2}}\int_0^T s(g, x)\left[\tilde{\Pi}(g, dx) - \beta(dx)\right],\end{aligned}$$

where $\tilde{\Pi}$ is a counting process with the properties of Π.

4 Discrete time computation of the likelihood

Computations of the likelihood will be illustrated with the simplest of examples. It is posited that the Z process has the form

$$Z(\omega, t) = \int_0^t s(Z(\omega, x))\, dx + B(\omega, t),$$

where B is a standard Wiener process. One supposes furthermore that values of an observed waveform $x(t)$ are available at times

$$0 = t_0 < t_1 < \cdots < t_{n-1} < t_n = T$$

"summarized" into a vector $\underline{t}_n$. The corresponding values of x are denoted $x(\underline{t}_n)$. Then, using standard approximations of integrals,

$$\begin{aligned}\ln\left[\frac{dP_{S+N}}{dP_N}\right](x) &\approx \sum_{i=0}^{n-1} s(m(x, t_i))\left[m(x, t_{i+1}) - m(x, t_i)\right] \\ &- \frac{1}{2}\sum_{i=1}^{n-1} s^2(m(x, t_i))(t_{i+1} - t_i).\end{aligned}$$

Let $\Sigma_N(\underline{t}_n)$ denote the covariance matrix of the noise variables

$$N(\cdot, t_0), N(\cdot, t_1), \ldots, N(\cdot, t_n),$$

and let L_n denote the matrix that transforms the vector $\underline{u}$ into a vector $L\underline{u}$ whose ith component is

$$[L\underline{u}]_i = \sum_{k=1}^{i} u_k.$$

$\Sigma_N(\underline{t}_n)$ can be decomposed as $\Sigma_N(\underline{t}_n) = F_n F_n^\star$, where F_n is lower triangular. One then sets $m(x, t_i) \approx \left[L_n F_n^{-1} x(\underline{t}_n)\right]_i$. One thus obtains, when $t_{i+1} - t_i = \tau$, $1 \le i \le n-1$, an approximation $\mathcal{L}_n(x, \underline{t}_n)$ to the logarithm of the likelihood $\ln\left[\frac{dP_{S+N}}{dP_N}\right](x)$ in the form

$$\mathcal{L}_n(x, \underline{t}_n) = \sum_{i=0}^{n-1} s\left(\left[L_n F_n^{-1} x(\underline{t}_n)\right]_i\right)\left[F_n^{-1} x(\underline{t}_n)\right]_{i+1} - \frac{\tau}{2}\sum_{k=0}^{n-1} s^2\left(\left[L_n F_n^{-1} x(\underline{t}_n)\right]_i\right),$$

which can be given the following recursive form (when a new observation $x(t_{n+1})$ is registered):

$$\begin{aligned} \mathcal{L}_{n+1}\left(x, \underline{t}_{n+1}\right) &= \mathcal{L}_n(x, \underline{t}_n) \\ &+ s\left(\left[L_n F_n^{-1} x(\underline{t}_n)\right]_n\right)\left[F_{n+1}^{-1} x\left(\underline{t}_{n+1}\right)\right]_{n+1} \\ &- \frac{\tau}{2} s^2\left(\left[L_n F_n^{-1} x(\underline{t}_n)\right]_n\right). \end{aligned}$$

The only new term to compute at step $n+1$ is the term $\left[F_{n+1}^{-1} x\left(\underline{t}_{n+1}\right)\right]_{n+1}$ which can be obtained from the cross correlation of $x\left(\underline{t}_{n+1}\right)$ with the $(n+1)$th row of F_{n+1}^{-1}.

Finally one must estimate s from the data. Some details can be found in [3, 5], and comments about practice follow.

Remark: As

$$N(\cdot, i\tau) = \sum_{k=1}^{i-1} F(i\tau, k\tau)\left[B(\cdot, (k+1)\tau) - B(\cdot, (k)\tau)\right] + \epsilon_i, \ 1 \le i \le n,$$

one could possibly either compute the values of F or validate its estimation using independent components analysis [8].

4.1 Computational implementations

The key problem in developing effective signal detection algorithms based on the above theory is the determination of a good approximation to the functionals

$\{s_i, \ 1 \leq i \leq M\}$ appearing in the expression for the vector functional $\underline{Z}$ having components

$$Z_i(\omega, t) = \int_0^t s_i(\omega, x)\beta_i(dx) + B_i(\omega, dx).$$

One can write each functional s_i as a causal functional of the vector process $\underline{Z}$:

$$s_i(\omega, x) = \sigma_i(\underline{Z}(\omega, v), 0 \leq v \leq x),$$

and it is this implementation that has been pursued in studies of algorithm development for sonar. Since the linear filters F_i appearing in the definition of the N process can generally be determined, the essence of the problem for implementation is then to determine the functionals σ_i. Various strategies have been used in studies performed on experimental data. A "bullet-proof" general approach is still to be determined and seems unlikely to exist. However, in limited studies to date, algorithms based on such representations have generally outperformed classical methods using both active and passive sonar data. These representations have for the most part been determined by a combination of examining representative data and consideration of the physical properties of the noise, the target, and the environment.

As can be seen, the theory developed here does not provide, as do approaches such as matched filtering, a defined algorithm that can be implemented in a straightforward manner for a given system. Rather, it provides a likelihood-ratio-based (LRB) framework within which one can seek an effective implementation. As such, the further development of effective algorithms based on the theory is far more dependent on serious analysis of data properties and representations than is generally performed in development of contemporary signal processing methods. This lack of explicit solutions and the need for a problem-by-problem approach to developing a solution are the negatives of the theoretical results so far as computational implementations are concerned. The positives are the existence of an LRB approach within which one can proceed with confidence and which applies to very general $S + N$ processes.

Bibliography

[1] Ash, R.B. (1965). Information Theory. Interscience, New York.

[2] Baker, C.R. and Gualtierotti, A.F. (1986) Discrimination with respect to a Gaussian process. Probability Theory and Related Fields 71, pp. 159-182.

[3] Baker, C.R. and Gualtierotti, A.F. (1986). Likelihood ratios and signal detection for NonGaussian processes, *Stochastic Processes in Underwater Acoustics,* Baker, C.R., Editor, Lecture Notes in Control and Information Sciences 85, Springer, Berlin, pp.154-180.

[4] Baker, C.R. and Gualtierotti, A.F. (1992). Absolute continuity and mutual information for Gaussian mixtures. Stochastics and Stochastics Reports 39, pp. 139-157.

[5] Baker, C.R. and Gualtierotti, A.F. (1993). Likelihood-Ratio detection of Stochastic signals, *Advances in Statistical Signal Processing 2,* Poor, H.V. and Thomas, J.B., Editors, JAI Press, Greenwhich, CT, pp. 1-33.

[6] Climescu-Haulica, A. (1999). Calcul stochastique appliqué aux problèmes de détection des signaux aléatoires. Ph.D. dissertaion, Federal Polytechnic, Lausanne.

[7] Climescu-Haulica, A. and Gualtierotti, A.F. (2001). Theory of likelihood-ratio detection of random signals in dependent, causally filtered Wiener-plus-Poisson noise. CRC Report No. CRC-RP-01-01, Communications Research Center Canada, Ottawaa, Ontario.

[8] Comon, P. (1994). Independent component analysis, a new concept? Signal Processing 36, pp. 287-314.

[9] Cramér, H. (1961). On some classes of non-stationary stochastic processes. *Proceedings of the Fourth Berkeley Symposium on Mathematical Statistics and Probability,* 2, pp. 57-77.

[10] Helstrom, C.W. (1968). Statistical Theory of Signal Detection, 2nd ed. Pergamon, Oxford.

[11] Hida, T. (1960). Canonical representations of Gaussian processes and their applications. *Memoirs of the College of Science 33, University of Kyoto,* pp. 109-155.

[12] Horton, C.W., Sr. (1969). Signal Processing of Underwater acoustic Waves. U.S. Government Printing Office, Washington, D.C.

[13] Huang, S.T. and Cambanis, S. (1979). Spherically invariant processes: their nonlinear structure, discrimination and estimation, Journal of Multivariate Analysis 9, pp. 59-83.

[14] Kailath, T. and Poor, H.V. (1998). Detection of stochastic processes. IEEE-IT 44, pp.2230-2259.

[15] Middleton, D. (1999). Non-Gaussian noise models in signal processing for telecommunications. IEEE-IT 45, pp. 1129-1149.

[16] Schatten, R. (1960). Norm Ideals of Completely Continuous Operators. Springer, New York.

[17] Slepian, D. (1958). Some comments on the detection of Gaussian signals in Gaussian noise. IEEE-IT 4, pp.65-69.

[18] Sobczyk, K. (1985). Stochastic Wave Propagation. Elsevier, Amsterdam.

[19] Todorovic, P. (1992). An Introduction to Stochastic Processes and Their Applications. Springer, New York.

Solution Of The General Harmonic Estimation Problem (High-Resolution Sinusoid Parameter Estimation)

M. D. Macleod
Department of Engineering, University of Cambridge, Trumpington Street, Cambridge, CB2 1PZ, UK

Abstract

This paper presents a method for the analysis of harmonic processes with discrete or mixed spectra, that is, signals which consist of a sum of sinusoids (or complex sinusoids) and additive white or coloured noise. This is a much studied problem, with many important applications. Nevertheless, existing approaches have significant limitations. In many cases, the model order (number of sinusoids) is assumed known, and in most cases Additive White Gaussian Noise (AWGN) is assumed. We present a new method for jointly determining the model order and estimating the sinusoid parameters in white or coloured noise. It is an iterative detection and estimation algorithm. At each iteration, deflation based on orthogonal projection (also known as the 'notch periodogram') is used to remove already-detected tones; a detection test is applied and if a new tone is detected the model order is increased and Maximum Likelihood tone frequency re-estimation is carried out. In the mixed spectrum case the detection test requires an estimate of the noise PSD, which is obtained by smoothing the logarithm of the notch periodogram.

1 Introduction

Let $\mathbf{x} = [x(0), x(1), \ldots, x(N-1)]^T$ be the signal consisting of M complex sinusoids (cisoids) in noise:

$$\mathbf{x} = \sum_{m=1}^{M} a_m \mathbf{e}(\omega_m) + \mathbf{v} = \mathbf{E}(\Omega)\,\mathbf{a} + \mathbf{v} \tag{1.1}$$

where a_m and ω_m are the complex amplitude and normalized frequency of the mth cisoid, $\mathbf{e}(\omega_m) = \left[e^0, e^{j\omega_m}, e^{j2\omega_m}, \ldots, e^{j(N-1)\omega_m}\right]^T$, $\mathbf{v} = [v(0), v(1), \ldots, v(N-1)]^T$ is the noise vector, $\mathbf{E}(\Omega) = [\mathbf{e}(\omega_1), \mathbf{e}(\omega_2), \ldots, \mathbf{e}(\omega_M)]^T$, $\Omega = [\omega_1, \omega_2, \ldots, \omega_M]^T$ and $\mathbf{a} = [a_0, a_1, \ldots, a_{N-1}]^T$.

The problem considered in this paper is to jointly estimate the Power Density Spectrum of the noise (which may be coloured), determine M, and estimate the signal parameter vectors $\mathbf{a}$ and Ω. We allow the minimum difference between frequencies to be less than $\omega_0 = 2\pi/N$; this is known as the 'high-resolution' estimation problem.

If $\mathbf{v}$ is white and Gaussian and M is known, the Maximum Likelihood Estimates of $\mathbf{a}$ and Ω are found [1] by solving the least squares minimization

$$\min_{\Omega,\mathbf{a}} J(\Omega,\mathbf{a}) \equiv \|\mathbf{x}-\mathbf{E}(\Omega)\,\mathbf{a}\|^2 . \tag{1.2}$$

For given Ω, the optimum $\mathbf{a}$ is given [1] by

$$\mathbf{a} = \left(\mathbf{E}(\Omega)^H\,\mathbf{E}(\Omega)\right)^{-1}\mathbf{E}(\Omega)^H\,\mathbf{x}. \tag{1.3}$$

Hence substituting (1.3) into (1.2) yields an equivalent minimization of an objective function of Ω only,

$$\min_{\Omega} J_1(\Omega) \equiv \left\|\mathbf{x}-\mathbf{P}_{\mathbf{E}(\Omega)}\mathbf{x}\right\|^2 = \left\|\mathbf{P}^{\perp}_{\mathbf{E}(\Omega)}\mathbf{x}\right\|^2 , \tag{1.4}$$

where $\mathbf{P}_{\mathbf{E}(\Omega)} = \mathbf{E}(\Omega)\left(\mathbf{E}(\Omega)^H\,\mathbf{E}(\Omega)\right)^{-1}\mathbf{E}(\Omega)^H$ is the projector onto span $(\mathbf{E}(\Omega))$ and $\mathbf{P}^{\perp}_{\mathbf{E}(\Omega)} = \mathbf{I}-\mathbf{P}_{\mathbf{E}(\Omega)}$ is the orthogonal projector. Assume that we *know* the values of the first $M-1$ frequencies which minimize $J_1(\Omega)$ in (1.4), and form the vector $\Omega_V = [\omega_1,\ldots,\omega_{M-1}]^T$, then it is proved in [2] that

$$J_1(\Omega) = J_1(\Omega_V) - P_V(\omega;\Omega_V),$$

where

$$P_V(\omega;\Omega_V) = \frac{\left|\mathbf{e}^H(\omega)\,\mathbf{P}^{\perp}_{\mathbf{E}(\Omega_V)}\mathbf{x}\right|^2}{\left\|\mathbf{P}^{\perp}_{\mathbf{E}(\Omega_V)}\mathbf{e}(\omega)\right\|^2} \tag{1.5}$$

is called the *notch periodogram* of the data vector $\mathbf{x}$ with respect to the *notch set* Ω_V. Hence the optimum estimate of the single unknown frequency $\hat{\omega}_M$ when all other frequencies are known is that which maximizes the univariate function $P_V(\hat{\omega}_M;\Omega_V)$.

2 Existing detection tests

For white noise, there are several classical detection tests [3], all based on the periodogram, including Schuster's and Fisher's tests for a single sinusoid, and Siegel's test for multiple sinusoids. These tests are satisfactory [3] when the sinusoid frequencies are Fourier frequencies ('on bin'), and they implicitly provide estimates of the frequencies of the detected sinusoids. However, in the case of a finite blocklength N (which is of the greatest practical importance) the frequencies cannot be assumed to be Fourier frequencies and the resulting actual leakage causes a loss of detection power [4]. Also the power of Siegel's test reduces as the number of sinusoids increases.

Three recent publications derive optimal detectors. The Matched and Adaptive Subspace detectors presented in [5, 6] are Generalized Likelihood Ratio Tests and are shown to be optimal (UMP invariant) detectors. The application of Likelihood Ratio Tests to the problem of estimating the number of sinusoids present in data is also analysed in [7]. However, the optimality of these tests depends on the sinusoid frequencies being known. Because the frequencies are unknown they have to be estimated before they can be detected. This is a fundamental problem, because frequency estimation requires non-linear minimisation, which requires good initial frequency estimates to avoid becoming trapped in erroneous local minima. These initial frequency estimates are intimately linked to the detection process, especially in the high-resolution case, as will be illustrated below. Therefore it is essential to consider detection and estimation jointly.

Over the past two decades, a popular approach for determining the order k of a model (which in this case is the number of cisoids) has been to use an information-theoretic test such as the Akaike Information Criterion (AIC), Minimum Description Length (MDL) [7], or EDC [8]:

$$\mathrm{EDC}\,(k) = N \ln J_1\,(\Omega_k) + \frac{3k}{2} \alpha \ln N \tag{2.1}$$

in which the first term is (ignoring a constant term) the log-maximum-likelihood (Ω_k being the parameter vector which maximizes the likelihood) and the second a penalty term. k is incremented only if it results in a reduction in EDC(k). In [9], the value $\alpha = 2$ in (2.1) is empirically found to be optimum. Interestingly, a more recent Bayesian model order criterion for this problem [10, 7] (for AWGN) corresponds to the choice $\alpha = 5/3$ in (2.1).

However, the information-theoretic approach has suboptimal performance, as reported in [11] and confirmed below.

Li and Djurić recently proposed a detection test [11] for closely spaced sinusoids in white noise. We will describe it in detail, as background to the new detection test presented in Section 4. Consider the case where there is in fact only one tone (M=1) and its frequency ω_1 is correctly estimated, and the variance of the white noise $\mathbf{v}$ is σ^2. Then with notch frequency ω_1, the *peak* of the notch periodogram in the range $(\omega_1 - \pi/N) < \omega < (\omega_1 + \pi/N)$, that is, one 'bin' width centred on ω_1, is located; assume its frequency is ω_P. Then Li and Djurić show that (in the tails) the normalized height of the peak, $2P_1\,(\omega_P;\omega_1)/\sigma^2$, is approximately distributed as χ^2_2.

Where σ^2 is not known, they propose estimating it by averaging the notch periodogram over the frequency 'bins' *not* close to the notch frequencies. Thus a suitable threshold γ_1 can be defined, so that if $P_1\,(\omega_P;\omega_1) < \gamma_1$ we should conclude the present periodogram peak is caused by only one sinusoid; otherwise we should conclude that there is at least one more sinusoid close in frequency.

Li and Djurić then describe an extension of this idea to test between the two hypotheses 'two sinusoids in the current peak' and 'more than two sinusoids'. A new notch set is formed either from the frequencies of the two largest peaks in $P_1(\omega;\omega_1)$ in the same frequency range $(\omega_1 - \pi/N) < \omega < (\omega_1 + \pi/N)$ or, if there is only one peak in that range at ω_P, as $\Omega_2 = [\omega_1, \omega_P]^T$. The new two-notch periodogram is computed over the same frequency range, and its peak is again tested; if it is less than a second threshold γ_2 we conclude there are only two sinusoids. (Note that $\gamma_2 < \gamma_1$ to give the same false alarm probability; we comment further on this below.) Otherwise the number of notches is increased in the same way and the test is repeated until it fails.

A practical problem is that the test imposes no control over the separation between the notch frequencies. Since coincident notches cause singularity in the notch periodogram algorithm, and very close notches cause numerical problems, suitable (minor) algorithm modifications are required to prevent these problems arising.

[11] proposes that each distinct peak in the original periodogram is processed in the way described above, but that while the kth peak is being analysed notches are also placed at the other peaks of the original periodogram, and the procedure for estimating σ^2 is modified to exclude those peaks. Detection of the distinct peaks in the periodogram requires a separate detection test, and therefore a separate initial estimate of σ^2.

2.1 Limitations of existing tests

We found that the Li and Djurić test performs satisfactorily for deciding between the hypotheses 0, 1, and more-than-1 tone in a given periodogram peak. However, if the number of tones is greater than one, it has a limitation which may be illustrated using the (often used [1]) test signal used in [11], that is

$$y(n) = \exp(j2\pi 0.5n) + b\exp(j2\pi 0.51n + \phi) + v(n), \tag{2.2}$$

where $n = 0, \ldots, 24$ and $\phi = \pi/4$. We use the signal-to-noise ratio definition $\mathrm{SNR} = 10\log_{10}(1/\mathrm{var}(v(n)))$ dB, as in [1, 11]. For this signal we find that the first notch is placed at the frequency ω_1 of the single periodogram peak (near the mean of the two true frequencies), and the second notch is placed very close to the first. Thus the two notches are not close to the two true frequencies. We believe that this is one of the reasons why it is necessary to have $\gamma_2 < \gamma_1$ to give the same false alarm probability in [11].

Any notch frequency errors also give rise to spurious peaks in the notch periodogram, and this gives rise to false detections if the SNR is sufficiently high. At a SNR of 20 dB, which is the maximum SNR considered in [11], this residual error is not sufficiently large to cause erroneous detection of a third tone. However, at higher SNR (for example 30 dB) the test erroneously detects 3 or 4 tones.

Next, we note that the signal (2.2) has a 'best case' phase relationship [14]. The 'worst case' phase in (2.2), in the sense that it makes both detection and estimation more difficult (for *any* algorithm), is [14] $\phi = -\pi/4$ rather than $+\pi/4$. After the first notch is placed (again near the mean of the two true frequencies), the residual notch periodogram has smaller total energy in this case; hence (for *any* detection test) a higher SNR is needed to detect two tones. However, we also note that although the residual has two significant peaks, they are well outside the one-bin frequency range of the test. This reduces the detection power of the test.

A further limitation of [11] concerns its method for noise floor estimation, and is commented on at the start of Section 4.3.

The detection test based on the EDC (2.1) has lower detection power than that in [11]. For example, if the signal (2.2) is used with $\phi = -\pi/4$ at SNR 40 dB, only one tone is detected by (2.1), whereas the test in [11] correctly detects two. This may be in part because the peak frequencies found in the initialization phase are biased due to other tones nearby in frequency. However, for $\phi = \pi/4$ the same bias is observed but two tones are successfully detected.

3 Existing joint detection-estimation algorithms

The literature on harmonic estimation is very extensive; for partial reviews, see [1, 3, 7]. In this section we review only certain algorithms which are direct precursors of the new algorithm presented in Section 4.

The idea of a deflationary solution to this problem is an old one. Priestly [22] and Walker [21] both suggest estimating the first frequency by finding the largest peak in the periodogram, then subtracting the corresponding estimated sinusoid from the data, estimating the second frequency by finding the largest peak in the periodogram of the residual, and so on. At each stage they propose the use of a classical single sinusoid detection test; if this test is passed, the model order (number of sinusoids) is incremented. Those authors observed that this procedure is valid asymptotically (as $N \to \infty$), because for sufficiently large N the periodogram at frequency ω_1 is negligibly affected by the presence of a sinusoid in the data at frequency ω_2.

However, for finite N, the influence of a sinusoid in the data at frequency ω_2 on the periodogram at frequency ω_1 ('leakage') is not neglible, resulting in a biased estimate of ω_1. Bloomfield [12] proposed that after each new sinusoid is detected, the whole set of frequency estimates should be reoptimized to minimize the sum-squared residual. Note that this corresponds to directly minimizing $J_1(\Omega)$ in (1.4). Bloomfield described how this multivariate nonlinear optimization can be carried out by an optimization procedure called 'cyclic descent'. In one cycle of this procedure, each tone in turn is re-estimated, with all the *other* estimated tones subtracted from the data. The whole cycle is repeated until a convergence criterion is satisfied. Macleod [13] presented computationally efficient frequency-domain algorithms for carrying out this procedure, and analysed their performance. He showed that, provided the separation of frequencies is sufficient, it converges quickly and results in nearly ML estimates.

In [14], Macleod presented a high-resolution algorithm, in which 'clusters' of closely spaced tones are optimized by multivariate optimization, within an overall 'cyclic descent' algorithm in which each cluster is optimized in turn until convergence is achieved. This algorithm again minimizes (1.4) directly, by subtraction of the estimated tones.

Hwang and Chen [9] described a combined detection and estimation algorithm based on iterative use of the notch periodogram. It has two phases. The initialization phase starts by identifying the frequency ω_{01} of the peak in the periodogram. Next, with $\Omega_1 = \omega_{01}$ as the notch set, the new peak frequency ω_{02} of the notch periodogram is found. Then with $\Omega_2 = [\omega_{01}, \omega_{02}]^T$ as the notch set, the frequency ω_{03} of the peak of the notch periodogram is determined, and so on. The model order k is incremented for as long as it results in a reduction in EDC(k), as given by (2.1).

The second phase of the algorithm in [9] refines the frequency estimates using the Alternating Notch Periodogram Algorithm (ANPA) [2] (which is again a 'cyclic descent' algorithm), repeated until convergence.

In [15, 16] Clarke presented the 'IMP' (Incremental Multi-Parameter) algorithm, and illustrated its application to the more complicated problem of sensor array processing where, in general, multiple blocks ('snapshots') of data are processed. At each stage of IMP, a new signal component is detected and its parameters are estimated. Clarke suggested that an appropriate single-source detection test should be applied to detect each new component (as in [12]). Following detection of a new component, the parameters of all the estimated signal components are then re-estimated, using 'cyclic descent'. The removal of all components other than the one to be re-estimated is achieved by normalized orthogonal projection, as in ANPA. To apply IMP to the problem we are considering, the number of data blocks would be set to one, and the 'bearing" and 'signature function' to ω and $\mathbf{e}(\omega)$. The objective function in IMP is then identical to (1.5). Hence the second stage of Hwang and Chen's ANPA algorithm is a special case of the IMP algorithm.

Successful operation of ANPA, and hence IMP, requires sufficiently large tone frequency separations, as explained in the next section; they are therefore not general 'high-resolution' algorithms (although it is true that when used in a 'cyclic descent' optimizer, deflation using orthogonal projection [9] allows smaller tone frequency separation than deflation by subtraction [13]).

3.1 Limitations of existing algorithms

The key disadvantage of the subtractive approach [14, 13] is that it is difficult to devise a statistically sound detection test, because of the difficulty of characterizing the pdf of the residual spectrum in the vicinity of the subtracted tones. If instead orthogonal projection is used, then there is a sound statistical basis [11, 5, 6] for the design of a detection test.

However, the detection test based on the EDC (2.1), as used in ANPA, has lower detection power than that in [11], which in turn has limitations, as explained in Section 2.1. In Section 4.1 we describe a new detection test which outperforms both.

For the high-resolution problem, the existing algorithms based on orthogonal projection [2, 16] have further limitations.

First, there are problems associated with the computation of (1.5). If the frequency of interest ω is very close to one of the notch frequencies in Ω_V, then the numerator and denominator of (1.5) are both small values computed as differences of larger values, so computational errors become significant. If ω is equal to one of the notch frequencies then (1.5) is 0/0. The solution adopted in [9, 11] is to compute the notch periodogram on a discrete grid of frequencies, $\omega = q\omega_\Delta$, where $\omega_\Delta = 2\pi/L$, $L = ZN$, Z integer. The notch frequencies are also quantized to this grid, and the notch periodogram is simply not computed at notch frequencies, since it is not in fact required at those frequencies.

However, the error resulting from quantizing the notch frequencies causes false peaks in the notch periodogram, and this gives rise to false detections if the SNR is sufficiently high, as mentioned in Section 2.1. For an isolated sinusoid at frequency ω_m, the height of the false peak caused by quantizing frequency ω_m to $\tilde{\omega}_m$ is found empirically to be

$$P_{\max} \approx \frac{a_m^2}{12} N^3 (\omega_m - \tilde{\omega}_m)^2$$

and the probability of a false detection becomes significant if $P_{\max} > \sigma^2$.

Second, the convergence speed of the ANPA or IMP algorithms is acceptable for 'good" phase relationships, such as $\phi = \pi/4$ in (2.2), which was one of the test cases used in [9]. However, like all univariate optimization methods, they converge much more slowly when the Hessian matrix of the objective function has a high eigenvalue ratio, as happens for example with $\phi = -\pi/4$ in (2.2). For example, at 40 dB SNR, with $\phi = \pi/4$, ANPA converges completely within fige iterations; but with $\phi = -\pi/4$ (and forced detection of two tones) convergence is slower, and still incomplete in 100 iterations. If notch frequencies are quantized, convergence is further impaired (it 'stalls').

The problems caused by quantization of notch frequencies can be reduced (at a computational cost) by increasing the zero padding factor Z, but this has to be increased very substantially to overcome the stalling problem. A better solution is to avoid quantization when *estimating* the frequency of a peak by using interpolation, as in [16], and to allow arbitrary notch frequencies when *computing* the notch periodogram by using a suitable algorithm [18]. Further details are given in Section 4.2.

4 New detection-estimation algorithm

In this section we describe a new high-resolution joint detection–estimation algorithm which: (a) overcomes the limitations of the detection test [11], and (b) overcomes the limitations of the ANPA (or IMP) optimization algorithm for high-resolution estimation. We then extend the detection test to accommodate coloured noise.

4.1 New detection algorithm

We assume initially that σ^2 is known. Instead of starting by detecting separate peaks in the periodogram and then carrying out tests in restricted frequency ranges around each peak, as in [11], we detect tones one at a time, over the whole frequency range, by computing the notch periodogram with existing tones notched, and comparing its peak with a threshold based on the χ_2^2 distribution. We re-optimize the frequency estimates after each new tone is detected, as described below.

If there is in fact a further tone present, not yet detected, then the estimated frequencies at the time of the detection test are incorrect; they are biased by the so-far-undetected tone(s). This reduces detection power, particularly in the case of closely spaced tones. To overcome this problem, we use the idea [17] of a lowered initial detection threshold γ_0. If that threshold is exceeded, we increase the model order, optimize the frequencies, then compute the notch periodogram with each notch removed in turn, and retest the resulting peak against the unmodified threshold γ_1. To minimize false alarm probability, the lowered threshold could be applied only at frequencies close to existing tones.

4.2 Joint detection and estimation

Because the cyclic descent algorithm has very slow convergence under some conditions, we use a Nelder–Mead simplex method (Matlab routine fminsearch) to carry out multivariate optimization. In an earlier paper [19] we proposed that the objective function for this optimization should be the summed notch periodogram

$$\sum_{k=0}^{N-1} P_V\left(k\omega_0; \Omega_V\right). \tag{4.1}$$

However the real objective is to minimize $J_1\left(\Omega_V\right)$, the Sum Squared Residual after subtraction of the estimated tones. In this paper we therefore minimize the objective function $J_1\left(\Omega_V\right)$, and use the notch periodogram only for the detection of new tones. Evaluating $J_1\left(\Omega_V\right)$ also requires less computation than (4.1).

The sequence in which detection, frequency estimation and notch frequency optimization are carried out is crucial. The following sequence was found to be best:

(1) compute notch periodogram, estimate noise floor and hence threshold;
(2) test peak of notch periodogram; if peak$< \gamma_0$, terminate;
(3) add new peak frequency to Ω and minimize $J_1\left(\Omega\right)$ using multivariate optimization;
(4) retest all peaks; if any peak$< \gamma_1$, remove it from Ω and terminate;
(5) go to 1.

The reason for retesting all peaks in step 4 is that after optimization the size of all peaks can alter.

The notch periodogram (1.5) is computed on a discrete grid of frequencies (typically using zero-padding factor $Z = 16$). An efficient algorithm [18] may be used for this computation; it does not require quantization of the notch frequencies. Numerical problems are avoided by not computing the notch periodogram at grid frequencies very close to notch frequencies, and then not including those grid frequencies when searching for peaks in step 2.

The peak frequency estimate in step 2 is obtained by interpolation [19]. It is possible (for example if the signal is given by (2.2) with $\phi = \pi/4$) for the second peak frequency to be identical to the first. This has not caused problems yet, but it may be prudent to enforce a minimum separation between each new frequency estimate and all existing frequency estimates, to give a better initial value for the optimizer.

4.3 Noise floor estimation

Estimation of σ^2 by averaging the notch periodogram (excluding frequency bands close to detected tones) [11] results in a positive bias during initial detection, due to the not-yet-detected tones. This may in certain cases reduce detection performance. To reduce the bias we propose estimation of σ^2 using the mean of the logarithm of the notch periodogram instead:

$$\hat{\sigma}^2 = 1.776 \exp\left(N^{-1}\sum_{k=0}^{N-1}\log P_V\left(k\omega_0;\Omega_V\right)\right) \tag{4.2}$$

where $\omega_0 = 2\pi/N$. Although this estimator has a higher variance than the (unrealizable) sample-mean of the periodogram of noise alone, it has a far lower bias from any undetected tones. Thus the probability of failed detection of new peaks is reduced.

We may extend this approach to accommodate coloured noise. If the PSD of $\mathbf{v}$ (the 'noise floor'), defined as

$$S_N\left(\omega\right) = \frac{1}{N}E\left\{\left|\mathbf{e}^H\left(\omega\right)\mathbf{v}\right|^2\right\}, \tag{4.3}$$

(where E signifies Expectation), varies only 'slowly' with frequency, then over a small frequency range it is approximately constant. Hence, instead of σ^2 in the detection test we may use an estimate of $S_N\left(\omega\right)$ obtained by replacing the global average in (4.1) with a weighted local average in the frequency domain

$$\hat{S}_N\left(m\omega_0\right) = 1.776 \exp\left(\sum_{k=-K}^{K} w_k \log P_V\left(\left(m+k\right)\omega_0;\Omega_V\right)\right) \tag{4.4}$$

where $w_k, k = -K, \ldots, K$ is a unit sum window function (such as a scaled Hanning window). Such an approach has been proposed [20] as a smoothed estimator of a broadband spectrum, but its attenuation of spectral peaks (previously viewed as a disadvantage [3]) has not previously been exploited for its ability to reduce bias.

Table 1 Two tones, worst case phase, 40 dB SNR

True freq (bins)	12	12.25
Mean result (bins)	11.996	12.253
rms error (bins)	0.033	0.031

Table 2 Two tones, best case phase, 40 dB SNR

True freq (bins)	12	12.25
Mean result (bins)	11.999	12.2497
rms error (bins)	0.0057	0.0059

5 Results

Example 1: two equal amplitude tones, equation (2.2) using *worst case* phase $\phi = -\pi/4$, white noise, SNR = 40 dB. In 200 trials, exactly two tones were detected in all cases. The true and mean estimated frequencies and rms estimation error were as in Table 1.

Example 2: two equal amplitude tones, equation (2.2) using *best case* phase $\phi = \pi/4$, white noise, SNR = 40 dB. In 200 trials, exactly two tones were detected in all cases. The results were as in Table 2.

Example 3: blocklength $N = 64$, four tones of different amplitudes in two close pairs, using worst-case phase relationships [14] between *all* tones, white noise. In 200 trials, exactly four tones were detected in all cases. The results were as in Table 3.

Example 4: blocklength $N = 64$, four isolated tones of different amplitudes, white noise. In 200 trials, exactly four tones were detected in all cases. The results were as in Table 4.

5.1 Computation issues

We have developed fast frequency domain algorithms for computing the notch periodogram [18], although the Matlab implementation we actually used does not use all the techniques in [18] (some of which are more advantageous for a C implementation or embedded processor). The fast algorithms in [18] give even greater relative speed-ups for larger values of N. Numerical problems due to ill-conditioning in computing (1.5) were avoided as explained in Section 4.2.

Average execution times to completion were as follows, on a Pentium II processor:

Example 1: (two tones, worst case phase) 1.2 secs.
Example 2: (two tones, best case phase) 0.8 secs.
Example 3: (four tones, two close pairs, N=64) 4.2 secs.
Example 4: (four isolated tones, N=64) 3.2 secs.

Note that although multivariate optimization is more complex than univariate optimization as used in ANPA and IMP, the number of function evaluations is actually reduced in some cases, including examples 1 and 2 above, and the evaluation of (1.4) is simpler than (1.5), so computation time is reduced.

Table 3 Four tones in two close pairs, worst-case phase

Tone SNR (dB)	44	53	53	58
True freq (bins)	7.5156	7.8156	15.0156	15.3156
Mean result	7.5158	7.8156	15.0155	15.3156
rms error	0.002	0.0007	0.0028	0.0017

Table 4 Four isolated tones

Tone SNR (dB)	54	40	50	50
True freq (bins)	3.2156	7.5156	11.8156	15.0156
Mean result	3.2156	7.5157	11.8156	15.0156
rms error	0.000083	0.00046	0.00017	0.00017

5.2 Further issues

When the data are real, it is only necessary to detect peaks over the positive frequency range $0 < \omega < \pi$, and the least squares minimization (1.2) must be replaced by

$$\min_{\Omega,\mathbf{a}} J(\Omega, \mathbf{a}) \equiv \|\mathbf{x} - \mathrm{Re}(\mathbf{E}(\Omega)\,\mathbf{a})\|^2 .$$

One technique for possible further computation reduction, as reported in [19], is to use the concept of 'clusters'. Clusters are sets of tones with the property that any tone in a given cluster differs in frequency from any tone in any other cluster by at least $\omega_{\min}$, a defined limit value, while any tone in a given cluster differs in frequency from at least one other tone in the same cluster by less than $\omega_{\min}$. Investigation of this technique is continuing.

5.3 Conclusions

This paper has presented a new and general practical algorithm for joint detection and high-resolution frequency estimation. It includes developments subsequent to those reported in [19], of which the most important is the use of (1.4) as the objective function for frequency optimization. This improves performance and reduces computation load.

Bibliography

[1] Kay, S.M. *Modern Spectral Estimation*, Prentice-Hall, Englewood Cliffs, NJ, 1988.

[2] Hwang, J.-K. and Chen, Y.-C. Superresolution Frequency Estimation by Alternating Notch Periodogram, *IEEE Trans. SP*, **41**, 1993, 727–741.

[3] Percival, D.B. and Walden, A.T. *Spectral Analysis for Physical Applications*, Cambridge University Press, UK, 1993.

[4] Whittle, P. The simultaneous estimation of a time series harmonic components and covariance structure, *Trabajos de Estadistica*, **13**, 43–57.

[5] Scharf, L. and Friedlander, B. Matched Subspace Detectors, *IEEE Trans. SP*, **42**, 1994, 2146–2157.

[6] Kraut, S., Scharf, L. and McWhorter, L.T. Adaptive Subspace Detectors, *IEEE Trans. SP*, **49**, 2001, 1–16.

[7] Quinn, B.G. and Hannan, E.J. *The Estimation and Tracking of Frequency*, Cambridge University Press, 2001.

[8] Zhao, L. C., Krishnaiah, P.R. and Bai, Z.D. On detection of the number of signals in presence of white noise, *J. Multivariate Anal.*, **20**, 1986, 1–25.

[9] Hwang, J.-K. and Chen, Y.-C. A Combined Detection-Estimation Algorithm for the Harmonic Retrieval Problem, *Signal Processing*, **30**, 1993, 177–197.

[10] Djurić, P.M. A Bayesian model order determination rule for harmonic signals, Proc. ISCAS-1995, Vol. 3, pp. 2273–2276.

[11] Li, H.T. and Djurić, P.M., A novel approach to detection of closely spaced sinusoids, *Signal Processing*, **51**, 1996, 93–104.

[12] Bloomfield, P. *Fourier Analysis of Time Series: An Introduction*, Wiley, New York, 1976.

[13] Macleod, M.D. Fast nearly-ML estimation of the parameters of real or complex single tones or resolved multiple tones, *IEEE Trans SP*, **46**, 1998, 141–148.

[14] Macleod, M.D. High resolution nearly-ML estimation of sinusoids in noise using a fast frequency domain approach, *Proc EUSIPCO 98*, (Sept 98), pp. 1849–1852.

[15] Clarke, I.J. Robustness of eigen-based techniques versus iterative adaptation, *RADAR 87*, IEE Conference Publication No. 281, (1987), London, pp. 84–88.

[16] Clarke, I.J. Supervised Interpretation of Sampled Data Using Efficient Implementations of Higher-Rank Spectrum Estimation, Chapter 2 of *Advances in Spectrum Analysis and Array Processing, Vol II*, (ed. S Haykin), Prentice-Hall, Englewood Cliffs, NJ, 1991.

[17] Clarke, I.J. Personal Communication.

[18] Macleod, M.D. A Fast Frequency Domain Notch Periodogram, *Signal Processing*, **81**, 2001, 1449–1463.

[19] Macleod, M.D. Solution Of The General Harmonic Estimation Problem By Iterated Orthogonal Projection And Residue Detection, *5th IMA Conference on Mathematics in Signal Processing*, Conference Digest, IMA, Dec. 2001.

[20] Wahba, G. Automatic Smoothing of the Log Periodogram, *J. Am. Stat. Assoc.*, **75**, 1980, 122–132.

[21] Waler, A.M. On the estimation of a harmonic component in a time series with stationary independent residuals, *Biometrika*, **58**, 21–36.

[22] Priestley, M.B. *Spectral Analysis and Time Series.* Academic Press, London, 1981.